BIOMAT 2013

International Symposium on
Mathematical and Computational Biology

BIOMAT 2013

International Symposium on
Mathematical and Computational Biology

Toronto, Ontario, Canada 4 – 8 November 2013

edited by

Rubem P Mondaini

Federal University of Rio de Janeiro, Brazil

World Scientific

NEW JERSEY LONDON SINGAPORE BEIJING SHANGHAI HONG KONG TAIPEI CHENNAI

Published by

World Scientific Publishing Co. Pte. Ltd.
5 Toh Tuck Link, Singapore 596224
USA office: 27 Warren Street, Suite 401-402, Hackensack, NJ 07601
UK office: 57 Shelton Street, Covent Garden, London WC2H 9HE

British Library Cataloguing-in-Publication Data
A catalogue record for this book is available from the British Library.

BIOMAT 2013
Proceedings of the International Symposium on Mathematical and Computational Biology

ISBN 978-981-4602-21-1

Printed in Singapore by World Scientific Printers.

Preface

The BIOMAT 2013 International Symposium has been held in Toronto, Ontario, Canada, on November 04-08, 2013, and the host institution was the prestigious Fields Institute for Research in Mathematical Sciences. The institute is named after the mathematician John Charles Fields, the founder of the Fields Medal, the Nobel Prize for Mathematics. We had an immense pleasure of working with senior representatives of the Canadian Scientific Community. Their support was fundamental to the organization of the conference in Toronto: Prof. Walter Craig, the Director of the Fields Institute, Dr. Matheus Grasselli, the Fields Deputy Director; our friends and colleagues from York University, Prof. Jianhong Wu, the Director of the Centre for Disease Modelling and Co-chair of the BIOMAT 2013 conference, Dr. Seyed Moghadas, Director of the Agent-Based Modelling Laboratory, Dr. Huaiping Zhu, Director of the Laboratory of Mathematical Parallel Systems, both members of the BIOMAT 2013 Organizing Committee.

As the authors of accepted contributed papers for presentation in the conference are concerned, scientific practitioners, research students, young postdocs, we had people coming from Asia, Eastern Europe, Northern Europe, Africa, South America and North America. Many collaborators on joint papers have already been working in Canada or studying for a first doctoral degree there. It has been then relatively easy to succeed at assembling together these practitioners and their interested colleagues into five days of continuous scientific sessions, according to one of the BIOMAT Consortium fine traditions.

The administrative support given to the local organization of the BIOMAT 2013 International Symposium by the Administrative Staff of the Fields Institute, was essential to reach the observed high efficiency level. Particularly, we are very pleased to acknowledge the expert help from Mrs. Alison Conway, the Manager of Scientific Program of the Fields Institute and from Ms. Natasha Hanif, Program Coordinator at Fields. Their professional dedication has always been beyond all expectatives (and the BIOMAT Consortium has some very rigorous ones) for a successful organization of an international conference and we dare to take their extra effort as a sincere token of friendship.

The Editor of this BIOMAT series of books, is indebted to his wife Carmem Lucia for the usual dedication she gives to the editorial work of the series. He would like to register that he is indebted for her love and her patience with the occasional oddities of the organizational work, which is usually done under the stress of many parallel commitments. He would like to acknowledge Dr. Cecilia Mondaini from Federal University of Rio de Janeiro, UFRJ, Institute of Mathematics, for the help given to the local organization in Toronto on the week previous to the conference. Last but not least, special thanks are due to Dr. Leonardo Mondaini from Federal University of Rio de Janeiro State, UNIRIO, who has worked with the Fields Administrative Staff, in charge of the General Secretariat of the BIOMAT Consortium, as well as a member of the BIOMAT 2013 Organizing Committee.

Rubem P. Mondaini
President of the BIOMAT Consortium
Chairman of the BIOMAT 2013 Scientific Advisory Committee

Toronto, Ontario, Canada, November 2013

Editorial Board of the BIOMAT Consortium

Contents

Computational Biology

Modelling Physiological Disorders

Mathematical Modelling of Infectious Diseases

THE PRINCESS AND THE PEA: THE UNEXPECTED IMPORTANCE OF MOVEMENT ALGORITHMS

REBECCA C. TYSON

Mathematics & Statistics, University of British Columbia Okanagan, 3333 University Way, Kelowna, BC, V1V 1V7, Canada

Movement patterns are varied and complex, and so movement models necessarily are simplified representations of any real movement pattern. Many model results are robust to different levels of movement complexity, but not always. The algorithms underlying movement decisions by individuals or subgroups in a population can have profound effects on the distribution of the population and the interaction patterns between that population and the landscape in which it resides. In order to understand population dynamics, it is thus often crucial that the movement behaviours of individuals or subgroups in the population be modeled with the appropriate level of detail. In this paper, I present models that encompass a range of formulations for organism movement, and a range of levels of detail. I focus on the role played by the movement patterns in shaping the modelling results, and the changes that follow from modifications of the movement patterns. I find that movement patterns are important at all scales, and even in problems that do not intially present themselves as being chiefly about organism displacement.

1. Introduction

Nature has demonstrated remarkable ingenuity in solving the spatial problem that arises when organisms and the resources they need for survival are not located in the same place. In addition to the myriad forms of locomotion (consider, for example, birds, snakes, dogs, inchworms and slugs), there is an equally remarkable variety of movement patterns. In the category of searching patterns alone, scientists have reported near random motion [4], circular forays [55,28], long-distance displacement [59], and intermittent search patterns [52,86]. The fine details of any search pattern are highly complex, and so models including organism movement have involved the same general simplifying approach that is so fundamental to modelling.

As a general rule, when building a model, small details can be ignored, rare events can be ignored and reasonable approximations should produce results that are generally correct. For the most part, this general approach works well, and there are a host of examples of models that are built on simplifying approximations and yet that yield results with surprising explanatory power [76,10,62,68,53,13,78]. In certain cases however, the rare events and simplifying approximations have a disproportionately large or unexpected effect on the results [19].

In the seminal paper by [40], it was shown that rare long-distance dispersal events are key factors in the spreading speed of an invasion wave. The presence of such events can turn a constant invasion wavespeed into an accelerating one, putting uninvaded areas at risk considerably sooner than anticipated. This work is a clear example of the unexpected importance of rare events in some situations.

In completely unrelated work on wolf hunting patterns, [48] sought to determine whether or not anthropogenic linear features (e.g. seismic lines) affected the movement patterns of wolves and thus the predation pressure they exerted on ungulate prey. Not only were these subtle changes found, as suspected, but the altered movement patterns of the predator also affected the predator's functional response to the prey. Traditionally, functional response is viewed as the per-capita response of a predator to prey density. The work by [48] however, shows that the functional response is also fundamentally a function of the predator's movement pattern.

These two examples serve to illustrate that the fine details of a movement pattern can sometimes be much more important than one would suspect, or can profoundly affect other dynamics that aren't traditionally thought of as arising from movement behaviours. This finding suggests that if movement is involved in a model, the details of the organism's movement strategy should be considered carefully. Similarly, aspects of the results may depend quite heavily on the movement pattern used.

Understanding the importance and effect of movement algorithms is increasingly important in a world dealing with the constant pressure of unintentional invasions. Increased global trade is moving a staggering number of species annually through ballast water and shipping crates. Most organisms transplanted in this drastic way do not survive the journey, or are not successful at their destination. Those few that do survive however, include some that are currently wreaking economically costly environmental havoc. What has been so surprising to the unwary, is the rapidity with which these invasive species can spread across the landscape once introduced.

Movement is also a critical component of the management tactic called "biocontrol". In this technique, a species foreign to a given region is brought in as an intentional invader, with the express purpose of controlling the explosive growth and spread of an unintended invader. These agents are introduced at particular sites, and then allowed to spread on their own across the landscape. How fast this spread occurs, and whether or not it will traverse international boundaries, are issues of very practical concern.

Climate warming is also enforcing movement of populations. As range boundaries shift northward and upward, the species inhabiting those biogeoclimatic zones must move also. Some seem well able to keep up with this enforced migration, while others appear less adaptable. There is real concern that there may be significant species losses simply through the inability of some to keep up with a shifting climate. Even among those more plastic species, there is always the possibility that their migration may run up against an insurmountable obstacle (large body of water, mountain range, etc) that will trap them on the wrong side of the shifting range and thus lead to their extirpation.

The patterns of spread of populations are fundamentally linked to understanding how organisms move in heterogeneous habitat. In this paper, I will present four models involving very different formulations of movement and movement patterns. In each case, I will focus on the role that the movement algorithm played in the results obtained from the modelling study. Each model was motivated by the study of a different biological organism, and so I will briefly describe each biological problem to provide context for the associated model. The organisms in each case exhibit a different set of movement behaviours, and these are modeled in differing levels of detail. These projects did not all initially present themselves as being about movement, but we will see that in each case, the movement pattern plays a key role in the results.

2. Movement Behaviour in an Agent-Based Model

Agent-based models (ABMs) have increased in popularity with modern advances in computing [25] and the advent of agent-based modelling software such as NetLogo [71]. These models are very useful when studying population dynamics at a scale where the discrete nature of individuals and the stochastic aspects of their behaviour are germane to the question being studied. In the ABM framework, a modeler has the freedom to include as much complexity in the movement algorithm as is warranted by the biological question and the movement data available. Movement "decisions" are often based on features of an interaction neighbourhood that can vary from the very localized nearest-neighbour community, out to the entire simulation domain. These interaction neighbourhoods can be strictly controlled, and so it is possible to tease apart nearest- and distant-neighbour effects [42,72]. Note that the agent-based modeler can also create movement algorithms where movement decisions are affected by features that have nothing to do with the environment or the agents, but this is not our focus here.

Here, I present a model in which the agents are males and females of a single species, moving stochastically in a manner consistent with biological observation. Individual interact at a nearest-neighbour level for mating, and also at a larger neighbourhood level through pheromone plumes emitted by females calling for a mate. The project was motivated by the question of control of codling moth, *Cydia pomonella* pest management in apple orchards.

2.1. *The Biological Context: Cydia pomonella*

The codling moth, *C. pomonella* is the iconic "worm in the apple". Originally from Asia Minor [5], it is now a globally distributed and highly damaging agricultural pest wherever pome fruits (chiefly apples and pears) are grown [5,21]. In recent years, scientists have developed an alternative to chemical sprays called the Sterile Insect Technique (SIT) [39]. Under an SIT program, pest insects are mass-reared, sterilized, and then released in target areas. The intent is to overwhelm the wild population with sterile insects, so that wild-wild matings are prevented and the wild population collapses. Key to the success of any SIT program, is determining how many sterile insects are needed to ensure a sufficiently high sterile-wild:wild-wild mating ratio. The necessary sterile:wild ratio is determined theoretically, and then monitored using pheromone traps.

The world's first fully operational SIT program against *Cydia pomonella* was established in the Okanagan region in southern British Columbia, Canada in 1994 [12], and the stated goal was to eliminate codling moth from the agricultural region. Over the first half decade of the SIT program in the Okanagan, appropriate numbers of sterile insects were released, and the data from monitoring traps indicated that there were plenty of sterile moths in the orchard. Nonetheless, the wild population decreased much less quickly than expected [36,54], and by 2004, there were signs that the wild moth population was increasing [77].

We were interested in using a modelling approach to determine whether suspected variations in moth behaviours in springtime could explain the lack of success of the Okanagan SIT program. These behaviours included female mate selectivity, wild population clumping at emergence from diapause, female movement between calls and sterile male uncompetitiveness, that is, propensity for taking "resting breaks" during the search for a mate. Half of these behaviours, the second two, are directly related to individual movement.

2.2. *The Codling Moth Movement Model*

Moth movement in our model had a random walk component, and also a biased response to pheromone attractants emitted by female moths and traps. We assumed that all moths not trapped or actively mating had some probability of engaging in movement. The direction and distance of each movement step depended on the wind velocity and, in the case of males, on the presence or absence of pheromone in the moth's immediate vicinity.

Females flew to a new position by choosing a flight vector. The distance d_i traveled by a given moth is given by

$$d_i = v \times 1 \text{ minute} \tag{1}$$

where the flight speed v is between 0 and 20 m/min, and was chosen at random from a uniform distribution at the beginning of each timestep. The direction ϕ_i of female flight was also chosen at random, from a uniform distribution between 0° and 360°. For a single moth, given the location $P_i = (x_i, y_i)$, distance d_i, and angle of flight ϕ_i at the beginning of a timestep, we determined the new position $P_{i+1} = (x_{i+1}, y_{i+1})$ of the moth at the end of the timestep using

$$x_{i+1} = x_i + d_i \cos(\phi_i), \tag{2a}$$

$$y_{i+1} = y_i + d_i \sin(\phi_i). \tag{2b}$$

Males not mating or resting flew in search of females. Each new position P_{i+1} was calculated in the same way as for females, except that the direction and speed of flight were affected by the presence or absence of pheromone. In the absence of pheromone, male moths engaged in "casting" behaviour, flying at speeds less than maximum and mainly in a cross-wind direction, attempting to intersect a pheromone plume drifting downwind. Thus the direction of flight, ϕ_i, was chosen from a nonuniform probability distribution as illustrated in Figure 1. Once a pheromone plume was detected, the male moth would fly directly upwind at his maximum flight speed until finding the source or losing the plume. This upwind flight direction appears as the gray area in Figure 1. If the plume was lost, casting behaviour resumed at the next timestep.

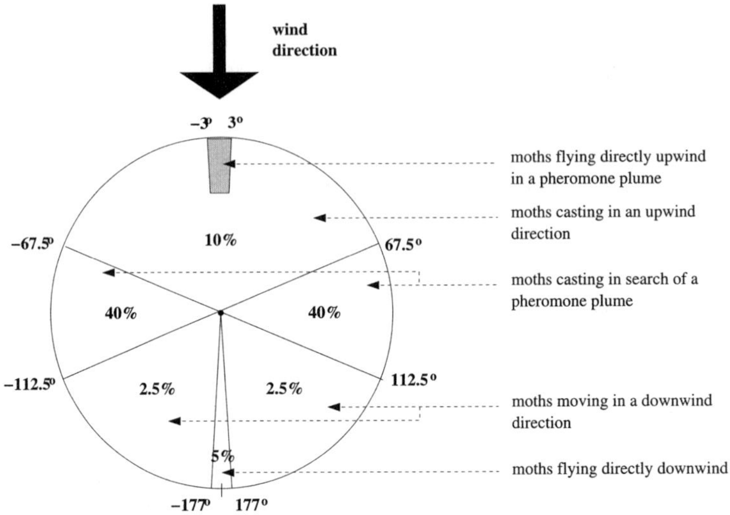

Figure 1. Male moth flight direction probability map. The circle is divided up into flight directions between -180° and 180°, where 0° is directly upwind. The percentage value given inside each wedge is the probability that a male moth's flight direction falls within that wedge. The only except to that rule is the gray area: When a male moth detects a pheromone plume, it flies directly upwind (with flight direction as indicated by the gray region) with 100% probability.

At each timestep, moths not actively mating could stay in the current location (resting behaviour) or move to another location according to the algorithm described above. In the model, lack of competitiveness on the part of sterile males corresponded to an increased propensity of engaging in resting behaviour.

2.3. *The Effect of Movement on Mating Rates*

With this model, we were able to look in detail at the effects of variations in specific moth behaviours on the sterile-wild and wild-wild mating rates. These specific behaviours were female mate selectivity, wild population clumping at emergence, female propensity to move between calls for a mate, and sterile male uncompetitiveness or propensity to take resting breaks. While population clumping could be related to movement, in this case it is the distribution of the population shortly after emergence from the overwintering state. These emergent moths did move as normal in our simulations: it was only the initial distribution that was more or less clumped.

The four behaviours we investigated therefore included two directly part of the moth movement algorithms (female movement and sterile male uncompetitiveness), and two not related to movement (female selectivity and population clumping). The effects of the four factors were compared using one-way anova, and the results are shown in Figure 2.

We see that all four behaviours have strong effects on the wild:wild mating rates, but largely very little effect on the sterile:wild trap catch ratios. The general lack of any effect by these behaviours on the sterile:wild trap catch ratios means that current monitoring methods (chiefly, regular assessment of trap catch ratios) may often not give any indication of problems in the implementation of the SIT. This is a significant issue for managers.

With regard to mating rates, it is not surprising that female selectivity and population clumping each are very important. Females that are dissatisfied with sterile male spermatophores and therefore continue searching for a mate clearly will increase the wild:wild mating rate. Population clumping (Figure 3) affects mating rates by giving the wild males, which emerge near the wild females, an advantage over the sterile males which have been subject to considerable handling and then blown into the orchard to find wild females.

What is perhaps less intuitively clear, is the comparison between the non-movement behaviours and the movement ones. We see that the effects of both sets of behaviours on the wild:wild mating rates are equally strong. The problem of determining an optimal movement strategy for finding a resource (e.g. mate, prey, nesting site), is a nontrivial one [20,63,74], and it is therefore not obvious at the outset how more or less movement by certain individuals will affect the success rate.

The cost or benefit of more or less movement may depend on the population density. If we consider a stationary calling female within a population of flying males (Figure 4), at high densities it should be unnecessary for the female to move in order to have a high probability of attracting a mate. At lower male population densities however, relocation after a certain period without success could potentially increase the female's mate-finding chances. Similarly, if sterile males are doing more resting than their wild counterparts, they spend less time actively seeking a mate. Whether or not this reduces their chances of finding a mate during each bout of flying however, is not immediately clear, and may also depend on the spatial density and distribution of calling females.

The theoretical underpinnings of the SIT are largely based on non-spatial ordinary differential equation models, that is, modelers assumed

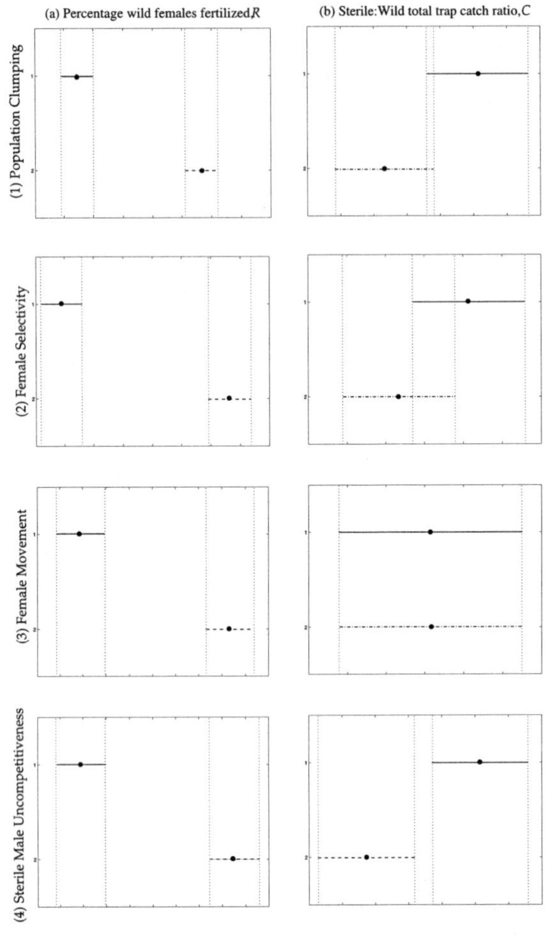

Figure 2. One-way anova results for (a) the percentage of wild females fertilized and (b) the sterile:wild trap catch ratio under (1) Clumping, (2) Female Selectivity, (3) Female Movement and (4) Sterile Male Uncompetitiveness. Results are based on nine replicates for each data point. Null scenario data are shown with solid lines. Dashed data are significantly different from solid line data. Dash-dot data are not significantly from any of the other data.

that the wild and sterile populations are well-mixed and uniformly distributed throughout the crop area [15,87,84]. This should be largely true for the sterile populations, as they are manually delivered to the crop area in a manner that should ensure a relatively uniform distribution of

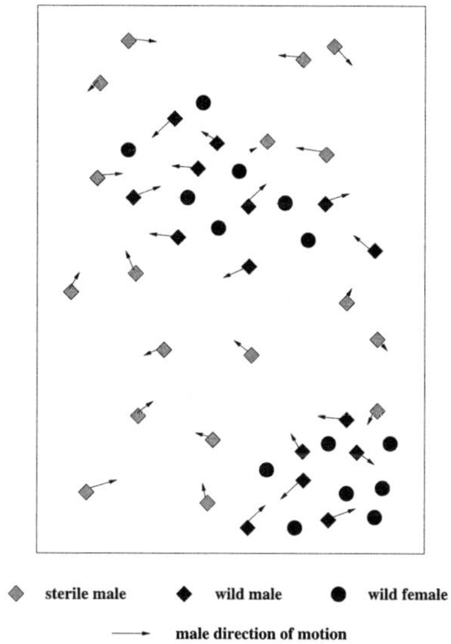

Figure 3. Schematic showing how wild population clumping can affect sterile-wild mating rates. Sterile moths are delivered to the orchard via a blower mounted on a tractor, and so are distributed throughout the orchard. In contrast, the wild population may be highly clumped at emergence, in which case the local ratio of wild to sterile males is quite low. In this scenario, the sterile moths are relatively ineffective at preventing wild:wild matings.

sterile insects throughout the crop. The distribution of the wild population however, is not under the control of management. In the summertime, Okanagan codling moths are likely fairly evenly distributed throughout orchards as temperatures are high and the moths are very active. Okanagan springtime weather however, is considerably cooler and wetter, and moth behaviours may be subject to the variations investigated here. Our results indicate that SIT models and monitoring activities must take spatial effects into account.

3. Movement Behaviour in a Patch Model

Another modelling paradigm often used in ecology is that of the patch-level model [61,37,83]. The landscape is divided up into habitat patches that may or may not be contiguous, and the population within each patch is generally considered homogeneous and well-mixed; individuals within these patches are indistinguishable. In contrast to the ABM approach, movement within each patch is assumed to occur quickly enough so that temporary

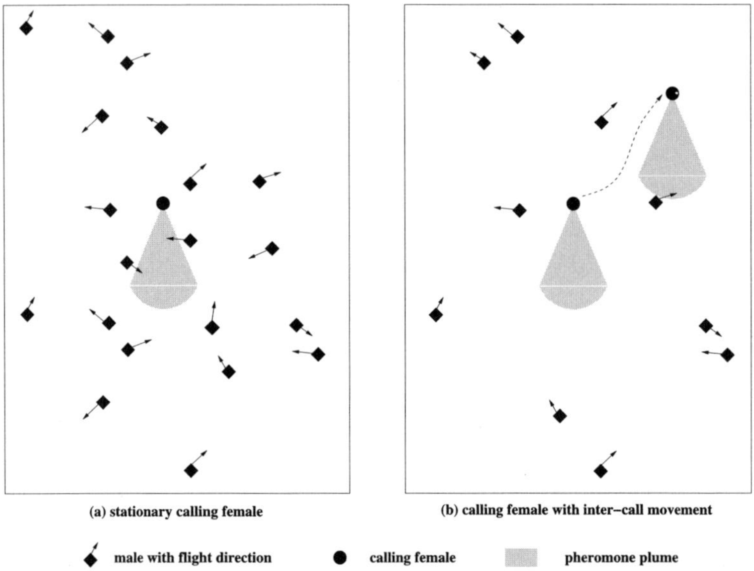

Figure 4. Schematic showing how female movement between calls could affect mating rates. In the left panel, the density of male moths is relatively high, and several male moths are in the near vicinity of the female pheromone plume. In the right panel, the male moth density is much lower. By relocating from the original position to another one (in this sketch, up and to the right), the female moves her pheromone plume from an isolated location with no males nearby to a location where now there is a male in the near vicinity.

differences in the resident population density across the patch are irrelevant. The population dynamics within each patch can thus be modeled using ODEs.

The movement aspect of such models occurs in the transfer of individuals between separate patches. This movement is a function of various factors which often include the inter-patch distance, and the relative size or quality of the two patches [64]. Here, movement is a population-level process that connects habitat patches through dispersal. This view of landscapes and the movement across them is useful when there are easily identifiable regions of distinct habitat quality, such as woodlots in a matrix of agricultural land, though the nature of the matrix also can have an effect on movement behaviours [29]. The exact trajectories followed by dispersers is not of concern here, instead it is the amount of immigration and emigration at each patch that are most relevant.

Here I present a model where the patches are areas of forest harvested at different times. The project was motivated by questions surrounding the process whereby tree squirrels, *Tamiasciurus hudsonicus*, recolonize forest patches that are regrowing after a harvesting event.

3.1. *The Biological Context: Tamiasciurus hudsonicus*

The North American red squirrel (*T. hudsonicus*) is a small, tree-dwelling rodent that ranges throughout much of the coniferous forests in North America [57]. Adults occupy individual, non-overlapping territories that provide resources essential for survival: squirrels that do not possess a territory and food storage site (midden) are unable to forage or cache sufficient food to live through winter [43].

When a forest patch is destroyed through harvesting or wildfire, the middens and cone-bearing trees are destroyed along with it and the area becomes uninhabitable for tree squirrels. As the forest regrows, either naturally or through replanting, the trees are initially too small for cone production to occur, and it is some years before the first cone harvest is produced and the forest can once again begin to support a squirrel population [2,44]. Additional years of squirrel occupancy are needed for middens to be rebuilt. Thus, the ongoing turnover of forest stands (by harvesting, or by wildfire), creates a shifting mosaic of habitat patches, ranging in quality from suitable to completely unsuitable.

This scenario of shifting, contiguous patches is different from the usual metapopulation paradigm of physically isolated populations connected via dispersal [45,22,27], though there has been some work on metapopulation persistence in landscapes where suitable habitat patches are temporary [46,73]. We were interested in examining the interaction dynamics between the squirrel population and coniferous forest. In particular, we sought to determine when harvested patches would cease to be population sinks and become population sources, and how this time compares with forest maturity as measured by tree age.

3.2. *The Patch-Scale Movement Model*

The model we used in this work is a system of ODEs following the straightforward framework

$$\boxed{\begin{array}{c}\text{change in}\\ \text{adult}\\ \text{squirrel}\\ \text{population}\\ \text{in patch } i\end{array}} = \boxed{\begin{array}{c}\text{local}\\ \text{settlers}\\ \text{in patch } i\end{array}} + \boxed{\begin{array}{c}\text{long-range}\\ \text{settlers}\\ \text{in patch } i\end{array}} + \boxed{\begin{array}{c}\text{deaths}\\ \text{in patch } i\end{array}}. \qquad (3)$$

Squirrels are not born as adults, but as juveniles who become adults when they claim a territory of their own. The number of juveniles in each patch is $\rho\ u$ where u is the number of adults in the patch. Growth of the adult population occurs through settlement of juveniles via dispersal (Figure 5). Juvenile dispersal is divided into local and long-range settlement. Here, instead of modelling the dispersal trajectories of individuals, as was done in the individual-based codling moth model, we simply assume that a fixed fraction β of dispersers will seek vacant territories locally in the current patch, while the remaining $(1-\beta)$ fraction of dispersers will seek vacant territories long-distance in other patches. This assumption is consistent with the view that squirrels are genetically "programmed" to search locally for a vacant territory, but a few others are genetically inclined to take off and search long-distance.

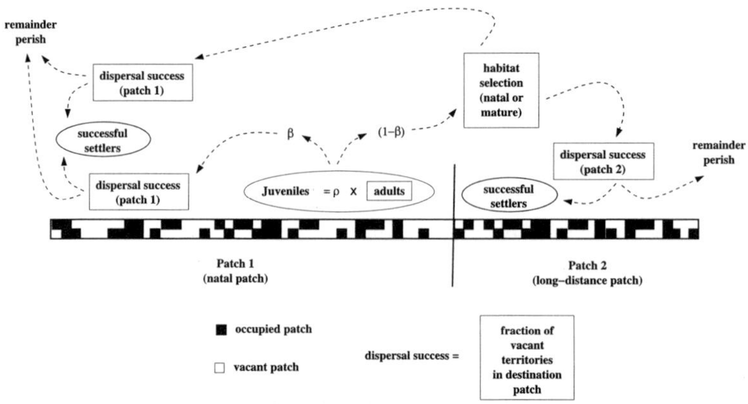

Figure 5. Schematic indicating how juveniles are redistributed across the landscape through dispersal. The algorithm works for multiple patches. For simplicity, we show only two. Juveniles in a patch are split into two populations: one that disperses locally and one that disperses long-distance (to another patch). Within these two subgroups, some dispersers are successful in finding a vacant territory, while the remainder perish. The dispersal success in a given patch is a function of the fraction of vacant territories in that patch, as well as the habitat quality there. See the text for details.

The success of long-distance and local dispersal events depends upon the fraction of vacant territories in the destination patch and the squirrels' bias towards the absolute or relative habitat quality there (see Figure 5). Dispersal success is an increasing function of both of these two factors (vacancy rate and habitat bias).

We investigated two possible habitat selection functions describing habitat bias: Juveniles preferred either high-quality habitat patches, or habitat patches with the same quality as their natal patch. In cases where the natal patch was of inferior quality, this second selection function meant that juveniles preferred lower-quality patches. While this may not be true, it appears that juveniles from low quality patches in proximity to high quality patches tend to settle in patches of similar quality to their natal patch, even though they spend time exploring the higher-quality patches [28]. The fraction of vacant territories in each patch depends on the population currently inhabiting the patch, and the habitat quality (which determines the carrying capacity). The distance between the natal and destination patches did not affect resettlement succes in the model, and there is some evidence to suggest that this assumption may be reasonable [34].

After having successfully located and claimed a vacant territory, juvenile squirrels must survive the winter. Survival is highest in mature forest patches, and lowest in young second-growth patches. The growth of the adult population thus depends on successful squirrel dispersal, and survival through the first winter.

3.3. *The Effect of Movement Behaviour on the Global Population*

Our model showed that as a harvested patch regrows, it acts as a population sink for the neighbouring patches. This sink behaviour occurs through the unequal balance between the movement of juveniles from the two patches. While the same fraction $(1 - \beta)$ of squirrels in each patch disperses, the total number dispersing from each mature forest patch is much greater than the total number dispersing from the inferior quality second-growth patch. In addition, the squirrels dispersing from the mature forest to the second-growth patch experience very low survival rates in the second-growth patch while the habitat quiality is recovering. Thus, the population in the second-growth patch provides little or no input to the mature forest population, which must at the same time support the growth of a new population in the second-growth patch.

A young second-growth patch, with a large fraction of vacant territories, represents additional forest where juveniles can obtain territories. In this case, the habitat selection function predicts that a pool of long-range settlers develops. In a two-patch system, with the parameters we used in our study [56], the fraction of juveniles settling in the mature forest patch gradually dropped from 1 to 1/2 as the second-growth patch regrew. This put pressure on the populations within each of the mature patches which then had a smaller pool of juveniles from which to maintain their numbers. At the same time, the resettlement success of juveniles in the second-growth patch was less than in the mature forest patches, and so for a time juveniles were attempting to settle in the second-growth patch rather than the mature forest patch and were perishing in the process.

For rapid recolonization of the second-growth patch, we found that it was important that $(1 - \beta)$ be large, so that there was a sufficient number of juveniles dispersing to the regrowing patch to nucleate a successful population there. At the same time, β needed to be sufficiently large to maintain the population in the source patches. Indeed, if the local dispersal fraction, β, was decreased, we found that the population in the second growth patch increased at the expense of the mature forest patch. At the same time, the second growth patch remained a sink due to the low survival probability there while the trees and middens were still immature. Thus, we drew the conclusion that inter-patch movement is necessary, but too much is detrimental to the overall population size on the landscape as a whole.

We found that the dispersal bias, toward patches of habitat similar in quality to the natal patch or toward high quality habitat, had little effect on the squirrel populations in either the mature or second-growth forest patches. On the other hand, the geometry of the managed forest, that is, the connectivity between mature and second-growth patches, did have an important effect on the recolonization process. Forest patches with low connectivity (such as the end patches in a linear arrangement) took longer to recolonize than more highly connected patches and had lower steady-state population levels.

4. Movement Behaviour in a Diffusion-Based Model

The third modelling framework we consider is that of diffusion-advection equations. This modelling approach originated in physics, where it works very well to describe the spread of inorganic quantities (such as, for example, molecules or particulate pollutants [75]). The approach has also been extremely useful for describing movement of organisms that range from the microscopic, such as bacteria [79], to the very large, such as wolves [49]. Many examples can be found in the books by [53] and [58].

In this framework, individual movement choices are viewed at the population level, where the deterministic, average behaviour can be observed.

If individuals simply move to the left or right with equal probability, and discrete steps in space and time scale appropriately, at the movement at the population level is described by Fickian diffusion

$$\frac{\partial u}{\partial t} = D\nabla u, \tag{4}$$

where u is the diffusing population, t is time, ∇ denotes the second partial derivative with respect to space and D is the constant diffusion coefficient. Other movement patterns at the individual level lead to modifications of the diffusion equation [30,48]. Simple examples are movement left and right with a bias towards one of the two directions, or including no movement as an option during a single timestep. The possible range of movement rules is limited only by the creativity of the modeler.

The project I present here is one that was motivated by questions surrounding the spread of transgenes via insect-mediated pollen dispersal. We are chiefly interested in the dispersal of pollen, but the pollen grains are carried by bees, largely honeybees. Thus, in order to understand the spread of transgenic pollen, we must first understand honeybee movement during foraging.

4.1. *The Biological Context: Dispersal of Apple Pollen*

Land planted with genetically modified crops has increased rapidly in recent years [16]. Simultaneously, widespread concern about the unintended spread of transgenes remains a public policy issue [65,8,51], with some agricultural regions lobbying for GMO-free status [23,50,7]. Plant pollen is dispersed either by wind or insects, with bees being the most important insect pollinator [38]. Most of the existing research on pollen gene flow from transgenic crops has been carried out with oilseed rape [32], which is both wind and insect-pollinated, and with exclusively wind-pollinated plants such as corn [3]. Nonetheless, crops that rely entirely on insect pollination have recently seen the development of GM varieties [82], and so models predicting bee-mediated outcrossing rates between transgenic and conventional plantings are needed [67]. We were thus interested in developing a mechanistic model for bee-mediated pollen movement, that could eventually be used to determine the propensity for transgene flow from transgenic to conventional plantings, or determine appropriate isolation distances between them.

4.2. *Diffusion-Based Movement Model*

Bee behaviour is complex and involves direct communication between conspecifics, return trips to the hive, and complex responses to the environment [6,60,51]. We opted for a simple diffusion-based description of bee movement. Traditionally, a reaction-diffusion model for dispersal of a population is a single equation of the following form [26]:

$$\boxed{\begin{array}{c}\text{change in}\\ \text{population}\\ \text{density at } (\vec{x}, t)\end{array}} = \boxed{\begin{array}{c}\text{spatial}\\ \text{movement}\\ \text{of}\\ \text{organisms}\end{array}} + \boxed{\begin{array}{c}\text{reaction terms}\\ \text{(growth, death,}\\ \text{emigration and}\\ \text{immigration)}\end{array}}. \tag{5}$$

The "spatial movement" box is actually a sum of terms, each one describing a different aspect of organism motion. Organism movement can be described as a composition of several different movement components, that might include, for example, diffusion (random motion), advection (directed motion in some vector flow field), chemotaxis (random motion biased up a chemical gradient) and central-place foraging (return trips to a central location such as a nest or hive). In this case, the "spatial movement term" would be written as the linear sum

$$\boxed{\begin{array}{c}\text{spatial}\\ \text{movement}\\ \text{of}\\ \text{organisms}\end{array}} = \boxed{\text{diffusion}} + \boxed{\text{advection}} + \boxed{\text{chemotaxis}} + \boxed{\begin{array}{c}\text{central-}\\ \text{place}\\ \text{foraging}\end{array}}. \tag{6}$$

The notion that all of these components operate simultaneously and can be combined in a linear sum has worked extremely well for organisms such as bacteria [79] and even larger insects [76], and has been the acceptable approach in mathematical biology for the better part of a century [58,53,69]. We found however, that the model solutions could not be made to match the bee dispersal data we had at our disposal [80]. Using the traditional formulation, we could match either the central part of the dispersal data, or the tails, but not both. Since we were interested in determining the spread of transgenes, it was important that we be able to capture the shape of the entire distribution, and the traditional approach was clearly inappropriate.

Motile organisms in search of resource, are faced with heterogeneous environments in which resource is patchily distributed [41]. Furthermore, within each patch, resource levels will vary spatially, depending on a number of factors including time since a previous forager visited the patch [24]. Consequently, foragers must constantly evaluate their environment, and decide whether they should continue foraging in a given patch, or relocate to a different patch altogether. If we consider the movement of individual foragers, rather than the entire foraging population, we find that the patchy distribution of resource results in two main types of movement patterns: foragers exhibit distinct periods of movement that are mainly diffusive (intensive search mode - ISM), followed by movement that is mainly advective (extensive search mode - ESM) [17]. For example, foraging bees will move diffusively from flower to neighbouring flower as they gather nectar, but occasionally will tire of a patch and then switch to a long-distance mode of flight in search of another flower patch [85]. These two movement modes are illustrated in Figure 6. Other examples of bimodal movement behaviour include leatherback turtles [35], eagle owls [11] and ungulates [18,33]. The rate at which organisms switch between the two different search modes, and the cues they use to inform this switching, vary widely across species and have a strong effect on the population distribution [41,14,17]. We identified ISM movement with diffusion, and ESM movement with advection, and were thus motivated to separate the diffusion and advection portions of the movement model (5), arriving at the "harvester-scout" model [80]

$$\boxed{\begin{array}{c}\text{change in}\\ \text{harvester}\\ \text{population}\\ \text{density at } (\vec{x},t)\end{array}} = \boxed{\text{diffusion}} - \boxed{\begin{array}{c}\text{harvesters}\\ \text{switching}\\ \text{to scouting}\end{array}} + \boxed{\begin{array}{c}\text{scouts}\\ \text{switching}\\ \text{to harvesting}\end{array}} \tag{7}$$

$$\boxed{\begin{array}{c}\text{change in}\\ \text{scout}\\ \text{population}\\ \text{density at } (\vec{x},t)\end{array}} = \boxed{\text{advection}} - \boxed{\begin{array}{c}\text{scouts}\\ \text{switching}\\ \text{to harvesting}\end{array}} + \boxed{\begin{array}{c}\text{harvesters}\\ \text{switching}\\ \text{to scouting}\end{array}} \tag{8}$$

where "harvesters" are those bees engaged in diffusive ISM movement, and "scouts" are those bees engaged in advective ESM movement between resource patches.

Since apple pollen is moved from one flower to another exclusively by bees, pollen movement is forcibly the same as bee movement. There is thus diffusing pollen P_d attached to harvesting bees, and advecting pollen P_a attached to scouting bees. The diffusing and advecting pollen are part of the "motile pollen" pool. New motile pollen is produced throughout the flowering period as new flowers open at some rate $\alpha\ F(x)$. This pollen is loosely attached to anthers, and is easily transferred to the surface of any bees that visit the flowers. This pollen is then transported from the original flower to others, courtesy of the bees, and eventually sticks to the stamen of a subsequent flower. At this point, the pollen is no longer part of the motile pool. There is thus a third pollen population, P_s, which

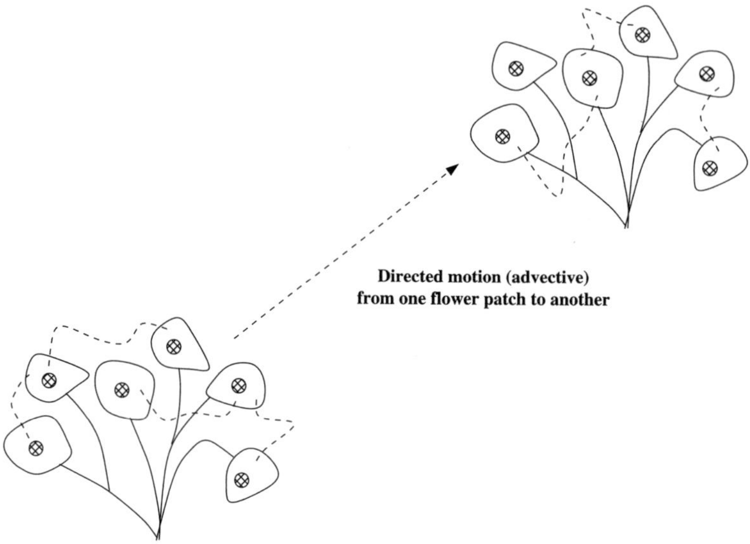

Figure 6. Movement of foraging bees is separated into movement that is chiefly diffusive, when the bee is moving from flower to flower within a patch, and movement that is chiefly advective, when the bee is moving in a directed path from one flower patch to another.

we called "stationary pollen". With these assumptions, we arrive at the mathematical equations for pollen movement:

$$\frac{\partial P_d(\vec{x},t)}{\partial t} = \overbrace{D\nabla^2 P_d}^{\text{diffusion}} + \overbrace{\alpha F(\vec{x})}^{\text{production}} - \overbrace{\beta P_d}^{\text{deposition}} - \overbrace{\gamma(P_d - P_a)}^{\text{movement mode switching}}, \tag{9a}$$

$$\frac{\partial P_{a\theta}(\vec{x},\theta,t)}{\partial t} = \overbrace{-v(\cos(\theta),\sin(\theta))\cdot\left(\frac{\partial P_{a\theta}}{\partial x},\frac{\partial P_{a\theta}}{\partial y}\right)}^{\text{advection in direction }\theta} + \overbrace{\gamma\left(\frac{P_d}{2\pi} - P_{a\theta}\right)}^{\text{movement mode switching}} \tag{9b}$$

$$\frac{\partial P_a(\vec{x},t)}{\partial t} = \overbrace{\int_0^{2\pi}\frac{\partial P_{a\theta}(\vec{x},\theta,t)}{\partial t}d\theta}^{\text{advection in all directions }\theta}, \tag{9c}$$

$$\frac{\partial P_s(\vec{x},t)}{\partial t} = \overbrace{\beta P_s}^{\text{deposition}}, \tag{9d}$$

where the parameters D and v are, respectively, the isotropic diffusion rate and the advection speed. The advection direction is given by θ. The resulting stationary pollen distribution $P_s(\vec{x}, t)$ determines the distribution of pollen from a source of flowers arranged as described by $F(\vec{x})$. We use the term "pollen shadow" to describe this distribution.

It is chiefly the distribution of transgenic seed, rather than pollen, that interests us. We must therefore determine how the transgenic pollen shadow relates to the distribution of transgenic seed. For simplicity, we assume that the ratio of transgenic to conventional seed produced at location $\vec{x}$ is equal to the ratio of stationary transgenic to conventional pollen at $\vec{x}$. Since only half of the pollen from the transgenic source carries the transgene, the amount of transgenic pollen deposited at $\vec{x}$ is half the total amount of pollen from the transgenic source. Mathematically, we have

$$\Phi_T(\vec{x}) = \frac{\frac{1}{2} P_{Ts}(\vec{x})}{P_{Ts}(\vec{x}) + P_{Ns}(\vec{x})} \tag{10}$$

where P_{Ts} and P_{Ns} are, respectively, the density of transgenic and non-transgenic viable stationary pollen at $\vec{x}$, and $\Phi_T(\vec{x})$ is the fraction of transgenic seed produced at $\vec{x}$. The spatial distributions $P_{Ts}(\vec{x})$ and $P_{Ns}(\vec{x})$ are obtained by solving equations (9). We call the distribution $\Phi_T(\vec{x})$ the transgenic "seed shadow".

4.3. *The Effect of Movement on Transgene Spread*

Using AIC, we showed that our harvester-scout model (8), in which ISM and ESM foraging behaviour are modeled separately, can provide a much better fit to real distribution data than the traditional diffusion or diffusion-advection models such as (5). Furthermore, this improved fit was obtained with only a slight increase in complexity: Diffusion and advection coefficients may be computed in the usual way, and only two new parameters, the behaviour switching rates, need to be determined. Using the harvester-scout model as the paradigm for bee movement, we derived a pollen dispersal model that gave seed shadows consistent with field data [81]

Our analytical work [80] indicated that the ISM-ESM framework is especially important for organisms where the advection speed, v, is much greater than the diffusion rate, D, or where time spent in the ESM movement mode is relatively large. In these cases, the forager distribution has long tails that decay much more slowly than the tails of the diffusion and diffusion-advection models. In fact, it can be rigorously shown that leptokurtic tails (fatter than gaussian) can be obtained from the harvester-scout model [70].

Such "fat-tailed" distributions have been shown to be critical in determining the spread rate of invasive species [69,40], or range expansion in response to climate change [9]. In retrospect, one can understand that rare long-distance dispersal events at the leading edge of an invasion can have an important effect on the spread rate of the invasion. That this process is also important at the scale of a single orchard however, was unexpected. Honeybees are rapid flyers, and are brought to orchards in sufficient numbers to ensure pollination of flowers throughout the orchard at a high density. Nonetheless, we found that it was absolutely necessary to separate ESM and ISM movement in order to match the pollen dispersal data at our disposal [81].

5. Discussion

In this paper, I focussed on the role played by movement paradigms in shaping the population processes observed. In each case, I found that the movement pattern was as important as other factors in shaping the results. That movement algorithms are influential in spread geometries, recolonization processes and invasion rates is self-evident. The extent to which local movement patterns affect and shape other processes not obviously connected with movement is a less intuitive result. One need only look at the extensive history of ordinary differential equation (ODE) models in mathematical biology [53,13] to see that it took some time before the importance of including movement in models was appreciated. Even then, a common paradigm was to keep reaction terms, those inherited from ODEs, the same and simply add movement terms. In diffusion-based modelling, this approach led to "reaction-diffusion" models [53]. In patch modelling, this meant that population dynamics were modeled with previously-studied ODEs within each patch, and then some fraction of the population in each patch was allowed to disperse to other patches [47,1]. This modelling progression is natural: each new feature is added onto existing ones that are understood.

Now, however, researchers are beginning to notice that those well-studied ODE functions are actually at least partly determined by individual movement patterns [48]. In addition, the stunning rise in easily-accessible computational power has made it possible for researchers to explore the population-level consequences of an enormous variety of movement algorithms. Modellers can thus explore the connection between detailed movement decisions and population dynamics at all levels from the IBM to the diffusion-based system. When we furthermore take the time to confront our models with data, and pay attention to the rare as well as the common events, we open ourselves to the opportunity to learn something new about movement and how it drives other population dynamics not traditionally associated with movement.

In the models presented here, movement had a fundamental effect on population distributions, as expected, but also on the population dynamics or the basic structure of the model equations. In the individual-based codling moth model, the individual movement algorithm was detailed and depended both on the state of the agents and the state of the local environment. When we simply ran this movement algorithm, we found that mating rates were strongly affected by behaviours obviously connected with mating (as in, female mate selectivity), and just as strongly affected by movement behaviours. In the patch-based squirrel model, the fraction of settlers at each point in time depended on environmental factors rather than simply being a fixed fraction of the juveniles in each patch. This more complex movement algorithm led to long-distance dispersal rates that varied with not just the population size in each patch but also with the habitat quality in neighbouring patches. In the diffusion-based pollen dispersal problem, we found that a traditional advection-diffusion model could not match both the centre and tails of the pollen distribution data, even on so small a scale as a single orchard. We were thus led to re-think individual movement behaviours and propose a new partial differential equation model.

Taken together, these observations suggest that a problem that initially does not present itself as being about movement, may in fact be fundamentally about the movement algorithms of individuals. Indeed, the large-scale consequences of small differences in movement decisions at the individual level is also receiving notice by ecologists [31,66]. It is thus worth thinking carefully about the details of the real and model movement algorithms, and to be cognizant of the possibility that the simplifications of movement processes in the model may be having a larger effect than expected on some of the results. This view of movement opens up a host of new ideas for research. There are innumerable ways in which one can think about movement, and just as many ways to assemble static and shifing heterogeneous environments. Considerable work is needed to develop a broad-based understanding of the way movement decisions interact with the environment to create patterns in dispersal and population dynamics.

6. Acknowledgements

The author would like to thank her writing group, Adriana Dawes, Elissa Schwartz and Christina Cobbold for candid and constructive feedback on the manuscript. Steve Krone's insights and pointed questions were invaluable throughout the writing process. The support of NSERC through the Discovery Grant Program is also gratefully acknowledged, as well as the support of UBC.

References

1. P. Amarasekare and R.M. Nisbet. Spatial heterogeneity, source-sink dynamics, and the local coexistence of competing species. *American Naturalist*, 158(6):572–584, 2001.
2. E. Bayne and K. Hobson. Relative use of contiguous and fragmented boreal forest by red squirrels (*tamiasciurus hudsonicus*). *Canadian Journal of Zoology*, 78(3):359–365, 2000.
3. H.J. Beckie, K.N. Harker, L.M. Hall, S.I. Warwick, A. Legere, P.H. Sikkema, J.Y. Leeson, G. Seguin-Swartz, and M.J. Simard. A decade of herbicide-resistant crops in canada. *Canadian Journal of Plant Science*, 86(4):1243–1264, 2006.
4. H.C. Berg. *Random Walks in Biology.* Princeton University Press, Princeton, NJ, 1993.
5. J.F. Brunner. Orchard pest management online: Codling moth. Washington State University Tree Fruit Research and Extension Center, 1993. accessed: Sep 17, 2013.
6. J.G. Burns and J.D. Thomson. A test of spatial memory and movement patterns of bumblebees at multiple spatial and temporal scales. *Behavioural Ecology*, 17(1):48–55, 2005.
7. K. Ceglarz. WATCH: Concerned residents demand G-M-O free zone. *Global News, British Columbia*, September 13, 2013.
8. S. Chandler and J. Dunwell. Gene flow, risk assessment and the environmental release of transgenic plants. *Critical Reviews in Plant Sciences*, 27:25–49, 2008.
9. J.S. Clark, M. Lewis, J.S. McLachlan, and J. HilleRisLambers. Estimating population spread: What can we forecast and how well? *Ecology*, 84(8):1979–1988, 2003.
10. A.T. Dawes and D. Iron. Cortical geometry may influence placement of interface between par protein domains in early caenorhabditis elegans embryos. *Journal of Theoretical Biology*, 2103.
11. M. del Mar Delgado and V. Penteriani. Behavioural states help translate dispersal movements into spatial distribution patterns of floaters. *The American Naturalist*, 172(4):475–485, 2008.
12. V.A. Dyck and M.G.T. Gardiner. Sterile-insect release programme to control the codling moth *cydia pomonella* (l.)(lepidoptera:olethreutidae) in british

columbia, canada. *Acta Phytopathologica et Entomologica Hungarica*, 27:219–222, 1992.

13. L. Edelstein-Keshet. *Mathematical Models in Biology.* SIAM, New York, 2005.
14. K.D. Farnsworth and J.A. Beecham. How do grazers achieve their distribution? A continuum of models from random diffusion to the ideal free distribution using biased random walks. *American Naturalist*, 153(5):509–526, 1999.
15. K.R. Fister, M.L. McCarthy, S.F. Oppenheimer, and C. Collins. Optimal control of insects through sterile insect release and habitat modification. *Mathematical Biosciences*, 244(2):201–212, 2013.
16. Council for Agricultural Science and Technology. Implications of gene flow in the scale-up and commercial use of biotechnology-derived crops: Economic and policy considerations. Issue Paper 37, CAST, Ames, Iowa, December 2007.
17. D. Fortin, J.M. Fryxell, and R. Pilote. The temporal scale of foraging decisions in bison. *Ecology*, 83(4):970–982, 2002.
18. J.M. Fryxell, M. Hazell, L. Börger, B.D. Dalziel, D.T. Haydon, J.M Morales, T. McIntosh, and R.C. Rosatta. Multiple movement modes by large herbivores at multiple spatiotemporal scales. *Proceedings of the National Academy of Sciences*, 105(49):19114–19119, 2008.
19. G.F. Fussmann and B. Blasius. Community response to enrichment is highly sensitive to model structure. *Biology Letters*, 1(1):9–12, 2005.
20. A. Gabel, S.N. Majumdar, N.K. Panduranga, and S. Redner. Can a lamb reach a haven before being eaten by diffusing lions? *Journal of Statistical Mechanics*, P05011:arXiv:1203.2985, 2012.
21. P.W. Geier. The codling moth, *cydia pomonella (l.)*: profile of a key pest. In R.L. Kitching and R.E. Jones, editors, *The Ecology of Pests*, pages 108–129, Melbourne, Australia, 1981. CSIRO.
22. M. Gilpin and I. Hanski. Metapopulation dynamcis: Brief history and conceptual domain. *Biological Journal of the Linnean Society*, 42(1–2):3–16, 1991.
23. GMO-free-Europe. 7th European Conference of GMO-free Regions, Sep 2012.
24. D Goulson. Foraging strategies of insects for gathering nectar and pollen, and implications for plant ecology and evolution. *Perspectives in Plant Ecology, Evolution and Systematics*, 2(2):185–209, 1999.
25. V. Grimm and S.F. Railsback. *Individual-based Modeling and Ecology.* Princeton University Press, Princeton, NJ, 2005.
26. D. Grünbaum. Using spatially explicit models to characterize foraging performance in heterogeneous landscapes. *American Naturalist*, 151(2):97–115, 1998.
27. I. Hanski and M. Gilpin. *Metapopulation Biology: Ecology, genetics and evolution.* Academic Press, London, 1997.
28. D.L. Haughland and K.W. Larsen. Exploration correlates with settlement: Red squirrel dispersal in contrasting habitats. *Ecology*, 73(6):1024–1034, 2004.
29. K.J. Haynes and J.T. Cronin. Interpatch movement and edge effects: The

role of behavioral responses to the landscape matrix. *Oikos*, 113(1):43–54, 2006.

30. T. Hillen and K. Painter. Transport models for movement in oriented habitats and anisotropic diffusion. In M.A. Lewis, P. Maini, and S. Petrovskii, editors, *Dispersal, individual movement and spatial ecology: A mathematical perspective*. Springer, Heidelberg, 2012.
31. B.T. Hirsch, M.P. Tujague, Y.E. Di Blanco, M.S. Di Bitetti, and C.H. Janson. Comparing capuchins and coatis: Causes and consequences of differing movement ecology in two sympatric mammals. *Animal Behaviour*, 86(2):331–338, 2013.
32. A. Hüsken and A. Dietz-Pfeilstetter. Pollen-mediated intraspecific gene flow from herbicide resistant oilseed rape (Brassica napus L.). *Transgenic Research*, 16:557–569, 2007.
33. C. Johnson, K. Parker, D. Heard, and M. Gillingham. Unrealistic animal movement rates as behavioural bouts: A reply. *Journal of Animal Ecology*, 75:303–308, 2006.
34. W.T. Jones. Dispersal distance and the range of nightly movements in Merriam's kangaroo rats. *Journal of Mammology*, 70(1):27–34, 1989.
35. I.D. Jonsen, R.A. Myers, and M.C. James. Identifying leatherback turtle foraging behaviour from satellite telemetry using a switching state-space model. *Marine Ecology Progress Series*, 337:255–264, 2007.
36. G.J.R. Judd, M.G.T. Gardiner, and H.M.A. Thistlewood. Seasonal variation in recapture of mass-reared sterile codling moth, *cydia pomonella* (l.)(lepidoptera:tortricidae): implications for control by sterile insect technique in british columbia. *Journal of the Entomological Society of British Columbia*, 101:29–44, 2004.
37. Y. Kang and N. Lanchier. Expansion or extinction: Deterministic and stochastic two-patch models with allee effects. *Journal of Mathematical Biology*, 62(6):925–973, 2011.
38. A.-M. Klein, B.E. Vaissiere, J.H. Cane, I. Steffan-Dewenter, S.A. Cunningham, C. Kremen, and T. Tscharntke. Importance of pollinators in changing landscapes for world crops. *Proceedings of the Royal Society B-Biological Sciences*, 274(1608):303–313, 2007.
39. E.F. Knipling. The basic principles of insect suppression and management. Agriculture Handbook 512, United States Department of Agriculture, Science and Education Administration, Washington, D.C., 1979.
40. M. Kot, M.A. Lewis, and P. van den Driessche. Dispersal data and the spread of invading organisms. *Ecology*, 77(7):2027–2042, 1996.
41. D.C. Krakauer and M.A. Rodriguez-Gironès. Searching and learning in a random environment. *Journal of Theoretical Biology*, 177:417–419, 1994.
42. S.M. Krone. Spatial models: Stochastic and deterministic. *Mathematical and Computer Modelling*, 40(3–4):393–409, 2004.
43. K. Larsen and S. Boutin. Movements, survival, and settlement of red squirrel (*tamiasciurus hudsonicus*) offspring. *Ecology*, 75(1):214–223, 1994.
44. K.W. Larsen. Dispersal in a gradient of habitats: Activity by juvenile North American red squirrels (Tamiasciurus hudsonicus) in varying-age for-

est stands. *Ecoscience*, 16(1):75–83, 2009. Symposium on Natural Hazards and Natural Disturbances in Mountain Forest - Challenges and Opportunities for Silviculture, Trento, ITALY.

45. R. Levins. Extinction. In M. Gerstenhaber, editor, *Some Mathematiacal Problems in Biology. Symposium on Mathematical Biology*, pages 77–107, Providence, RI, 1970. American Mathematical Society.
46. C. Loehle. Effect of ephemeral stepping stones on metapopulations on fragmented landscapes. *Ecological Complexity*, 4(1–2):42–47, 2007.
47. M. Marva, J.C. Poggiale, and R.B. De La Parra. Reduction of slow-fast periodic systems with applications to population dynamics. *Mathematical Models and Methods in Applied Sciences*, 22(10), 2012.
48. H.W. McKenzie, E.H. Merrill, R.J. Spiteri, and M.A. Lewis. How linear features alter predator movement and the functional response. *Interface Focus*, 2(2):205–216, 2012.
49. P.R. Moorcroft and M.A. Lewis. *Mechanistic Home Range Analysis*. Princeton University Press, Woodstock, 2006.
50. O. Moore. PEI won't bite on GMO crops. *Globe and Mail*, April, 2009.
51. W.F. Morris, P.M Kareiva, and P.L. Raymer. Do barren zones and pollen traps reduce gene escape from transgenic crops? *Ecological Applications*, 4(1):157–165, 1994.
52. Thomas Mueller, William F Fagan, and Volker Grimm. Integrating individual search and navigation behaviors in mechanistic movement models. *Theoretical Ecology*, 4(3):341–355, May 2010.
53. J. D. Murray. *Mathematical Biology*. Springer-Verlag, New York, NY, 3 edition, 2000.
54. J.H. Myers, D. Simberloff, A.M. Kuris, and J.R. Carey. Eradication revisited: dealing with exotic species. *Tree*, 15(8):316–320, 2000.
55. K.L. Nash, N.A.J. Graham, F.A. Januchowski-Hartley, and D.R. Bellwood. Influence of habitat condition and competition on foraging behaviour of parrotfishes. *Marine Ecology Progress Series*, 457:113–124, 2012.
56. B. Nyquist, R. Tyson, and K. Larsen. Modelling recolonization of second-growth forest stands by the North American red squirrel, Tamiasciurus hudsonicus. *Bulletin of Mathematical Biology*, 69:1311–1339, 2007.
57. M.E. Obbard. *Red Squirrel*, pages 265–281. Ministry of Natural Resources, Ontario, Canada, 1987.
58. Akira Okubo and Simon A. Levin. *Diffusion and Ecological Problems: Modern Perspectives*. Springer, New York, NY, 2001.
59. J.L. Osborne, S.J. Clark, R.J. Morris, I.H. Williams, J.R. Riley, A.D. Smith, D.R. Reynolds, and A.S. Edwards. A landscape-scale study of bumble bee foraging range and constancy, using harmonic radar. *Journal of Applied Ecology*, 36:519–533, 1999.
60. J.L. Osborne, A.P. Martin, N.L. Carreck, J.L. Swain, M.E. Knight, D. Goulson, R.J. Hale, and R.A. Sanderson. Bumblebee flight distances in relation to the forage landscape. *Journal of Animal Ecology*, 77:406–415, 2001.
61. O. Ovaskainen and I. Hanski. From individual behavior to metapopulation dynamics: Unifying the patchy population and classic metapopulation mod-

els. *American Naturalist*, 164(3):364–377, 2004.
62. N. Perony, C.J. Tessone, B. Koenig, and F. Schweitzer. How random is social behaviour? Disentangling social complexity through the study of a wild house mouse population. *PLOS Computational Biology*, 8(11), 2013.
63. GH Pyke. Optimal Foraging Theory: A Critical Review. *Annual review of ecology and systematics*, 1984.
64. E. Revilla and T. Wiegand. Individual movement behavior, matrix heterogeneity, and the dynamics of spatially structured populations. *Proceedings of the National Academy of Sciences*, 105(49):19120–19125, 2008.
65. M.A. Rieger, M. Lamond, C. Preston, S.B. Powles, and R.T. Roush. Pollen-mediated movement of herbicide resistance between commercial canola fields. *Science*, 296:2386–2388, 2002.
66. P. Romanczuk and L. Schimansky-Geier. Mean-field theory of collective motion due to velocity alignment. *Ecological Complexity*, 10(SI):83–92, 2012.
67. J. Rong, Z. Song, T.J. de Jong, X. Zhang, S. Sun, X. Xu, H. Xial, B. Liu, and B.-R. Lu. Modelling pollen-mediated gene flow in rice: risk assessment and management of transgene escape. *Plant Biotechnology Journal*, 8(4):452464, 2010.
68. E.J. Schwartz, E.N. Bodine, and S. Blower. Effectiveness and efficiency of imperfect therapeutic HSV-2 vaccines. *Human Vaccines*, 3(6):231–238, 2007.
69. N. Shigesada and K. Kawasaki. *Biological Invasions: Theory and Practice.* Oxford University Press, Oxford, 1997.
70. G.T. Skalski and James F. Gilliam. A diffusion-based theory of organism dispersal in heterogeneous populations. *The American Naturalist*, 161(3):441–458, 2003.
71. E. Sklar. Netlogo, a multi-agent simulation environment. *Artificial Life*, 13(3):303–311, 2007.
72. P.E. Smaldino, C.L. Pickett, J.W. Sherman, and J.C. Schank. An agent-based model of social identity dynamics. *Journal of Artificial Societies and Social Simulation*, 15(4):7, 2012.
73. T. Snall, J Pennanen, L. Kivisto, and I. Hanski. Modelling epiphyte metapopulation dynamics in a dynamic forest landscape. *Oikos*, 109(2):209–222, 2005.
74. DW Stephens, JS Brown, and RC Ydenberg. *Foraging: behavior and ecology.* University of Chicago Press, Chicago, 2007.
75. John M. Stockie. The Mathematics of Atmospheric Dispersion Modeling. *SIAM REVIEW*, 53(2):349–372, 2011.
76. S. Strohm, R.C. Tyson, and J.A. Powell. Pattern formation in a model for mountain pine beetle dispersal: Linking model predictions to data. *Bulletin of Mathematical Biology*, 2013.
77. H.M.A. Thistlewood, G.J.R. Judd, and M.E.O. Clodius. Sustainable management and nomitoring of codling moth, *cydia pomonella*, in an area-wide program employing sit. In International Atomic Energy Agency, editor, *Improvement of Codling Moth SIT to Facilitate Expansion of Field Application, Stellenbosch, South Africa, March 8–12, 2004*, Working Paper Series IAWA-314-D4-Rc876, pages 79–87. Atomic Energy Agency, Vienna, Austria, 2004.
78. P. Turchin. *Complex Population Dynamics: A Theoretical/Empirical Synthe-*

sis. Princeton University Press, Princeton NJ, 2003.

79. R. Tyson, S.R. Lubkin, and J.D. Murray. A minimal mechanism for bacterial pattern formation. *Proceedings of the Royal Society of London Series B: Biological Sciences*, 266(1416):299–304, 1999.
80. R.C. Tyson, J.B. Wilson, and W.D. Lane. Beyond diffusion: Modelling local and long-distance dispersal for organisms exhibiting intensive and extensive search modes. *Theoretical Population Biology*, 79(3):70–81, 2011.
81. R.C. Tyson, J.B. Wilson, and W.D. Lane. A mechanistic model to predict transgenic seed contamination in bee-pollinated crops validated in an apple orchard. *Ecological Modelling*, 222(13):2084–2092, 2011.
82. United States Department of Agriculture. Biotechnology: Petitions for determination of nonregulated status, Sep 2013.
83. D. Vanpeteghem and B. Haegeman. An analytical approach to spatio-temporal dynamics of neutral community models. *Journal of Mathematical Biology*, 61(3):323–357, 2010.
84. M.J.B. Vreysen, H.J. Barclay, and J. Hendrichs. Modeling of preferential mating in areawide control programs that integrate the release of strains of sterile males only or both sexes. *Annals of the Entomological Society of America*, 99(3):607–616, 2006.
85. K. Waddington. Flight patterns of foraging bees relative to density of artificial flowers and distribution of nectar. *Oecologia*, 44:199–204, 1980.
86. Henri Weimerskirch. Are seabirds foraging for unpredictable resources? *Deep Sea Research Part II: Topical Studies in Oceanography*, 54(3-4):211–223, February 2007.
87. S.M. White, P. Rohani, and S.M. Sait. Modelling pulsed releases for sterile insect techniques: fitness costs of sterile and transgenic males and the effects on mosquito dynamics. *Journal of Applied Ecology*, 47(6):1329–1339, 2010.

PLANKTON NUTRIENT INTERACTION MODEL WITH HARVESTING UNDER CONSTANT ENVIRONMENT

SAMARES PAL* AND ANAL CHATTERJEE+
Department of Mathematics,
University of Kalyani,
Kalyani- 741235, India
E-mail:samaresp@yahoo.co.in, chatterjeeanal172@gmail.com

This paper deals with a nutrient-phytoplankton-zooplankton ecosystem model consisting of dissolved limiting nutrient with nutrient uptake functions. We use a Holling type-II harvest function to model density dependent plankton population. It is assumed that phytoplankton release toxic chemical for self defense against their predators. The model system is studied analytically and the threshold conditions for the existence and stability of various steady states are worked out. It is observed that if the rate of toxin produced by phytoplankton population crosses a certain critical value, the system enters into Hopf bifurcation. We have derived the direction of Hopf-bifurcation. Our observations indicate that constant nutrient input and the maximal zooplankton conversion rate influence the nutrient-plankton ecosystem model and maintain stability around the coexistence equilibrium in the presence of toxic chemical release by phytoplankton for self defense. It is observed that harvesting rates of the plankton population play a vital role in changing the stability criteria. Computer simulations have been carried out to illustrate different analytical results.

Key words: Nutrient, Phytoplankton, Zooplankton, Toxin, Equilibria.

1. Introduction

Marine plankton ecosystems play an important roles in earth ecosystem. Phytoplankton species are an important food source for organisms in an aquatic environment. The phytoplankton occupies a central position in the food chain. The basic trophic levels in an ecological food chain comprise of (i) primary producers, and (ii) primary consumers. The primary producers, also known as phytoplankton, are capable of producing their organic food requirement from inorganic materials either be photosynthesis or by inorganic oxidation. The study of interactions between nutrients and food webs is a complex matter even without taking into account the additional effects of toxin in marine ecosystem.

* *Corresponding author:* Tel.:+91-33-25666571 ; e-mail: samaresp@yahoo.co.in (Samares Pal)

However, an understanding of the additional effects of toxin play a vital role in marine plankton ecology.

The role of competition in phytoplankton population for the occurrence and control of plankton bloom and effects of toxin on NPZ model have been studied in [1],[2], [3]. Recently the authors in [4] have observed that due to the toxicity present in the zooplankton biomass the system is adversely affected and hence the equilibrium level of zooplankton biomass goes down. In marine plankton ecosystem, the toxic phytoplankton do not drive the zooplankton population towards extinction [5]. Again this has been discussed in [6] that the growth rates of both plankton populations are diminished due to uptake of toxin by phytoplankton and zooplankton in two different models. But the occurrence of toxin producing phytoplankton may not be always harmful, rather be helpful to maintain stable equilibrium in trophodynamics through coexistence of all the species [7]. The author in [8] suggested that the functional response of the whole zooplankton population in mean-field models might be described by type III response provided the depth of the column is sufficiently large.

The study of population dynamics with harvesting is a subject of mathematical bioeconomics and is mainly concerned with the optimal management of renewable resources [9]. The most obvious way to increase the harvest of a culture species is to lower oxygen demand by reducing plankton biomass. Greater oxygen availability would permit higher stocking densities. Plankton could be harvested either mechanically or biologically. Mechanical harvest would involve pumping water through filters and collecting the plankton retained. Because the mesh or screen size determines the size of the particle harvested, filter screen selection and placement would be critical. Screen mesh must be small enough to retain the size of plankton desired but large enough to allow smaller plankton and particles to pass through unobstructed. Larger particles such as zooplankton must be removed before filtering smaller particles like phytoplankton and minute zooplankton. Otherwise, the small mesh screens for phytoplankton would become clogged rapidly by the large zooplankton, and filtering would be disrupted. While mechanical harvest of plankton may be technologically feasible, it is likely that economic obstacles and the current lack of markets for plankton products would make this approach impractical [10].

The species of zooplankton which are being harvested in the world have been studied by many researcher. Calanus finmarchicus and C. plumchrus are the most abundant copepods in the northern part of the North Atlantic and North Pacific Oceans, respectively. C. finmarchicus is collected mainly with large stationary nets set in sounds where tidal currents are of moderate strength. The catches are usually best during late evening and night when the plankton rises to the surface. Also C. plumchrus has been taken in the vicinity of the Fraser River estuary, on the west coast of Canada [11]. Recently phytoplankton-zooplankton model with harvesting is proposed and investigated for the existence of bionomic equilibria and the optimal harvesting policy [12].

A two-zooplankton one-phytoplankton model with harvesting has been investigated and studied the impact of harvesting on the coexistence and exclusion of competitive predators [13]. Several kinds of planktonic algae can be harvested from laboratory cultures by continuous-flow centrifugation [14]. Three different harvesting techniques, plankton net trawling, traveling screen, and screw pump filtration, were compared in terms of energy consumption and harvesting efficiency [15]. The author studied the net energy balance and economic benefits of harvesting detrimental environmental algal blooms and conversion of the harvested biomass into biofuels. A method for harvesting large quantities of zooplankton from hatchery ponds has been discussed in [16].

In this present research, we have taken Holling Type II and III function in embodying the effect of saturation. Here an open system is considered with three interacting components consisting of phytoplankton (P), zooplankton (Z) and dissolved limiting nutrient (N). In this paper a nutrient-phytoplankton-zooplankton model with the effect of toxin producing phytoplankton and harvesting rates of both plankton population is described. The stability of equilibrium point is described. We have also derived the conditions for instability of the system around the interior equilibrium and Hopf bifurcation. Numerical simulations under a set of parameter values have been performed to support our analytical results.

2. The mathematical model

Let $N(t)$ be the concentration of the nutrient at time t. Let P(t) and Z(t) be the concentration of phytoplankton and zooplankton population respectively at time t. Let N^0 be the constant input of nutrient concentration, and D be the dilution rate of nutrient [17]. The constant $\frac{1}{D}$ has the physical dimension of a time and represents the average time that nutrient and waste products spend in the system [18]. Let α_1 and α_2 be the nutrient uptake rate for the phytoplankton population and conversion rate of nutrient for the growth of phytoplankton population, respectively ($\alpha_2 \leq \alpha_1$). Let μ_1 be the mortality rate of the phytoplankton population and μ_2 be the mortality rate of the zooplankton population. Let μ_3 ($\mu_3 \leq \mu_1$) be the nutrient recycle rate after the death of phytoplankton population and μ_4 ($\mu_4 \leq \mu_2$) be the nutrient recycle rate after the death of zooplankton population. Let β_1 be the maximal zooplankton ingestion rate and β_2 ($\beta_2 \leq \beta_1$) be the maximal zooplankton conversion rate. We choose Holling type II and type III functional forms to describe the grazing phenomena with K_1 and K_2 as half saturation constant. Here harvesting is represented by a Holling type II function with h_1 and h_2 as the harvest rates (also known as catchability constant) of phytoplankton and zooplankton respectively. Here E_1 and E_2 are the measures of the effort required to harvest the phytoplankton and zooplankton population respectively. This type of harvest function implies that when fewer plankton are available, it is harder to find them and so the daily catch drops. On the other hand, when there are sufficiently many plankton, then $lim_{H_1\to\infty} \frac{h_1 H_1}{E_1+H_1} = h_1$ and $lim_{H_2\to\infty} \frac{h_2 H_2}{E_2+H_2} = h_2$, so that the harvesting level is close to h_1 and h_2 respectively. It is assumed θ is the rate of zooplankton decay due to toxin producing phytoplankton.

With these assumptions our model is given by :

$$\begin{aligned}
\frac{dN}{dt} &= D(N^0 - N) - \frac{\alpha_1 PN}{K_1 + N} + \mu_3 P + \mu_4 Z \equiv F_1(N,P,Z),\\
\frac{dP}{dt} &= \frac{\alpha_2 PN}{K_1 + N} - \frac{\beta_1 P^2 Z}{K_2 + P^2} - \mu_1 P - \frac{h_1 P}{E_1 + P} \equiv F_2(N,P,Z),\\
\frac{dZ}{dt} &= \frac{(\beta_2 - \theta)P^2 Z}{K_2 + P^2} - \mu_2 Z - \frac{h_2 Z}{E_2 + Z} \equiv F_3(N,P,Z). \qquad (1)
\end{aligned}$$

The system (1) will be analyzed with the following initial conditions,

$$N(0) > 0,\ P(0) > 0,\ Z(0) > 0. \qquad (2)$$

3. Some preliminary results

3.1. *Positive invariance*

By setting $X = (N,\ P,\ Z)^T \in \mathbf{R}^3$ and $F(X) = [F_1(X),\ F_2(X),\ F_3(X)]^T$, with $F : \mathbf{R_+}^3 \to \mathbf{R}^3$ and $F \in C^\infty(\mathbf{R}^3)$, equation (1) becomes

$$\dot{X} = F(X), \tag{3}$$

together with $X(0) = X_0 \in \mathbf{R_+}^3$. It is easy to check that whenever $X(0) \in \mathbf{R_+}^3$ with $X_i = 0$, for i=1, 2, 3, then $F_i(X) \mid_{X_i=0} \geq 0$. Then any solution of equation (3) with $X_0 \in \mathbf{R_+}^3$, say $X(t) = X(t; X_0)$, is such that $X(t) \in \mathbf{R_+}^3$ for all $t > 0$.

3.2. *Boundedness of the System*

Theorem 1. *All the solutions of (1) are ultimately bounded.*
Proof. We define a function

$$w = N + P + Z.$$

The time derivative of w along the solutions of (1) is

$$\frac{dw}{dt} = D(N^0 - N) - [\frac{\alpha_1}{K_1 + N} - \frac{\alpha_2}{K_1 + N}]NP - [\frac{\beta_1}{K_2 + P^2} - \frac{(\beta_2 - \theta)}{K_2 + P^2}]P^2 Z -$$

$$(\mu_1 - \mu_3)P - (\mu_2 - \mu_4)Z - \frac{h_1 P}{E_1 + P} - \frac{h_2 Z}{E_2 + Z},$$

and therefore $\frac{dw}{dt} \leq -D_0(N + P + Z) + DN^0$,
where

$$D_0 = \min\{D, \mu_1 - \mu_3, \mu_2 - \mu_4\}.$$

We can find a constant $m > 0$ such that

$$\frac{dw}{dt} + D_0 w \leq m. \tag{4}$$

Using the variation of constants formula, this inequality is transformed into

$$w(N(t), P(t), Z(t)) \leq \frac{m}{D_0}(1 - e^{-D_0 t}) + w(N(0), P(0), Z(0))e^{-D_0 t},$$

from which we get

$$(N + P + Z)(t) \leq \max\left[(N(0) + P(0) + Z(0)), \frac{m}{D_0}\right], \forall t \geq 0,$$

and for $t \to \infty$, we have

$$(N + P + Z)(t) \leq \frac{m}{D_0}.$$

Hence, all the solutions of (1) which initiate in R^3_+ are eventually confined in the region

$$B = \left\{(N, P, Z) \in R^3{}_+ : N + P + Z = \frac{m}{D_0} + \epsilon, \forall \epsilon > 0\right\}.$$

3.3. *Equilibria*

The system (1) possesses the following equilibria:
(i) The phytoplankton free equilibrium $E_0 = (N^0, 0, 0)$.
(ii) The zooplankton free equilibrium $E_{11} = (N_1, P_1, 0)$ with

$$P_1 = \frac{(h_1 - E_1\alpha_2 + E_1\mu_1)N_1 + (h_1 + E_1\mu_1)K_1}{(\alpha_2 - \mu_1)N_1 - K_1\mu_1},$$

and N_1 is given from cubic equation $aN_1^3 + bN_1^2 + cN_1 + d = 0$, where
$a = D(\alpha_2 - \mu_1)$,
$b = (\alpha_1 - \mu_3)(h_1 - E_1\alpha_2 + E_1\mu_1) - D(N^0 - K_1)(\alpha_2 - \mu_1) - DK_1\mu_1$,
$c = (\alpha_1 - \mu_3)(h_1 + E_1\mu_1)K_1 - \mu_3 K_1(h_1 - E_1\alpha_2 + E_1\mu_1) + D(N^0 - K_1)K_1\mu_1$
$- DK_1N^0(\alpha_2 - \mu_1)$,
$d = DK_1^2\mu_1 N^0 - \mu_3 K_1^2(h_1 + E_1\mu_1)$,
which exists if

$$\frac{DN^0\mu_1}{\mu_3} < (h_1 + E_1\mu_1) < \min\{\frac{D(N^0-K_1)(\alpha_2-\mu_1)+DK_1\mu_1+E_1\alpha_2(\alpha_1-\mu_3)}{\alpha_1-\mu_3},$$
$$\frac{\mu_3K_1(h_1-E_1\alpha_2+E_1\mu_1)+DN^0K_1(\alpha_2-\mu_1)-D(N^0-K_1)K_1\mu_1}{\alpha_1-\mu_3}\}.$$

(iii) The positive interior equilibrium $E^* = (N^*, P^*, Z^*)$
where $Z^* = \frac{(K_2+P^{*2})h_2}{(\beta_2-\theta-\mu_2)P^{*2}-K_2\mu_2} - E_2$, $\quad N^* = \frac{(\frac{\beta_1P^*Z^*}{K_2+P^{*2}}+\mu_1+\frac{h_1}{E_1+P^*})K_1}{\alpha_2-\frac{\beta_1P^*Z^*}{K_2+P^{*2}}-\mu_1-\frac{h_1}{E_1+P^*}}$.
Putting the value of N^* and Z^* in

$$D(N^0 - N^*) - \frac{\alpha_1 N^* P^*}{K_1 + N^*} + \mu_3 P^* + \mu_4 Z^* = 0,$$

we get the positive value of P^*.

Thus the condition for the existence of the interior equilibrium point $E^*(N^*, P^*, Z^*)$ is given by, $N^* > 0$, $P^* > 0$, $Z^* > 0$, and

$$h_2 > \frac{E_2[(\beta_2 - \theta - \mu_2)P^{*2} - K_2\mu_2]}{K_2 + P^{*2}} \tag{5}$$

and

$$\alpha_2 > \mu_1 + \frac{\beta_1 P^* Z^*}{K_2 + P^{*2}} + \frac{h_1}{E_1 + P^*}. \tag{6}$$

4. Stability analysis of the system (1)

Let $\overline{E} = (\overline{N}, \overline{P}, \overline{Z})$ be any arbitrary equilibrium. Then the variational matrix about $\overline{E}$ is given by

$$\overline{V} = \begin{bmatrix} -D - \frac{\alpha_1K_1\overline{P}}{(K_1+\overline{N})^2} & -\frac{\alpha_1\overline{N}}{K_1+\overline{N}} + \mu_3 & \mu_4 \\ \frac{K_1\alpha_2\overline{P}}{(K_1+\overline{N})^2} & \frac{\alpha_2\overline{N}}{K_1+\overline{N}} - \frac{2K_2\beta_1\overline{Z}\ \overline{P}}{(K_2+\overline{P}^2)^2} - \mu_1 - \frac{E_1h_1}{(E_1+\overline{P})^2} & -\frac{\beta_1\overline{P}^2}{K_2+\overline{P}^2} \\ 0 & \frac{2K_2(\beta_2-\theta)\overline{P}\ \overline{Z}}{(K_2+\overline{P}^2)^2} & \frac{(\beta_2-\theta)\overline{P}^2}{K_2+\overline{P}^2} - \mu_2 - \frac{E_2h_2}{(E_2+\overline{Z})^2} \end{bmatrix}.$$

4.1. *Criterion for the extinction of phytoplankton*

Theorem: 2. *Let the inequality*

$$R_0 = \frac{\alpha_2 N^0}{(\mu_1 + \frac{h_1}{E_1})(K_1 + N^0)} < 1 \tag{7}$$

hold. Then $\lim_{t\to\infty}(N(t), P(t), Z(t)) = E_0$.

Theorem (2) demonstrates that if the ratio of maximal nutrient conversion rate for the growth of phytoplankton and the loss rate of the phytoplankton is less than unity, then the phytoplankton population will eliminate from the system. The proof is obvious.

4.2. *Eigenvalue analysis and Hopf-bifurcation*

In this section, local stability analysis of the system around the biologically feasible equilibria is performed.

Lemma 1: *If* $R_0 > 1$, *then the plankton free steady state* E_0 *of the system (1) is unstable.*

Lemma 2: *There exists a feasible zooplankton free steady state* E_{11} *of the system (1) which is unstable if*

$$R_{11} = \frac{(\beta_2 - \theta)P_1^2}{(K_2 + P_1^2)(\mu_2 + \frac{h_2}{E_2})} > 1. \tag{8}$$

The analytical results are summarized in the Table 1.

Table 1: The table representing thresholds and stability of steady states

Thresholds (R_0, R_{11})	$(N_0, 0, 0)$	$(N_1, P_1, 0)$	(N^*, P^*, Z^*)
$R_0 < 1$	Asymptotically stable	Not feasible	Not feasible
$R_0 > 1,\ R_{11} < 1$	Unstable	Asymptotically stable	Not feasible
$R_{11} > 1$	Unstable	Unstable	Asymptotically stable

Stability analysis of the positive interior equilibrium of the system (1). The variational matrix of system (1) around the positive equilibrium $E^* = (N^*, P^*, Z^*)$ is

$$V^* = \begin{bmatrix} m_{11} & m_{12} & m_{13} \\ m_{21} & m_{22} & m_{23} \\ m_{31} & m_{32} & m_{33} \end{bmatrix},$$

where $m_{11} = -D - \frac{K_1\alpha_1 P^*}{(K_1+N^*)^2} < 0$,
$m_{12} = -\frac{\alpha_1 N^*}{K_1+N^*} + \mu_3 < 0$,
$m_{13} = \mu_4 > 0$,
$m_{21} = \frac{K_1\alpha_2 P^*}{(K_1+N^*)^2} > 0$,
$m_{22} = \frac{\beta_1 P^* Z^*[P^{*2} - K_2]}{(K_2+P^{*2})^2} + \frac{h_1 P^*}{(E_1+P^*)^2}$,
$m_{23} = -\frac{\beta_1 P^{*2}}{K_2+P^{*2}} < 0$,

$m_{31} = 0,$
$m_{32} = \frac{2K_2(\beta_2-\theta)P^*Z^*}{(K_2+P^{*2})^2} > 0,$
$m_{33} = \frac{h_2Z^*}{(E_2+Z^*)^2} > 0.$
The characteristic equation is $y^3 + A_1y^2 + A_2y + A_3 = 0$, where
$A_1 = -(m_{11} + m_{22} + m_{33}),$
$A_2 = m_{11}m_{22} - m_{23}m_{32} + m_{11}m_{33} + m_{22}m_{33} - m_{12}m_{21},$
$A_3 = (m_{11}m_{23} - m_{13}m_{21})m_{32} - m_{33}(m_{11}m_{22} - m_{12}m_{21}).$

Case 1: When $m_{22} > 0$, then $A_1 = -(m_{11} + m_{22} + m_{33}) > 0$
if $D + \frac{K_1\alpha_1P^*}{(K_1+N^*)^2} > \frac{\beta_1P^*Z^*[P^{*2}-K_2]}{(K_2+P^{*2})^2} + \frac{h_1P^*}{(E_1+P^*)^2} + \frac{h_2Z^*}{(E_2+Z^*)^2}.$
$A_2 = m_{11}m_{22} - m_{23}m_{32} + m_{11}m_{33} + m_{22}m_{33} - m_{12}m_{21} > 0$ if
$m_{22}m_{33} - m_{23}m_{32} - m_{12}m_{21} > -(m_{11}m_{22} + m_{11}m_{33}).$
Also $A_3 = (m_{11}m_{23} - m_{13}m_{21})m_{32} - m_{33}(m_{11}m_{22} - m_{12}m_{21}) > 0$ if
$m_{11}m_{23}m_{32} - m_{33}m_{11}m_{22} > m_{13}m_{21}m_{32} - m_{33}m_{12}m_{21}.$

Case 2: When $m_{22} < 0$, then $A_1 = -(m_{11} + m_{22} + m_{33}) > 0$
if $D + \frac{K_1\alpha_1P^*}{(K_1+N^*)^2} - \frac{\beta_1P^*Z^*[P^{*2}-K_2]}{(K_2+P^{*2})^2} - \frac{h_1P^*}{(E_1+P^*)^2} > \frac{h_2Z^*}{(E_2+Z^*)^2}.$
$A_2 = m_{11}m_{22} - m_{23}m_{32} + m_{11}m_{33} + m_{22}m_{33} - m_{12}m_{21} > 0$ if
$m_{11}m_{22} - m_{23}m_{32} - m_{12}m_{21} > -(m_{11}m_{33} + m_{22}m_{33}).$
Also $A_3 = (m_{11}m_{23} - m_{13}m_{21})m_{32} - m_{33}(m_{11}m_{22} - m_{12}m_{21}) > 0$ if
$m_{11}m_{23}m_{32} - m_{33}m_{12}m_{21} > m_{13}m_{21}m_{32} + m_{33}m_{11}m_{22}.$

Therefore according the Routh-Hurwitz criteria, all roots of the above cubic equation have negative real parts satisfying $A_1 > 0$, $A_3 > 0$ and $A_1A_2 - A_3 > 0$. Then the system become locally asymptotically stable around E^*. Thus depending upon system parameters, the system may exhibit stable or unstable behavior in this case.

Theorem: 3. *When the rate of zooplankton decay due to toxin production phytoplankton θ crosses a critical value, say θ^*, the system (1) enters into a Hopf-bifurcation around the positive equilibrium, which induces oscillations of the populations.*

Proof. The necessary and sufficient conditions for the existence of the Hopf-bifurcation for $\theta = \theta^*$ if it exist are (i) $A_i(\theta^*) > 0$, $i = 1, 2, 3$ (ii)$A_1(\theta^*)A_2(\theta^*) - A_3(\theta^*) = 0$ and (iii) the eigenvalues of above characteristic equation should be of the form $\lambda_i = u_i + \mathbf{i}v_i$, and $\frac{du_i}{d\theta} \neq 0$, $i = 1, 2, 3$.
We will now verify the Hopf-bifurcation condition (iii), putting $\lambda = u + \mathbf{i}v$ in the above equation, we get

$$(u + \mathbf{i}v)^3 + A_1(u + \mathbf{i}v)^2 + A_2(u + \mathbf{i}v) + A_3 = 0. \quad (9)$$

On separating the real and imaginary parts and eliminating v, we get

$$8u^3 + 8A_1u^2 + 2u(A_1^2 + A_2) + A_1A_2 - A_3 = 0. \quad (10)$$

It is clear from the above that $u(\theta^*) = 0$ iff $A_1(\theta^*)A_2(\theta^*) - A_3(\theta^*) = 0$. Further, at $\theta = \theta^*$, $u(\theta^*)$ is the only root, since the discriminate $8u^2 + 8A_1u + 2(A_1^2 + A_2) = 0$ if $64A_1^2 - 64(A_1^2 + A_2) < 0$. Again, differentiating (10) with respect to θ, we have
$24u^2\frac{du}{d\theta} + 16A_1u\frac{du}{d\theta} + 2(A_1^2 + A_2)\frac{du}{d\theta} + 2u[2A_1\frac{dA_1}{d\theta} + \frac{dA_2}{d\theta}] + \frac{dS}{d\theta} = 0$, where $S = A_1A_2 - A_3$.
Since at $\theta = \theta^*$, $u(\theta^*) = 0$ we get
$\left[\frac{du}{d\theta}\right]_{\theta=\theta^*} = \frac{-\frac{dS}{d\theta}}{2(A_1^2+A_2)} \neq 0.$

This ensures that the above system has a Hopf-bifurcation around the positive interior equilibrium E^*.

Lemma 3: The Hopf-bifurcation will be of supercritical (subcritical) nature and there will be emergence of stable (unstable) limit cycles if the first lyapunov coefficient is negative (positive).

5. Numerical simulations

In this section, we focus our attention on the occurrence and termination of the fluctuating plankton population. We begin with a parameter set (cf. Table 2, reference [19]) for which the existence criterion at equilibrium of coexistence E^* is satisfied. In this case coexistence equilibrium point $E^* = (1.4690, 0.2510, 0.3967)$ is locally asymptotically stable in the form of a stable focus with eigenvalues $-0.9684, -0.1453 \pm \mathrm{i}0.7984$ (cf. Fig. 1(a)).

Table 2: A set of parameter values

Parameter	Definition	Value	Unit
N^0	Constant input of nutrient	2	$mgml^{-1}$
D	Dilution rate of nutrient	1	day^{-1}
α_1	Nutrient uptake rate for the phytoplankton	3.2	day^{-1}
α_2	Nutrient conversion rate for the phytoplankton	2.4	day^{-1}
β_1	Phytoplankton uptake rate for the zooplankton	2.1	day^{-1}
β_2	Phytoplankton conversion rate for the zooplankton	1	day^{-1}
μ_1	Mortality rate of phytoplankton	0.1	day^{-1}
μ_2	Mortality rate of zooplankton	0.1	day^{-1}
μ_3	Nutrient recycle rate due to the death of phytoplankton	0.06	day^{-1}
μ_4	Nutrient recycle rate due to the death of zooplankton	0.06	day^{-1}
K_1	Half saturation constant for phytoplankton	0.6	$mgml^{-1}$
K_2	Half saturation constant for zooplankton	0.1	$mgml^{-1}$
h_1	Harvesting rate of phytoplankton population	0.4	day^{-1}
E_1	Effort required to harvest the phytoplankton	1	day^{-1}
h_2	Harvesting rate of zooplankton population	0.4	day^{-1}
E_2	Effort required to harvest the zooplankton	1	day^{-1}
θ	Rate of zooplankton decay due to toxin production phytoplankton	0	day^{-1}

5.1. *Effects of θ*

For $\theta = 0.3$, leaving all other parameter values unaltered, the system exhibits oscillations around the positive interior equilibrium E^* with eigenvalues $-1.0022, 0.0542 \pm \mathrm{i}0.7063$ (cf. Fig. 1(b)).

5.2. *Combined effect of θ and β_2*

It has already been shown that the system depicts oscillatory behavior around the positive interior equilibrium E^* for $\theta = 0.3$. If the maximal zooplankton conversion rate, β_2, is increased from 1 to 1.2, the system becomes asymptotically stable at positive interior equilibrium $E^* = (1.420, 0.2755, 0.3872)$ in the form of a stable focus with eigenvalues $-.9753, -0.0903 \pm \mathrm{i}0.7823$ (cf. Fig. 1(c)).

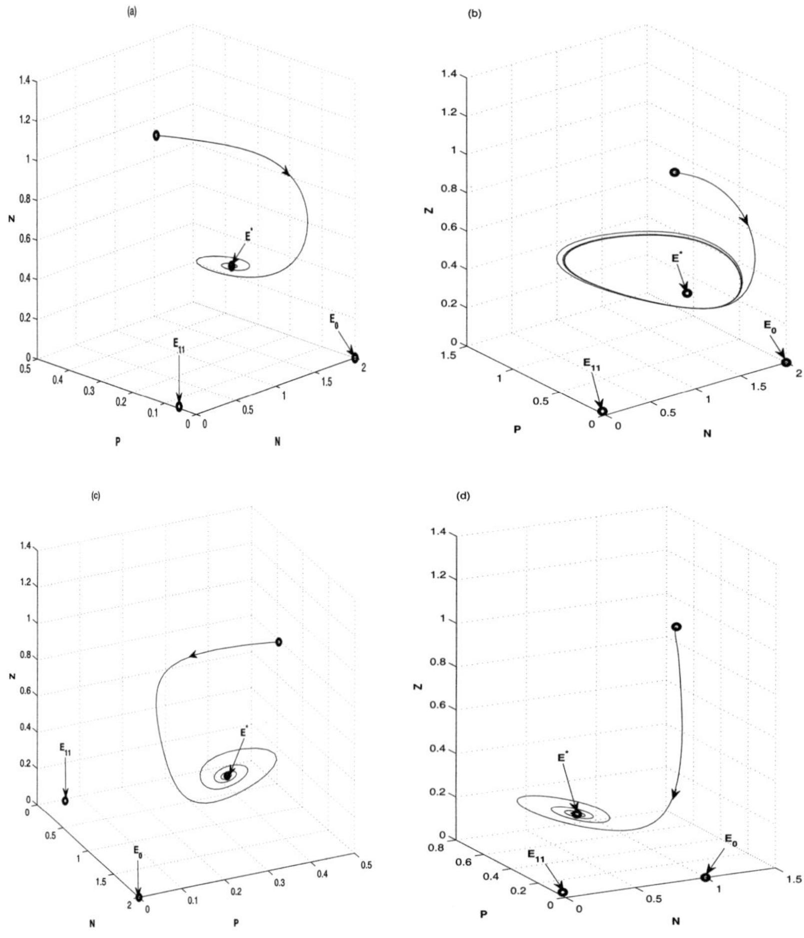

Figure 1. **(a)** The equilibrium point E^* is asymptotically stable for the parameter values as given in the Table 2. **(b)** The figure depicts oscillatory behavior around the positive interior equilibrium E^* for $\theta = 0.3$ with other parameter values as given in Table 2. **(c)** The figure depicts asymptotically stable behavior at E^* of the system (1) for $\theta = 0.3$ and $\beta_2 = 1.2$ with same set of parameter values as given in Table 2. **(d)** The figure depicts asymptotically stable behavior at E^* of the system (1) for $\theta = 0.3$ and $N^0 = 1$ with same set of parameter values as given in the Table 2.

5.3. *Combined effect of θ and N^0*

If the constant nutrient input, N^0, decrease from 2 to 1, the system becomes asymptotically stable at positive interior equilibrium $E^* = (0.4721, 0.4005, 0.2078)$ in the form of a stable focus with eigenvalues $-1.2604, -0.0571 \pm \mathrm{i}0.5237$(cf. Fig. 1(d)).

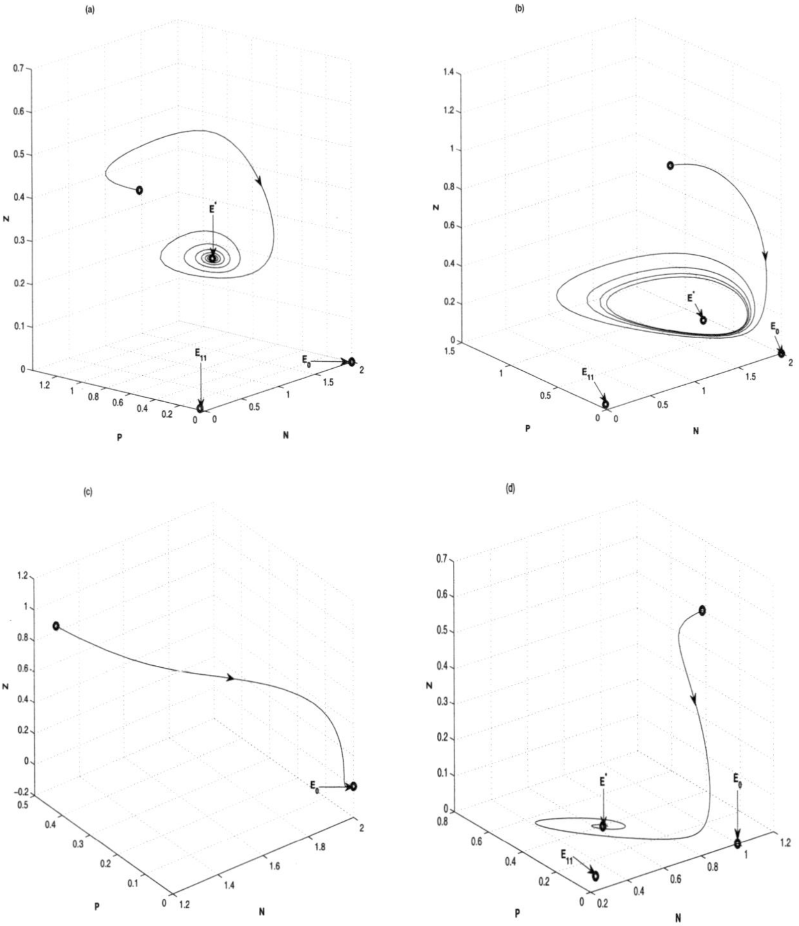

Figure 2. **(a)** The figure depicts asymptotically stable behavior at E^* of the system (1) for $\theta = .3$ and $D = .5$ with same set of parameter values as given in the Table 2. **(b)** The figures depicts oscillatory behavior around the positive interior equilibrium E^* of the system (1) for increasing h_1 from .4 to 1.2 and other parameter values as given in the Table 2. **(c)** The figure depicts stable behavior at E_0 of the system (1) for increasing h_1 from .4 to 1.8 with same set of parameter values as given in the Table 2. **(d)** The figure depicts asymptotically stable behavior at E^* of the system (1) for $h_1 = 1.2$ and $N^0 = 1$ with same set of parameter values as given in the Table 2.

5.4. *Combined effect of θ and D*

Decreasing the value of dilution rate of nutrient, D from 1 to 0.5, the system swifts to a stable behavior at positive interior equilibrium $E^* = (0.7442, 0.3769, 0.2869)$ in the form of a stable focus with eigenvalues $-0.4993, -0.0447 \pm \mathbf{i}0.7225$(cf. Fig. 2(a)).

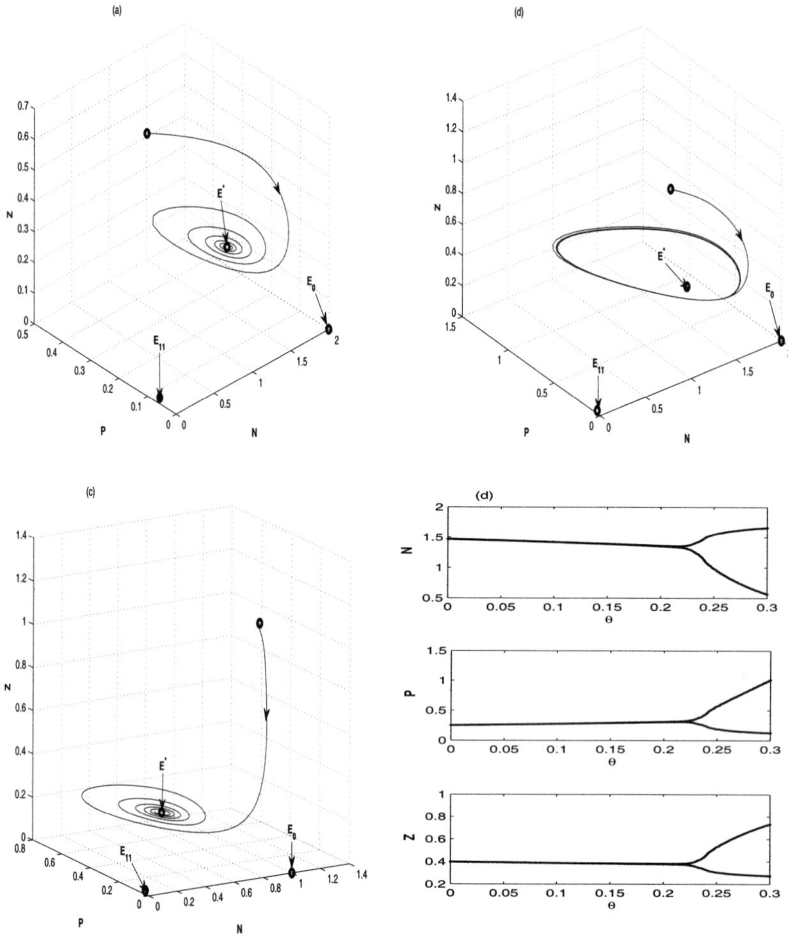

Figure 3. **(a)** The figure depicts asymptotically stable behavior at E^* of the system (1) for $h_1 = 1.2$ and $\beta_2 = 1.3$ with same set of parameter values as given in Table 2. **(b)** The figures depicts oscillatory behavior around the positive interior equilibrium E^* of the system (1) for increasing h_2 from 0.4 to 0.6 and other parameter values as given in Table 2. **(c)** The figure depicts stable behavior at E^* of the system (1) for $h_2 = 0.6$ and $N^0 = 1$ with same set of parameter values as given in the Table 2. **(d)** The bifurcation diagram for θ with all parameter values as given in Table 2.

5.5. *Effects of h_1*

Keeping the other parameter values fixed as in Table 2 and increasing harvesting rate of phytoplankton (h_1) from 0.4 to 1.2, the system swifts to oscillatory behavior around the positive interior equilibrium E^* with eigenvalues $-0.9765, -0.0094 \pm \mathbf{i}0.5997$ (cf.

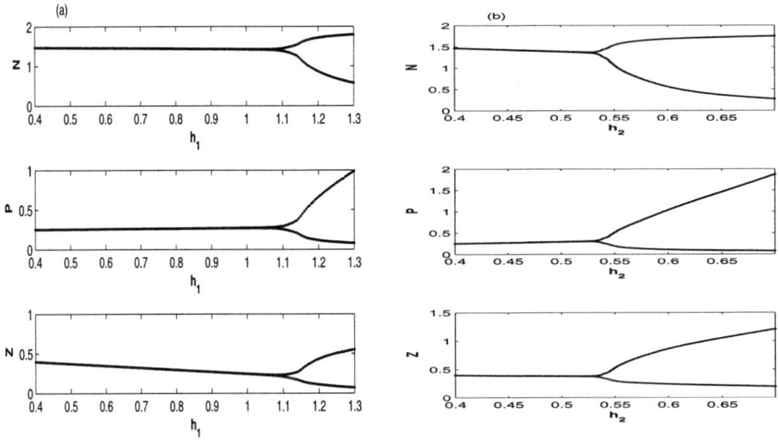

Figure 4. **(a)** The bifurcation diagram for h_1; **(b)** The bifurcation diagram for h_2 with all parameter values as given in Table 2.

Fig. 2(b)). Further increasing the value of harvesting rate of phytoplankton (h_1) from 1.2 to 1.8 with the same set of parameter values, the system again swifts to plankton free equilibrium $E_0 = (2, 0, 0)$ (cf. Fig. 2(c)).

5.6. *Combined effect of h_1 and N^0*

We have already seen (cf. Fig. 2(b)) that the system shows oscillatory behavior around the positive interior equilibrium E^* for $h_1 = 1.2$. But decreasing the value of constant nutrient input (N^0) from 2 to 1, the system again swifts back to a stable behavior at positive interior equilibrium $E^* = (0.5508, 0.3068, 0.0393)$ in the form of a stable focus with eigenvalues $-1.0808, -0.0689 \pm \mathbf{i}0.2805$ (cf. Fig. 2(d)).

5.7. *Combined effect of h_1 and β_2*

The oscillatory behavior around the positive interior equilibrium E^* shown in Fig. 2(b) swifts to a stable behavior at $E^* = (1.5141, 0.2232, 0.2039)$ in the form of a stable focus with eigenvalues $-0.9702, -0.0521 \pm \mathbf{i}0.6318$ when the value of maximal conversion rate (β_2) increased from 1 to 1.3 (cf. Fig. 3(a)).

5.8. *Effects of h_2*

The other parameter values remaining fixed as in Table 2 and increasing harvesting rate of zooplankton (h_2) from 0.4 to 0.6, the system exhibits to oscillatory behavior around the positive interior equilibrium E^* with eigenvalues $-1.0111, 0.0575 \pm \mathbf{i}0.8388$ (cf. Fig. 3(b)).

5.9. *Combined effect of h_2 and N^0*

We have already observed (cf. Fig. 3(b)) the system shows oscillatory behavior around the positive interior equilibrium E^* for $h_2 = 0.6$. But decreasing the value of constant nutrient input (N^0=1), the system again swifts to a stable behavior at positive interior equilibrium $E^* = (0.4859, 0.3839, 0.2102)$ in the form of a stable focus with eigenvalues $-1.2443, -0.0417 \pm \mathbf{i}0.6343$ (cf. Fig. 3(c)).

5.10. *Hopf-bifurcation*

For a clear understanding of the dynamical changes as a result of change in the rate of zooplankton decay due to toxin producing phytoplankton, θ, a bifurcation diagram is plotted with this parameter as the bifurcation parameter with other three species (cf. Fig. 3(d)). Here we see a Hopf bifurcation point at θ=0.229127 with the first Lyapunov coefficient -0.3780012. Clearly the first Lyapunov coefficient is negative. This means that a stable limit cycle (super critical bifurcation) bifurcates from the equilibrium, when it looses stability. Another bifurcation diagram is plotted with harvesting rate of phytoplankton h_1, as the bifurcation parameter with other three species (cf. Fig. 4(a)). Here the value of the first Lyapunov coefficient is -0.05101716 at bifurcation point for h_1 =1.078385. This implies that a stable limit cycle (super critical bifurcation) bifurcates from the equilibrium, when it looses stability. Finally we have plotted a bifurcation diagram with harvesting rate of zooplankton h_2, as the bifurcation parameter with other three species (cf. Fig. 4(b)). Here we see a Hopf bifurcation point at $h_2 = 0.537536$ with the first Lyapunov coefficient -0.3662218. This means that a super critical bifurcation occurs when it looses stability.

6. *Discussion*

A nutrient-plankton interaction model in the presence of harvesting is considered with nutrient uptake functions and instantaneous nutrient recycling. Firstly, The model is studied analytically and the threshold conditions for the existence and stability of various steady states are worked out, see Table 1. It is observed that there is a chance for the fluctuating population for a high value of the zooplankton decay rate due to toxin producing phytoplankton. Thus our analysis shows that if the rate of zooplankton decay due to toxin producing phytoplankton θ crosses a critical value, the system (1) enters into a Hopf bifurcation that induces stable limit cycle periodic solutions around the positive equilibrium.

But in the presence of high value of the zooplankton decay rate due to toxin producing phytoplankton, there is no chance for the fluctuating population at low values of constant nutrient input and dilution rate of nutrient simultaneously. A similar case arises for high values of the maximal zooplankton conversion rate.

Further, our observation indicates that there is a chance for oscillatory behavior in the presence of a high value of harvesting rate of phytoplankton population. This oscillatory behavior can be controlled by decreasing the value of constant nutrient input and increasing the value of maximal zooplankton conversion rate simultaneously. The oscillatory behavior has been seen again for a high value of harvesting rate of zooplankton population. But in the presence of a low value of constant nutrient, even at high harvesting rates of zooplankton population, there is no chance for the fluctuating population.

Thus our analysis suggests that either decreasing the value of constant nutrient or increasing the value of maximal zooplankton conversion rate may prevent the instability of system, even there is a high harvesting rate of plankton population.

Our study is to search a suitable mechanism to control and maintain a stable coexistence between all the species. It is observed that to control the fluctuating population and to maintain stability around the coexistence equilibrium, we need to control the rate of zooplankton decay due to toxin producing phytoplankton and harvesting rate of plankton population.

References

1. S. Pal, S. Chatterjee and J. Chattopadhyay, Role of toxin and nutrient for the occurrence and termination of plankton bloom. Results drawn from field observations and a mathematical model,*BioSystems* **90**: 87-100, (2007).
2. A.Chatterjee, S. Pal and S. Chatterjee, Bottom up and top down effect on toxin producing phytoplankton and its consequence on the formation of plankton bloom, *Applied Mathematics and Comp.* **218**: 3387-3398, (2011).
3. N. Bairagi, S. Pal, S. Chatterjee and J. Chattopadhyay, Nutrient, non-toxic phytoplankton, toxic phytoplankton and zooplankton interaction in the open marine system, in: R.J. Hosking, E. Venturino (Eds.), Aspects of Mathematical Modelling, *Mathematics and Biosciences in Interaction*, Birkhauser Verlag, Basel, Switzerland, pp. 4163, (2008).
4. S. Khare, O.P. Misra and J. Dhar, Role of toxin producing phytoplankton on a plankton ecosystem, *Nonlinear Anal. Hybrid Syst.* **4**: 496-502, (2010).
5. M. Banerjee and E. Venturino, A phytoplankton-toxic phytoplankton-zooplankton model, *Ecological Comp.* **8(3)**: 239-248, (2011).
6. S. R.-J. Jang, J. Baglama and J. Rick, Nutrient-phytoplankton-zooplankton models with a toxin, *Math. Comput. Model.* **43**: 105-118, (2006).
7. J. Chattopadhyay, R.R. Sarkar and S. Pal, Mathematical modelling of harmful algal blooms supported by experimental findings, *Ecol. Complexity* **1**: 225-235, (2004).
8. A.Y. Morozov, Emergence of Holling type III zooplankton functional response: bringing together field evidence and mathematical modelling, *J. Theor. Biol.* **265**: 4554, (2010).
9. C.W.Clark, The Optimal Management of Renewable Resources, *Mathematical Bioeconomics*, 2nd edition, Wiley Interscience, (1990).
10. W. A. Wurts, Sustainable Aquaculture in the Twenty-First Century, *Reviews in Fisheries Science* **8(2)**, 141-150, (2000).
11. M.Omori, Zooplankton Fisheries of the World: A Review, *Marine Biology* **48**: 199-205, (1978).
12. L.Yunfei, P.Yongzhena, G.Shujing and L.Changguo, Harvesting of a phytoplankton-zooplankton model, *Nonlinear Analysis: Real World Applications* **11**: 3608-3619, (2010).
13. P.Yongzhen, L.Yunfei and L.Changguo, Evolutionary consequences of harvesting for a two-zooplankton one-phytoplankton system, *Appl. Math. Modell.* doi:10.1016/j.apm.2011.09.015, (2011).
14. C. A. Price, L.R. Mendiola-Morgenthaler, M. Goldstein, E. N. Breden and R. R. L.Guillard, Harvest of Planktonic Marine Algae by Centrifugation into Gradients of Silica in the CF-6 Continuous -Flow Zonal Rotor, *Biol.Bull.***147**: 136-145, (1974).
15. C. T. Kuo, Harvesting natural algal blooms for concurrent biofuel production and hypoxia mitigation, *Agricultural & Biological Engineering* (2011).
16. K. G. Gravesa and J. C. Morrowa , Method for Harvesting Large Quantities of Zooplankton from Hatchery Ponds, *The Progressive Fish-Culturist* **50(3)**: 184-186, (1988).
17. A.G.Fredrickson and G.Stephanopoulos, Microbial competition, *Science.* **213(4511)**: 972-979, (1981).
18. H.L.Smith, Competitive coexistence in an oscillating chemostat, *SIAM J. Appl. Math.* **40**: 498-522, (1981).
19. B.Mukhopadhyay and R.Bhattacharyya, On a three -tier ecological food chain model with deterministic and random harvesting: A mathematical study, *Nonlinear Analysis: Modelling and Control* **16(1)**: 77-88 (2011).

Acknowledgements

The authors thank the referees very much for their valuable comments and suggestions. We are also thankful to editor, for his helpful comment. The first author is also thankful to INSA, CSIR for providing partial financial assistance to attend the conference.

TRAVELING WAVE SOLUTIONS FOR A CHEMOTAXIS SYSTEM*

F. CATRINA †

Department of Mathematics and Computer Science,
St. John's University,
Queens, New York 11439
E-mail: catrinaf@stjohns.edu

V. M. REYES G.

CUNY Graduate Center,
Mathematics Department, 365 Fifth Avenue,
New York, NY 10016
E-mail: Vreyes@gc.cuny.edu

In this article we study the existence of traveling wave solutions for a system of equations modeling one-dimensional movement of bacteria, driven by an existing substrate. We propose a class of *chemotactic* coefficients depending on two parameters m and p. According to the values of these coefficients, as well as to the diffusion parameters (corresponding to the bacteria and to the substrate), we prove existence and properties of traveling wave solutions, and even (in very special cases) provide explicit solutions.

1. Introduction

In this article we study the existence of traveling wave solutions for a system of the type

$$\begin{cases} \dfrac{\partial U}{\partial T} = D_U \cdot \dfrac{\partial^2 U}{\partial X^2} - \dfrac{\partial}{\partial X}\left(U \cdot \chi(V) \cdot \dfrac{\partial V}{\partial X}\right) \\ \dfrac{\partial V}{\partial T} = D_V \cdot \dfrac{\partial^2 V}{\partial X^2} - k \cdot U \end{cases} \tag{1}$$

*The authors would like to thank Prof. Marcello Lucia for suggesting this problem and for useful discussions.
†Work partially supported by the *2012 Summer Support for Research Program* at St. John's University.

where the independent variables $(T, X) \in (0, \infty) \times \mathbb{R}$ represent time and one spatial variable, respectively. We require that the unknown functions U, V are positive, and that for any T fixed,

$$\begin{aligned} &U(T,X) \to 0 \text{ for } |X| \to \infty, \quad V(T,X) \to 0 \text{ as } X \to -\infty, \\ &\text{and} \quad V(T,X) \to 1 \text{ as } X \to \infty. \end{aligned} \tag{2}$$

Keller and Segel[9] propose the system (1) as a model for bacteria moving in an infinitely long tube under the stimulus of a dissipating substrate. In experiments with *Escherichia Coli* it was observed that in the presence of certain types of supporting substrate, the bacteria advances in a migrating bands pattern. The system above is a possible model for this kind of cell motility. For variants of (1), motivations and explanations, we refer to Ch 11 in Murray[10], Ch1 and Ch 5 in Murray[11] and to Tyson [13].

The unknown functions $U(X,T)$ and $V(X,T)$ represent the density of bacteria and the concentration of substrate, respectively. The space variable X is the distance along the tube where migration takes place, and T is the time variable. The coefficients D_U and D_V are the diffusion constants for the bacteria and the substrate, respectively. The constant k represents the rate of consumption of substrate per unit of bacteria density. A very important role is played by *the chemotactic coefficient* χ. This is a function depending on the substrate density V, and this dependency is crucial in the behavior of the system.

Generalizations of the system (1) to one or several space dimensions have been proposed by numerous authors (see e.g. the survey by Horstmann[6] [7] and the references therein). In different models, the coefficients D_U, χ, k are either constants, or functions depending on one or both unknowns. Particular choices of these coefficients were made in[9] so that explicit *traveling wave* solutions were obtained. By *traveling wave* we mean solutions of the form $U(X,T) = u(X - cT)$ and $V(X,T) = v(X - cT)$.

One set of assumptions in Keller and Segel[9] is that D_U, k are positive constants, and $\chi(V) = \delta/V$ for some positive constant $\delta > D_U$. It is also assumed that the substrate diffusion is negligible, and therefore D_V is set equal to zero.

In this paper we discuss the existence of traveling wave solution for the system (1) with a very specific choice of the chemotactic coefficient, namely

$$\chi(V) = D_U \frac{m - pV^{p-1}}{mV - V^p} \tag{3}$$

for some $p \geq 1$ and $m > p$ (in the general case $D_V > 0$, $p > 1$, we will put more stringent requirements on m - see (28) in Remark 3 below-). We treat separately the cases $D_V = 0$ and D_V positive, and for $D_V = 0$ or $D_V = 2D_U$, as well as for $p = 1$, we are able to produce explicit families of solutions. Our results differ from [9] not only by the alternative choice of χ, but also because for the main part of our work the diffusion of the substrate D_V, is not neglected. Even when we consider $D_V = 0$ and $p = 1$, we have the case $\delta = D_U$ which surprisingly, was not discussed in [9]. We note that Horstmann and Stevens [8] find sufficient conditions on the coefficients so that the system admits traveling wave solutions. However our examples do not seem to fit in the classes of coefficients considered in [8].

As it turns out, in all the cases we consider we have conservation of mass in the amount of bacteria, that is

$$\int_{-\infty}^{\infty} U(T,X)\,dX = \int_{-\infty}^{\infty} U(0,X)\,dX < \infty \quad \text{for any } T.$$

Following [8] we call such solutions *traveling pulse* in U and *traveling wave front* in V. Refer to Figure (3.1) for the shape of solutions (incidentally, this shape is similar throughout all four cases in which we can write explicit solutions, even though the functions involved are significantly different).

We prove existence of traveling wave solutions of (1), (2) (with the chemotactic coefficient (3)) and we discuss properties of solutions in each of the cases in Table (1). The equation of the substrate V will be the crucial part of the system; it will become an ODE to which we refer in the third column below.

$D_V = 0$	$p > 1$	nonlinear first order, explicit solutions
$D_V = 0$	$p = 1$	linear first order, explicit solutions
$D_V > 0$	$p > 1$	nonlinear second order, fixed point argument
$D_V = 2D_U > 0$	$p > 1$	nonlinear second order, explicit solutions
$D_V > 0$	$p = 1$	linear second order, explicit solutions

The results will be presented in the form of a theorem at the beginning of the subsection corresponding to each case.

2. Preliminaries

Traveling wave solutions of (1) are solutions of the form

$$U(X,T) = u(\xi) \text{ and } V(X,T) = v(\xi), \text{ where } \xi = X - cT,$$

for a constant c (the speed of the wave) to be determined. We denote by u', v' the derivatives of u and v with respect to ξ. The system (1) becomes

$$\begin{cases} -cu' = (D_U \cdot u' - u \cdot \chi(v) \cdot v')' \\ -cv' = D_V \cdot v'' - k \cdot u, \end{cases} \tag{4}$$

with asymptotic conditions

$$\begin{aligned} &u(\xi) \to 0 \text{ as } |\xi| \to \infty, v(\xi) \to 0 \text{ as } \xi \to -\infty, \\ &\text{and } v(\xi) \to 1 \text{ as } \xi \to \infty. \end{aligned} \tag{5}$$

After integration, the first equation in (4) is of first order. Our approach is to integrate this first equation to find u in terms of v. After substituting u thus found into the second equation, we obtain an ODE for v. It is somewhat surprising that the Keller-Segel case, where the authors considered $D_V = 0$ and the system was first order in both equations, is sufficiently robust so that even for $D_V > 0$, for certain type of chemotactic coefficients $\chi(V)$, the system still admits explicit solutions satisfying the asymptotic solutions at $\pm\infty$ given by (5).

In the Horstmann-Stevens paper [8], the main goal is to find various types of diffusion coefficients D_U (not necessarily constant), chemotactic coefficients χ in the first equation, and functions $g(u, v)$ more general than $-ku$ in the second equation, so that the system admits solutions with the prescribed asymptotic behavior as $|\xi| \to \infty$. In this paper we will consider D_U to be constant throughout, but it is important to note that in [8], in the case it is not assumed constant, the authors take D_U to depend only on u rather than just on v or on both u and v (see discussion on page 237 in [9]).

We are interested in solutions that satisfy the asymptotic conditions

$$u(\xi) \to 0 \text{ as } |\xi| \to \infty, \; v(\xi) \to 0 \text{ as } \xi \to -\infty \text{ and } v(\xi) \to 1 \text{ as } \xi \to \infty. \tag{6}$$

These asymptotic conditions reflect the fact that the motion happens in an infinitely long tube, at the left end of the tube there is neither bacteria nor substrate, in advance of the wave the concentration of bacteria is almost zero, and the substrate concentration approaches a positive value, which we normalize to 1. Due to the "local" shape of u (it is exponentially small outside a bounded interval), such solutions are called *pulse* solutions.

Getting back to the system (4), after integrating the first equation from $-\infty$ to ξ we obtain

$$-cu = D_U \cdot u' - u \cdot \chi(v) \cdot v'. \tag{7}$$

Note that we pick the constant of integration to be zero because we want solutions to have u, u', $v' \to 0$ and $\chi(v)$ bounded as $\xi \to \infty$. Equation (7) can be written as

$$\frac{u'}{u} = -\frac{c}{D_U} + \frac{\chi(v)}{D_U} \cdot v'$$

and integrating again from 0 to ξ we get

$$\ln(u(\xi)) = \int_0^\xi -\frac{c}{D_U} + \frac{\chi(v(\zeta))}{D_U} \cdot v'(\zeta)\ d\zeta + \ln(u(0)).$$

Therefore

$$u(\xi) = u(0) \cdot \exp\left(-\frac{c}{D_U}\xi\right) \cdot \exp\left(\int_0^\xi \frac{\chi(v(\zeta))}{D_U} \cdot v'(\zeta)\ d\zeta\right). \tag{8}$$

From(8), it seems that the behavior of the substrate controls the dynamics of the system, but this is misleading because the specific form of u as a function of v (and more specifically of $\chi(v)$) determines the form of the third term in the second equation in (4).

The function $\chi(v)$ has to satisfy certain properties for u and v to satisfy the asymptotic conditions we require in (6). Note that if χ has no singularity at $v = 0$ then

$$\lim_{\xi \to -\infty} \int_0^\xi \chi(v(\zeta)) \cdot v'(\zeta)\ d\zeta = \lim_{v(\xi) \to 0} \int_{v(0)}^{v(\xi)} \chi(v)\ dv, \quad \text{finite.}$$

The equation (8) implies then that $\lim_{\xi \to -\infty} u(\xi) = \infty$ and this will violate the requirement that $\lim_{\xi \to -\infty} u(\xi) = 0$.

In the Keller and Segel paper[9] the authors also make the argument that the chemotactic coefficient χ must be *sufficiently singular* at $v = 0$ for the system to yield solutions with the given conditions at infinity. In that paper, the choice of a singular $\chi(v) = \delta/v$ is attributed to the "pervasiveness of the Weber-Fechner Law", (which roughly states that response rate is inverse proportional to stimulus) and it is referred to Weibull [15] for the justification of this law by laboratory experiments.

Let us denote by $K(v)$ a positive function that satisfies

$$\frac{d}{dv}(\ln K(v)) = \frac{\frac{d}{dv}K(v)}{K(v)} = \frac{\chi(v)}{D_U}.$$

Since

$$\int_0^\xi \frac{\chi(v(\zeta))}{D_U} \cdot v'(\zeta)\ d\zeta = \ln K(v(\xi)) - \ln K(v(0)),$$

with this new notation we have

$$u(\xi) = \frac{u(0)}{K(v(0))} K(v(\xi)) \exp\left(-\frac{c\xi}{D_U}\right).$$

We now introduce two positive parameters which are the values of the unknowns at $\xi = 0$

$$u(0) = a > 0 \text{ and } v(0) = b \in (0,1). \tag{9}$$

The second equation in (4) is

$$D_V v'' + cv' - ku = 0,$$

where u is replaced from (8).

We have that

$$u(\xi) = \frac{a}{K(b)} K(v(\xi)) \exp\left(-\frac{c\xi}{D_U}\right) \tag{10}$$

and the second equation in (4) becomes

$$D_V v''(\xi) + cv'(\xi) - \frac{ak}{K(b)} K(v(\xi)) \exp\left(-\frac{c\xi}{D_U}\right) = 0, \tag{11}$$

which depending on our choice of k, χ, D_U (and therefore of K) is a second order, possibly nonlinear, differential equation in v.

For the remaining of the paper, we choose

$$K(v) = mv - v^p, \text{ for some } p \geq 1 \text{ and some } m > p. \tag{12}$$

Thus,

$$\frac{d}{dv}(\ln K(v)) = \frac{m - pv^{p-1}}{mv - v^p} = \frac{\chi(v)}{D_U},$$

and therefore this special choice of K is equivalent to the special choice of

$$\chi(v) = D_U \frac{m - pv^{p-1}}{mv - v^p}. \tag{13}$$

Note that for $p = 1$ our choice of χ matches that of [9] ($\chi(v) = \delta/v$) with $\delta = D_U$ (see Subsection 3.2 below).

Therefore, with the ansatz (12), the function from (10) is

$$u(\xi) = \frac{a}{mb - b^p}(mv(\xi) - v^p(\xi)) \cdot \exp\left(-\frac{c\xi}{D_U}\right) \tag{14}$$

where v is solution of the equation

$$D_V v''(\xi) + cv'(\xi) - ku(0)\frac{mv(\xi) - v^p(\xi)}{mv(0) - v^p(0)} \cdot \exp\left(-\frac{c\xi}{D_U}\right) = 0.$$

Recalling the notation (9), let us denote $M = \dfrac{ak}{K(b)} = \dfrac{ak}{mb - b^p}$. Therefore, we are looking for functions v which solve

$$D_V v'' + cv' - M(mv - v^p) \cdot \exp\left(-\frac{c\xi}{D_U}\right) = 0. \tag{15}$$

At this point we make the following change of variables. Let

$$t = \exp\left(-\frac{c\xi}{2D_U}\right) \in (0, \infty) \quad \text{and} \quad v(\xi) = f(t). \tag{16}$$

Then as $\xi \to \infty$ we have $t \to 0$ and as $\xi \to -\infty$, $t \to +\infty$. Making the substitution in (15) we get

$$D_V \left(f_{tt} \left(\frac{dt}{d\xi}\right)^2 + f_t \frac{d^2 t}{d\xi^2} \right) + c\left(f_t \frac{dt}{d\xi} \right) - M\left(mf - f^p\right) \exp\left(-\frac{c\xi}{D_U}\right) = 0.$$

This reduces to

$$D_V f_{tt} + \frac{D_V - 2D_U}{t} f_t - \frac{4D_U^2 M}{c^2}\left(mf - f^p\right) = 0 \tag{17}$$

with boundary conditions

$$f(0) = 1 \text{ and } f(\infty) = 0, \text{ and the extra-requirement that } f(1) = b. \tag{18}$$

3. The Case $D_V = 0$

In this section we analyze the situation when the dissipation of the substrate is negligible, therefore we will assume $D_V = 0$. This case was considered originally in [9] and it is also treated in [12]. In this case the equation (17) is first order and separable

$$\frac{-2D_U}{t} f_t - \frac{4D_U^2 M}{c^2}\left(mf - f^p\right) = 0. \tag{19}$$

3.1. *The Case $p > 1$*

Theorem 1. *For $D_U > 0$, $D_V = 0$, $k > 0$, $m > p > 1$, the system* (1), (2), *admits traveling wave solutions given by $U(X,T) = u(X - cT)$ and $V(X,T) = v(X - cT)$, where $u(\xi) = a\frac{mv(\xi) - v^p(\xi)}{mb - b^p} \exp\left(-\frac{c\xi}{D_U}\right)$,*

$$v(\xi) = \left(\frac{m}{1 + (m-1)\left(\frac{m - b^{p-1}}{b^{p-1}(m-1)}\right)^{\exp\left(-\frac{c\xi}{D_U}\right)}} \right)^{\frac{1}{p-1}}$$

and the traveling speed is

$$c = \left(\frac{akD_U(p-1)}{b\left(1 - \frac{b^{p-1}}{m}\right) \ln\left(\frac{m - b^{p-1}}{(m-1)b^{p-1}}\right)} \right)^{\frac{1}{2}}.$$

Proof. Equation (19) takes the form $\frac{f_t}{f(m-f^{p-1})} = -\frac{2D_U M}{c^2}t$. Integrating, we get

$$\frac{1}{m}\int_{f(0)}^{f(t)} \frac{m-f^{p-1}}{f(m-f^{p-1})}\, df + \frac{1}{m}\int_{f(0)}^{f(t)} \frac{f^{p-1}}{f(m-f^{p-1})}\, df = -\frac{D_U M}{c^2}t^2 + C.$$

Therefore

$$\ln f(t) - \ln f(0) - \frac{1}{p-1}\ln(m-f^{p-1}(t)) + \frac{1}{p-1}\ln(m-f^{p-1}(0)) = -\frac{D_U mM}{c^2}t^2 + C.$$

Thus, considering the boundary condition $f(0)=1$ we obtain that $C=0$, and therefore

$$\ln\frac{(m-1)f^{p-1}(t)}{m-f^{p-1}(t)} = -\frac{D_U mM(p-1)}{c^2}t^2,$$

which implies

$$f(t) = \left(\frac{m}{1+(m-1)\exp\left(\frac{D_U mM(p-1)}{c^2}t^2\right)}\right)^{\frac{1}{p-1}}.$$

Since $f(1)=b>0$, by substituting in the formula for f above, we get

$$b = \left(\frac{m}{1+(m-1)\exp\left(\frac{D_U mM(p-1)}{c^2}\right)}\right)^{\frac{1}{p-1}}.$$

From this equation we solve for the speed of the wave in terms of the parameters in the model and the conditions $u(0)=a$ and $v(0)=b$. Recalling that $M = \dfrac{ak}{mb-b^p}$, we get the speed of the wave to be

$$c = \sqrt{\frac{kD_U(p-1)}{\left(1-\frac{b^{p-1}}{m}\right)\ln\left(\frac{m-b^{p-1}}{(m-1)b^{p-1}}\right)}}\cdot\sqrt{\frac{a}{b}}. \tag{20}$$

From $v(\xi) = f(t)$ after we substitute from (20) the value for c, we obtain the desired solution. □

Remark 1. Keeping all other constants involved fixed, the speed c of the wave has the properties:

(1) it is proportional to $\sqrt{a}$;
(2) increases to infinity as $b \searrow 0$ or as $b \nearrow 1$.

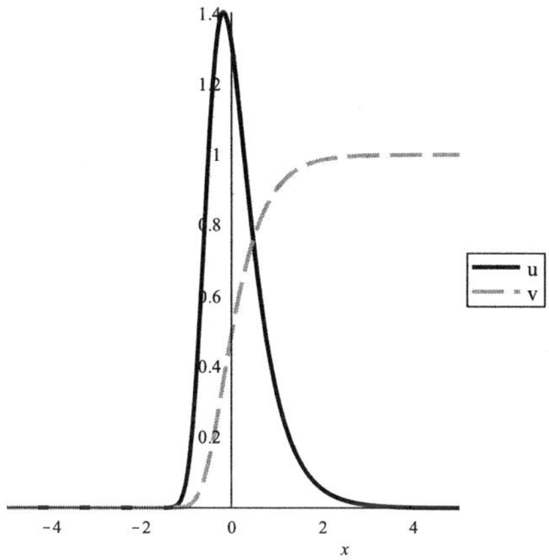

Figure 1. Solution (u, v) for parameters: $D_U = 1$, $D_V = 0$, $p = 2$, $a = 1.3$, $b = 0.5$, $k = 1$; $c = 1.845273097$

3.2. *The Case* $p = 1$

In this case $K(v) = (m-1)v$ and the chemotactic coefficient as given by (13) is $\chi(v) = \dfrac{D_U}{v}$. The chemotactic coefficients considered in [9] are $\chi(v) = \dfrac{\delta}{v}$. In the Keller-Segel paper [9] the authors only consider the situation $\delta > D_U$ (there, the dissipation coefficient of bacteria D_U is denoted by μ). It is stated on page 245 that *under the assumptions of their paper,* "steadily traveling waves cannot appear unless δ exceeds μ". As we will show here, traveling waves appear for $\delta = D_U$ also.

Theorem 2. *For* $D_U > 0$, $D_V = 0$, $k > 0$, $m > p = 1$, *the system* (1), (2), *admits traveling wave solutions given by* $U(X,T) = u(X - cT)$ *and* $V(X,T) = v(X - cT)$, *where*

$$u(\xi) = ab^{\exp\left(-\frac{c\xi}{D_U}\right)-1}\exp\left(-\frac{c\xi}{D_U}\right), \quad v(\xi) = b^{\exp\left(-\frac{c\xi}{D_U}\right)}$$

and the traveling speed is $c = \sqrt{-\dfrac{akD_U}{b\ln b}}$.

Proof. The equation (19) is now $\frac{f_t}{f} = -\frac{2(m-1)D_U M}{c^2}t$. Solving, we obtain

$$f(t) = \exp\left(-\frac{(m-1)\,D_U M}{c^2}t^2\right).$$

Substituting again $f(1) = b$ we obtain the speed of the wave to be

$$c = \sqrt{-\frac{akD_U}{b\ln b}}.$$

This gives

$$f(t) = b^{t^2}, \quad \text{and so} \quad v(\xi) = b^{\exp\left(-\frac{c\xi}{D_U}\right)}.$$

With $v(\xi)$ and c above, we obtain the explicit formula for $u(\xi)$. □

Remark 2. The solutions (u, v) we obtained for $p = 1$ in Theorem 2 can also be achieved by a limit process (even if tedious) either by

(1) letting the parameter $\overline{\delta} = \dfrac{\delta}{D_U} \to 1^+$ in the solution (12) in the Keller-Segel paper [9], or by
(2) letting the parameter $p \to 1^+$ in the solutions we obtained in Theorem 1.

4. The Case $D_V > 0$

The existence and stability of traveling waves for the case $D_V > 0$ with $\chi(v) = \delta/v$ and $\delta < D_U$ was discussed by Nagai and Ikeda[12]. For further references and generalizations we refer to Horstmann[6 7].

4.1. *The Case $p > 1$.*

In this case we only have qualitative results. It will be interesting to find what are the minimal conditions for the existence of traveling wave solutions as well as to obtain precise estimates for the speed of the wave.

Theorem 3. *For any $p > 1$, $D_U > 0$, $D_V > 0$ let $\nu = \frac{D_U}{D_V}$. There exists $m_0 = m_0(\nu, p) \geq p$, such that for any $m > m_0$ in (3), the system (1), (2), admits traveling wave solutions.*

Starting again from (17), we make further the notations and the change of variables $\nu = \dfrac{D_U}{D_V}$, $\quad t = \dfrac{c}{2D_U}\sqrt{\dfrac{D_V}{mM}}s$, $\quad f(t) = m^{\frac{1}{p-1}}g(s)$. By direct calculation we get that (17) becomes

$$-g_{ss} + \frac{2\nu - 1}{s}g_s + g = g^p, \tag{21}$$

and the boundary conditions (18) become

$$g(0) = m^{-\frac{1}{p-1}}, \quad g(\infty) = 0, \text{ with the extra-requirement that}$$
$$g\left(\frac{2D_U}{c}\sqrt{\frac{mM}{D_V}}\right) = \frac{b}{m^{\frac{1}{p-1}}}. \tag{22}$$

Theorem 3 is a consequence of the following

Theorem 4. *For any $p > 1$ and any $\nu > 0$ fixed, there exists a positive $\delta_0 = \delta_0(\nu, p)$ such that for any given $0 < \delta \le \delta_0$, the nonlinear ODE (21) admits a positive solution on $(0, \infty)$ with $g(0) = \delta$ and $g(s) \to 0$ as $s \to \infty$.*

Proof. We use a Fixed Point argument. We prove the existence of a positive function g which is a fixed point for the operator

$$\mathcal{T}g = \alpha\zeta + \mathcal{G}g^p, \quad \text{where} \quad \alpha = \frac{\delta}{2^{\nu-1}\Gamma(\nu)}$$

and ζ and $\mathcal{G}$ are described below. Let $\zeta(s) = s^\nu K_\nu(s)$, $\xi(s) = s^\nu I_\nu(s)$ be solutions of the homogeneous linear equation

$$\mathcal{L}\xi = -\xi_{ss} + \frac{2\nu - 1}{s}\xi_s + \xi = 0$$

(I_ν and K_ν are the Modified Bessel Functions of first and second kind, respectively) and $\mathcal{G} = \mathcal{L}^{-1}$ is the Green's operator

$$(\mathcal{G}h)(t) = \int_0^\infty \mathcal{K}(t,s)h(s)\,ds, \quad \text{where} \quad \mathcal{K}(t,s) = s^{-2\nu+1}\xi(s \wedge t)\zeta(s \vee t).$$

For notations and properties of $\mathcal{G}$ we refer to Appendix.

If g is a fixed point of

$$(\mathcal{T}g)(t) = \alpha\zeta(t) + \int_0^\infty \mathcal{K}(t,s)g^p(s)\,ds,$$

by applying $\mathcal{L}$ on both sides of $g = \mathcal{T}g$ we obtain

$$\mathcal{L}g = \mathcal{L}\mathcal{T}g = \alpha\mathcal{L}\zeta + \mathcal{L}\mathcal{G}g^p = g^p,$$

i.e. we obtain a solution of (21). To prove that $\mathcal{T}g = g$ has a solution with the desired properties we use an iteration scheme. Define $g_0(s) \equiv 0$, and for $n \ge 0$ let

$$g_{n+1} = \mathcal{T}g_n = \alpha\zeta + \mathcal{G}g_n^p, \tag{23}$$

so that $g_1 = \alpha\zeta$.

For $n \geq 1$ we have

$$g_{n+1}(t) - g_n(t) = (\mathcal{T}(g_n - g_{n-1}))(t) = \int_0^\infty \mathcal{K}(t,s)\left(g_n^p(s) - g_{n-1}^p(s)\right)\, ds.$$

Since $\mathcal{K} > 0$, by induction we obtain $g_{n+1}(t) > g_n(t)$ for all $t > 0$.

We now estimate $\varphi(s) := (\mathcal{G}\zeta^p)(s)$. We have that φ is solution of the ODE

$$-\varphi_{ss} + \frac{2\nu - 1}{s}\varphi_s + \varphi = \zeta^p, \quad \text{and} \quad \varphi(0) = \lim_{s\to\infty}\varphi(s) = 0.$$

Using the Green's Function we have that

$$\varphi(x) = \zeta(x)\int_0^x \frac{t^{2\nu-1}}{\zeta^2(t)}\int_t^\infty \frac{\zeta^{p+1}(s)}{s^{2\nu-1}}\, ds\, dt. \tag{24}$$

Let

$$C(\nu,p) = \int_0^\infty \frac{t^{2\nu-1}}{\zeta^2(t)}\int_t^\infty \frac{\zeta^{p+1}(s)}{s^{2\nu-1}}\, ds\, dt. \tag{25}$$

Alternatively, we have

$$C(\nu,p) = \int_0^\infty t^{-2\nu+1}\xi(t)\zeta^p(t)\, dt = \int_0^\infty t^{\nu(p-1)+1} I_\nu(t) K_\nu^p(t)\, dt.$$

We can write

$$C(\nu,p) = \int_0^\infty t\left(t^{-\nu} I_\nu(t)\right)\left(t^\nu K_\nu(t)\right)^p\, dt.$$

From (38) and (39) below, we have that the integral defining $C(\nu,p)$ is finite at zero. Also, from (40) and (41) we get that for $p > 1$ the integral defining $C(\nu,p)$ is finite at infinity, and so $C(\nu,p)$ is finite. We could find the exact value of $C(\nu,p)$ (using Maple) only for $p = 2$ or for $\nu = 1/2$. In these cases we have $C(\nu,2) = \frac{2^\nu\Gamma(\nu)}{3}\left({}_2F_1\left(\frac{1}{2},\nu-\frac{1}{2};\nu+\frac{1}{2};\frac{3}{4}\right) - \frac{1}{2}\right)$, and $C(1/2,p) = \frac{1}{p^2-1}\left(\frac{\pi}{2}\right)^{\frac{p-1}{2}}$. Define β to be the smaller of the two solutions of

$$\alpha + \beta^p C(\nu,p) = \beta. \tag{26}$$

Note that the maximum of the function $\beta \to (\beta - \beta^p C(\nu,p))$ happens for $\beta = \frac{1}{p^{\frac{1}{p-1}} C^{\frac{1}{p-1}}(\nu,p)}$ and it is $\alpha\text{max} = \frac{p-1}{p^{\frac{p}{p-1}}}\frac{1}{C^{\frac{1}{p-1}}(\nu,p)}$. So, for as long as $\alpha \leq \alpha\text{max}$, the equation (26) has two (possibly equal) solutions.

From (24) and (25) we have

$$\varphi(s) = (\mathcal{G}\zeta^p)(s) < C(\nu,p)\zeta(s) \text{ for all } s \in (0,\infty). \tag{27}$$

It follows now by Mathematical Induction that $g_n(s) \leq \beta\zeta(s)$. Indeed, $g_1 = \alpha\zeta < \beta\zeta$, and assume that $g_n(s) \leq \beta\zeta(s)$. Then

$$g_{n+1} = \mathcal{T} g_n = \alpha\zeta + \mathcal{G} g_n^p \leq \alpha\zeta + \beta^p \mathcal{G}\zeta^p \leq (\alpha + \beta^p C(\nu, p))\, \zeta = \beta\zeta.$$

The fact that the sequence $\{g_n\}$ is uniformly bounded, immediately implies from (23) that the sequence of derivatives $\{g_n'\}$ is uniformly bounded. We therefore have a monotonically increasing sequence $\{g_n\}_n$ bounded above by $\beta\zeta$ and therefore convergent. Since the derivatives are uniformly bounded it follows that the limit is differentiable and by repeating the argument for higher order derivatives, the limit is actually in $C^\infty(0, \infty)$. As the solutions obtained above tend to zero as $s \to \infty$, if we had an exact form solution, from the extra-condition in (22) one could solve for the speed of the wave c in terms of the parameters a, b, k, m, D_U and D_V. □

Remark 3. The condition $\alpha \leq \alpha_{\max}$ translates into

$$\delta \leq \delta_0(\nu, p) = \frac{(p-1)2^{\nu-1}\Gamma(\nu)}{p^{\frac{p}{p-1}} C^{\frac{1}{p-1}}(\nu, p)},$$

and since in Theorem 3 we need $\delta = m^{-\frac{1}{p-1}}$, this means that we need

$$m \geq \frac{p^p C(\nu, p)}{((p-1)2^{\nu-1}\Gamma(\nu))^{p-1}}. \tag{28}$$

As p tends to 1 from above, the right hand side tends to infinity and the condition $m > p$ is far from sufficient for (28) to hold. In this situation the existence of solutions is guaranteed only for sufficiently large m.

4.2. *Particular Case* $D_V = 2D_U$ *and* $p > 1$.

In this subsection we consider the special case $D_V = 2D_U > 0$. In this case the solutions u and v can be written explicitly because (21) becomes

$$-g_{ss} + g = g^p \tag{29}$$

which has a family of exponentially decreasing solutions as $|s| \to \infty$ given by

$$g(s) = \left(\frac{p+1}{2}\right)^{\frac{1}{p-1}} \left[\cosh\left(\frac{(p-1)(s-s_0)}{2}\right)\right]^{\frac{-2}{p-1}}. \tag{30}$$

Theorem 5. *For* $D_V = 2D_U > 0$, $k > 0$, $m > p > 1$, *the system* (1), (2), *admits explicit traveling wave solution and traveling speed.*

Proof. From (30) the equation (17) has the family of solutions

$$f(t)=\left(\frac{m(p+1)}{2}\right)^{\frac{1}{p-1}}\left[\cosh\left(\frac{p-1}{c}\sqrt{\frac{mMD_U}{2}}(t-t_0)\right)\right]^{\frac{-2}{p-1}}. \tag{31}$$

From (18) $f(0)=1$ we can solve for $\frac{t_0}{c}$ to get

$$\frac{t_0}{c}=\pm\frac{1}{p-1}\sqrt{\frac{2}{mMD_U}}\operatorname{arccosh}\left(\sqrt{\frac{m(p+1)}{2}}\right).$$

The extra-requirement $f(1)=b$ will then determine c (and at the same time t_0). We get that

$$\frac{1}{c}=\frac{1}{p-1}\sqrt{\frac{2}{mMD_U}}\left(\operatorname{arccosh}\left(\sqrt{\frac{m(p+1)}{2}}\frac{1}{b^{\frac{p-1}{2}}}\right)\pm\operatorname{arccosh}\left(\sqrt{\frac{m(p+1)}{2}}\right)\right)$$

Then

$$t_0=\pm\frac{c}{p-1}\sqrt{\frac{2}{mMD_U}}\operatorname{arccosh}\left(\sqrt{\frac{m(p+1)}{2}}\right),$$

where c is the value obtained above. Therefore we get

$$t_0=\pm\frac{\operatorname{arccosh}\left(\sqrt{\frac{m(p+1)}{2}}\right)}{\operatorname{arccosh}\left(\sqrt{\frac{m(p+1)}{2}}\frac{1}{b^{\frac{p-1}{2}}}\right)\pm\operatorname{arccosh}\left(\sqrt{\frac{m(p+1)}{2}}\right)},$$

where the two $\pm$ signs are not independent (i.e. plus with plus and minus with minus). Consequently, (15) has solutions

$$v(\xi)=\left(\frac{m(p+1)}{2}\right)^{\frac{1}{p-1}}\left[\cosh\left(\frac{p-1}{c}\sqrt{\frac{mMD_U}{2}}\left(e^{-\frac{c}{2D_U}\xi}-t_0\right)\right)\right]^{\frac{-2}{p-1}} \tag{32}$$

where t_0 and c have to be chosen as explained. However, if $t_0>0$, the maximum of $v(\xi)$ will be $v_{\max}=\left(\frac{m(p+1)}{2}\right)^{\frac{1}{p-1}}$ and our requirement that χ (as given by (3)) is positive means that $m>pv^{p-1}(\xi)$ for all ξ, but this will be violated. Therefore we can only take the case t_0 negative,

$$t_0=-\frac{\operatorname{arccosh}\left(\sqrt{\frac{m(p+1)}{2}}\right)}{\operatorname{arccosh}\left(\sqrt{\frac{m(p+1)}{2}}\frac{1}{b^{\frac{p-1}{2}}}\right)-\operatorname{arccosh}\left(\sqrt{\frac{m(p+1)}{2}}\right)} \tag{33}$$

and the corresponding wave speed

$$c = \frac{(p-1)\sqrt{\frac{mMD_U}{2}}}{\operatorname{arccosh}\left(\sqrt{\frac{m(p+1)}{2}}\frac{1}{b^{\frac{p-1}{2}}}\right) - \operatorname{arccosh}\left(\sqrt{\frac{m(p+1)}{2}}\right)}. \quad (34)$$

□

4.3. *The Case $p = 1$*

Theorem 6. *For any $m > 1$, $D_U > 0$, $D_V > 0$, the system* (1), (2), *admits explicit traveling wave solutions.*

Proof. In this case the equation (17) becomes $-f_{tt} + \frac{2\nu-1}{t}f_t + \frac{4\nu D_U M}{c^2}(m-1)f = 0$, where $\nu = \dfrac{D_U}{D_V}$ and $M = \dfrac{ak}{(m-1)b}$ and so

$$-f_{tt} + \frac{2\nu-1}{t}f_t + \frac{4ak\nu D_U}{bc^2}f = 0. \quad (35)$$

Considering that from (18) we want $f(0) = 1$, and from (39), the solution will have the form $f(t) = \frac{2}{\Gamma(\nu)}\left(\frac{1}{c}\sqrt{\frac{ak\nu D_U}{b}}\right)^{\nu} t^{\nu} K_{\nu}\left(\frac{2}{c}\sqrt{\frac{ak\nu D_U}{b}}t\right)$. By setting $f(1) = b$ (the extra-requirement in (18)) we obtain a unique value for the speed of the wave c in terms of the other parameters. Therefore we obtain from (14) that $u(\xi) = \frac{a}{b}v(\xi)\exp\left(-\frac{c\xi}{D_u}\right)$, where $v(x)$ is obtained from $f(t)$ by (16): $v(\xi) = \frac{2}{\Gamma(\nu)}\left(\frac{1}{c}\sqrt{\frac{ak\nu D_U}{b}}\right)^{\nu} \cdot \exp\left(-\frac{c\nu\xi}{2D_U}\right) \cdot K_{\nu}\left(\frac{2}{c}\sqrt{\frac{ak\nu D_U}{b}}\exp\left(-\frac{c\xi}{2D_U}\right)\right)$. □

5. Appendix

5.1. *Modified Bessel Functions*

In the main text we need certain properties of the modified Bessel functions I_{ν} and K_{ν} of the first and second kind. Most of these properties are obtained from [1], [14], [16], and [18]. The functions I_{ν} and K_{ν} are independent solution of the modified Bessel equation

$$y_{ss} + \frac{1}{s}y_s - \left(1 + \frac{\nu^2}{s^2}\right)y = 0.$$

We are only interested for s in the interval $(0, \infty)$, on which both I_{ν} and K_{ν} are real valued. By 3.7 (2) in [14] we have

$$I_{\nu}(s) = \sum_{k=0}^{\infty} \frac{(s/2)^{\nu+2k}}{k!\,\Gamma(\nu+k+1)}, \quad (36)$$

and by 3.7 (6), K_ν can be defined as

$$K_\nu(s) = \frac{\pi}{2}\frac{I_{-\nu}(s) - I_\nu(s)}{\sin \nu\pi}, \tag{37}$$

where for integer index $\nu = n$, the ratio is understood in the sense of limit as $\nu \to n$. From (36) we get for $\nu \geq 0$,

$$\lim_{s\to 0} s^{-\nu} I_\nu(s) = \frac{1}{2^\nu\,\Gamma(\nu+1)}, \tag{38}$$

and combining with (37) we get for $\nu > 0$,

$$\lim_{s\to 0} s^{\nu} K_\nu(s) = 2^{\nu-1}\,\Gamma(\nu). \tag{39}$$

From [18] we have that for $s >> 1$,

$$I_\nu(s) \sim \frac{e^s}{(2\pi s)^{\frac{1}{2}}}\sum_{k=0}^{\infty}(-1)^k\frac{a_k(\nu)}{s^k}, \tag{40}$$

and

$$K_\nu(s) \sim \left(\frac{\pi}{2s}\right)^{\frac{1}{2}} e^{-s}\sum_{k=0}^{\infty}\frac{a_k(\nu)}{s^k}, \tag{41}$$

where $a_0(\nu) = 1$ and for $k \geq 1$,

$$a_k(\nu) = \frac{(4\nu^2-1^2)(4\nu^2-3^2)\cdots(4\nu^2-(2k-1)^2)}{k!8^k}.$$

The weighted Wronskian corresponding to the modified Bessel operator applied to I_ν and K_ν is constant (by Abel's Theorem) and by formula 3.71 (19) in [14] we have

$$s\left(I_\nu(s)\frac{d}{ds}K_\nu(s) - K_\nu(s)\frac{d}{ds}I_\nu(s)\right) \equiv -1. \tag{42}$$

For $s > 0$ we have

$$\int_0^s tI_\nu^2(t)\,dt = \frac{s^2}{2}I_\nu^2(s) - \frac{s^2}{2}I_{\nu-1}(s)I_{\nu+1}(s) \tag{43}$$

and

$$\int_s^\infty tK_\nu^2(t)\,dt = -\frac{s^2}{2}K_\nu^2(s) + \frac{s^2}{2}K_{\nu-1}(s)K_{\nu+1}(s) \tag{44}$$

which both follow by integration by parts and by identities 3.71 (2), (3), and (4) in [14]. Combining (43) with (38), for $\nu > 0$ we have

$$\lim_{s\to 0+} s^{-2\nu-2}\int_0^s tI_\nu^2(t)\,dt = \frac{\nu+1}{2^{2\nu+1}\Gamma^2(\nu+2)}. \tag{45}$$

At infinity, for every $\nu \geq 0$ we have

$$\lim_{s\to\infty} e^{-2s} \int_0^s tI_\nu^2(t)\, dt = \frac{1}{4\pi}. \tag{46}$$

Combining (44) with (39), for $\nu > 1$ we have

$$\lim_{s\to 0+} s^{2\nu-2} \int_s^\infty tK_\nu^2(t)\, dt = \frac{2^{2\nu-3}}{\nu-1}\Gamma^2(\nu). \tag{47}$$

For $\nu = 1$ we have

$$\lim_{s\to 0+} \frac{1}{|\ln s|} \int_s^\infty tK_\nu^2(t)\, dt = 1. \tag{48}$$

For $0 \leq \nu < 1$ we have

$$\lim_{s\to 0+} \int_s^\infty tK_\nu^2(t)\, dt = \frac{\nu\pi}{2\sin(\nu\pi)}, \tag{49}$$

where it is understood that when $\nu = 0$ the value of the integral is $1/2$.

At infinity, it holds that for every $\nu \geq 0$ we have

$$\lim_{s\to\infty} e^{2s} \int_s^\infty tK_\nu^2(t)\, dt = \frac{\pi}{4}. \tag{50}$$

From (45), (46), (47), (48), (49), (50) it follows that for all $\nu \geq 0$,

$$\left(\int_s^\infty tK_\nu^2(t)\, dt \cdot \int_0^s tI_\nu^2(t)\, dt \right)\Big|_0^\infty = \frac{1}{16}. \tag{51}$$

From the same estimates we also get that

$$\int_0^\infty tI_\nu^2(t) \left(\int_t^\infty sK_\nu^2(s)\, ds \right) dt = A(\nu) < \infty. \tag{52}$$

5.2. *Green's Function*

In this section we discuss briefly the invertibility of the differential operator

$$\mathcal{L}g = -g_{ss} + \frac{2\nu - 1}{s} g_s + g.$$

We consider only the cases $\nu > 0$. The analysis can be done in the spirit of [2, 4, 5, 17] (whose notations and terminology we follow) for $\nu \in (0,1)$ but this setting is too restrictive for $\nu \geq 1$. We deal with the operator

$$\mathcal{L}g = \frac{-(pg')' + qg}{w}, \text{ with } p(s) = q(s) = w(s) = s^{-2\nu+1} \text{ on } (0,\infty).$$

Let $L^2((0,\infty);w)$ be the space of equivalence classes of Lebesgue measurable functions $g:(0,\infty)\to\mathbb{R}$ for which

$$\int_0^\infty g^2(s)w(s)\;ds<\infty.$$

As usual, we consider two functions in $(0,\infty)$ equivalent if they are equal almost everywhere. Then, $L^2((0,\infty);w)$ is a Hilbert space with inner product

$$\langle f,g\rangle_w=\int_0^\infty f(s)g(s)w(s)\;ds.$$

The maximal operator generated by $\mathcal{L}$ is the operator $\mathcal{T}_1$ with domain

$$D(\mathcal{T}_1)=\{f:(0,\infty)\to\mathbb{R}\;:\;f,\,pf'\in AC_{loc}(0,\infty),\;f,\,\mathcal{L}f\in L^2((0,\infty);w)\},$$

given by $\mathcal{T}_1 f=\mathcal{L}f$.
The notation $AC_{loc}(0,\infty)$ represents the set of functions which are absolutely continuous on any interval $[a,b]$ with a positive and b finite.

Define an antisymmetric bilinear form associated to $\mathcal{L}$ (which we will call *Wronskian*) of two functions f and g as

$$\begin{aligned}\mathcal{W}[f,g](s)&=f(s)\left(p(s)g_s(s)\right)-\left(p(s)f_s(s)\right)g(s)\\&=s^{-2\nu+1}\left(f(s)g_s(s)-f_s(s)g(s)\right).\end{aligned}$$

Our definition of the Wronskian is slightly different from the usual determinant $\begin{vmatrix} g & \xi \\ g_s & \xi_s\end{vmatrix}$ as it incorporates the weight $s^{-2\nu+1}$, but this makes for simpler notations.

Note that Green's formula on $0<a<b<\infty$:

$$\int_a^b\left(g(s)\mathcal{L}[f](s)-f(s)\mathcal{L}[g](s)\right)w(s)\;ds=\mathcal{W}[f,g](b)-\mathcal{W}[f,g](a),$$

implies that for any $f,g\in D(\mathcal{T}_1)$ the limits

$$\mathcal{W}[f,g](\infty)=\lim_{b\to\infty}\mathcal{W}[f,g](b)\text{ and }\mathcal{W}[f,g](0)=\lim_{a\to0^+}\mathcal{W}[f,g](a)$$

exist and are finite.

The minimal operator generated by $\mathcal{L}$ is the operator $\mathcal{T}_0$ with domain

$$D(\mathcal{T}_0)=\{f\in D(\mathcal{T}_1)\;:\mathcal{W}[f,g](\infty)=\mathcal{W}[f,g](0)=0\text{ for all }g\in D(\mathcal{T}_1)\},$$

and as before, $\mathcal{T}_0 f=\mathcal{L}f$.

The endpoint $a=0$ is a *regular endpoint* if and only if $\nu\in(0,1)$ (because only in this range all of p^{-1},q,w are integrable near 0). In this case $\xi(s)=s^\nu I_\nu(s)$ is *principal solution.*

In the case $\nu \geq 1$, the endpoint $a = 0$ is *limit point* because

$$\int_0^1 \zeta^2(s)w(s)\, ds = \int_0^1 K_\nu^2(s)s\, ds = \infty$$

where $\zeta(s) = s^\nu K_\nu(s)$ is solution of $\mathcal{L}g = 0$. For all $\nu > 0$ the endpoint $b = \infty$ is *limit point* because

$$\int_1^\infty \xi^2(s)w(s)\, ds = \int_1^\infty I_\nu^2(s)s\, ds = \infty$$

where $\xi(s) = s^\nu I_\nu(s)$ is solution of $\mathcal{L}g = 0$.

Even though ξ and ζ may not be in $L^2((0,\infty);w)$ (and in fact only ζ is in $L^2((0,\infty);w)$ and only for $0 < \nu < 1$), we have that $\xi \in L^2((0,1];w)$ and $\zeta \in L^2([1,\infty);w)$. We define the domain of the operator $\mathcal{L}$ to be

$$D(\mathcal{L}) = \{g \in D(\mathcal{T}_1)\ : \mathcal{W}[g,\xi](0) = \mathcal{W}[g,\zeta](\infty) = 0\}.$$

Also, we define the range of $\mathcal{L}$ to be

$$R(\mathcal{L}) = \Big\{\, h \in L^1_{loc}(0,\infty)\ : \int_0^1 \xi(s)\,|h(s)|\, w(s)\, ds < \infty, \\ \text{and } \int_1^\infty \zeta(s)\,|h(s)|\, w(s)\, ds < \infty \Big\}.$$

We state without proof the folowing

Theorem 7. *For $\nu \geq 0$ fixed, the differential operator $\mathcal{L} : D(\mathcal{L}) \to R(\mathcal{L})$ given by $\mathcal{L}g = -g_{ss} + \dfrac{2\nu - 1}{s} g_s + g$, is invertible.*

References

1. Andrews, George E.; Askey, Richard; Roy, Ranjan; *"Special functions"*. Encyclopedia of Mathematics and its Applications, 71. Cambridge University Press, Cambridge, 1999.
2. Bailey, P. B.; Everitt, W. N.; Zettl, A.; *Computing eigenvalues of singular Sturm-Liouville problems.* Results Math. 20 (1991), no. 1-2, 391173.
3. Bailey, P. B.; Everitt, W. N.; Zettl, A.; *The SLEIGN2 Sturm-Liouville code,* ACM Trans. Math. Software. 27 (2001), 143–192.
4. Everitt, W. Norrie; *A catalogue of Sturm-Liouville differential equations.* (English summary) "Sturm-Liouville theory", 271171, Birkhuser, Basel, 2005.
5. Hajmirzaahmad, Mojdeh; Krall, Allan M.; *Singular second-order operators: the maximal and minimal operators, and selfadjoint operators in between.* SIAM Rev. 34 (1992), no. 4, 614174.
6. Horstmann, Dirk; *From 1970 until present: the Keller-Segel model in chemotaxis and its consequences. I.* Jahresber. Deutsch. Math.-Verein. 105 (2003), no. 3, 103175.

7. Horstmann, Dirk; *From 1970 until present: the Keller-Segel model in chemotaxis and its consequences. II.* Jahresber. Deutsch. Math.-Verein. 106 (2004), no. 2, 5117.
8. Horstmann, D.; Stevens, A.; *A constructive Approach to Traveling Waves in Chemotaxis.* J. Nonlinear Sci. Vol. 14. 1–25 (2005).
9. Keller, Evelyn F. ; Segel, Lee A.; *Traveling Bands of Chemotactic Bacteria: A Theoretical Analysis.* Journal of Theoretical Biology (1971) Vol. **30**, 235–248.
10. Murray, J. D.; *Mathematical biology. I. An introduction.* Third edition. Interdisciplinary Applied Mathematics, 17. Springer-Verlag, New York, 2002.
11. Murray, J. D.; *Mathematical biology. II. Spatial models and biomedical applications.* Third edition. Interdisciplinary Applied Mathematics, 18. Springer-Verlag, New York, 2003.
12. Nagai, Toshitaka; Ikeda, Tsutomu; *Traveling waves in a chemotactic model.* J. Math. Biol. 30 (1991), no. 2, 169–184.
13. Tyson, Rebecca; *Pattern formation by E. coli - mathematical and numerical investigation of a biological phenomenon*, PhD Dissertation.
14. Watson, G. N.; *"A Treatise on the Theory of Bessel Functions".* Reprint of the second (1944) edition. Cambridge Mathematical Library. Cambridge University Press, Cambridge, 2005.
15. Weibull, Claes; *"The Bacteria".* Vol. **I**, (I. C. Gunsalus and R. Y. Steiner, eds.) 153–205, Academic Press, New York, 1960.
16. Whittaker E. T.; Watson, G. N.; *"A Course of Modern Analysis".* Reprint of the fourth (1927) edition. Cambridge Mathematical Library. Cambridge University Press, Cambridge, 1995.
17. Zettl, Anton; *"Sturm-Liouville Theory".* Mathematical Surveys and Monographs, 121. American Mathematical Society, Providence, RI, 2005.
18. http://dlmf.nist.gov/10.25

DYNAMICS ON A GENERAL STAGE STRUCTURED N PARALLEL FOOD CHAINS*

ISAM AL-DARABSAH† AND YUAN YUAN ‡

Department of Mathematics and Statistics
Memorial University of Newfoundland
St. John's NL A1C 5S7, Canada

We propose a general model with n parallel food chains through the stage structured maturation time delay, which can cover most of the prey-predator models in the literature. We discuss some basic dynamical properties of the system with single or multiple patches and with general or some particular functional responses, including the existence of equilibrium points and their local and global stabilities. Numerical simulations are given to compliment the theoretical analysis.

Keywords: Phytoplankton; Zooplankton; Prey-predator; Equilibrium points; Stability.

1. Introduction

Plankton is the productive base of both marine and freshwater ecosystems and provides food for larger animals. There are many mathematical models to describe the dynamics of plankton community structure in literature, we refer [3,4,5,6,11,12,14] and the references cited there. Most of the models are based on the so-called prey-predator model.

Armstrong [1] studied a model structure with n parallel food chains, each consisting of a phytoplankton species P_i and its dedicated zooplankton predator Z_i, proposed the following system

$$\begin{aligned} P_i' &= P_i\left[\mu_i(N) - Z_i h_i(P_i)\right] \\ Z_i' &= Z_i\left[\gamma_i P_i h_i(P_i) - \epsilon_i\left(\sum_{j=1}^{n} Z_j\right)\right], \end{aligned} \tag{1}$$

*the research is supported in part by the NSERC of canada.

†imaad6@mun.ca.

‡corresponding author, yyuan@mun.ca.

for $i = 1, \ldots, n$ and $N = T - \sum_{i=1}^{n} P_i - \sum_{i=1}^{n} Z_i$, with the growth rate $\mu_i(N)$ of P_i which is a function of nutrient concentration N, the per-phytoplankton-per-zooplankton harvest rate $h_i(P_i)$ of P_i by Z_i, predation on the zooplankton $\epsilon_i \left(\sum_{j=1}^{n} Z_j \right)$ by higher trophic levels. The constant parameter T is the nutrient supply, γ_i is the growth efficiency of Z_i. The diagram Fig. (1) shows the structure of three parallel food chains.

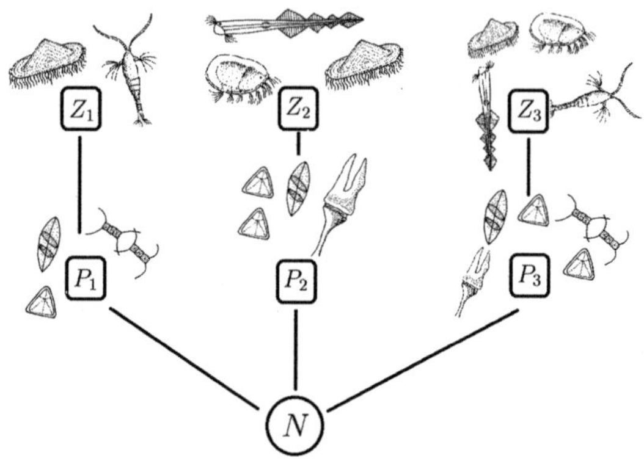

Figure 1.: Three parallel food chains.

As pointed out in [8], in reality, the growth of species is a combined result of birth and death processes, which is closely linked to the resource supply. The authors in [8] formulated a general prey-predator model with stage structure with constant maturation time delay. Adopting the idea of stage structure, we modify the model (1) to the following:

$$\begin{aligned} \frac{dP_i}{dt} &= P_i[\mu_i(\mathcal{N}) - Z_i h_i(P_i)], \\ \frac{dZ_i}{dt} &= b_i e^{-d_i \tau_i} P_i(t-\tau_i) Z_i(t-\tau_i) h_i(P_i(t-\tau_i)) - Z_i \epsilon_i \left(\sum_{k=1}^{n} Z_k \right), \end{aligned} \tag{2}$$

where $\mathcal{N} = T - \sum_{k=1}^{n} \alpha_k P_k - \sum_{k=1}^{n} \beta_k Z_k$ is the nutrient concentration with the coefficients $\alpha_k, \beta_k, (k = 1, \cdots, n)$ related to the efficiency of nutrient consuming for each species. We assume that, only the adult predators are capable of preying on the prey species, so Z_i is the adult zooplankton taxon feeding on the phytoplankton P_i, b_i denotes the adult predator's birth rate, d_i is the mortality death rate of the juvenile (through-stage death rate), τ_i is the unit of time to mature. We omit the equation of juvenile since it is decoupled. All the parameters are positive.

From the view points in biology and analysis, we assume that all the functions $\mu_i(\mathcal{N}), h_i(P_i)$ and $\epsilon_i(\sum_{k=1}^n Z_k)$ are continuous and differentiable, and satisfy the following hypothesis for $i = 1, \ldots, n$:

(C_1) $\mu_i(0) = 0$, $\frac{d\mu_i}{d\mathcal{N}} > 0$ when $\mathcal{N} \in [0, T]$;

(C_2) $h_i(P_i) \geq 0$, $\frac{d(P_i h_i(P_i))}{dP_i} > 0$, $0 \leq P_i h_i(P_i) \leq M_i$, M_i is a constant;

(C_3) $\frac{\partial \epsilon_i(Z)}{\partial Z_j} \geq 0$ for each Z_j in $Z = \sum_{k=1}^n Z_k$.

The general model (2) can cover almost all the prey-predator models in the literature, such as the well known Lotka-Volterra model and Rosenzweig-MacArthur model. In addition, (2) includes several general, partial general and specific models with some particular choices of the functions. For example, when $n = 1$, with $\mu(\mathcal{N}) = g(P), h(P) = \frac{p(P)}{P}, \epsilon(Z) = d(\text{constant})$ and $\tau = 0$, (2) becomes the general model in [10]; with $\mu(\mathcal{N}) = r(1 - P(t)/K), h(P) = \frac{p(P)}{P}, \tau \neq 0$ and constant $\epsilon(Z) = d_j$, (2) covers the partial general model in [8]; when we take $\mu(\mathcal{N}) = R(1 - P(t)/K)$ and the Holling type II and III functional responses $h(P) = \frac{A}{C_1 + P(t)}, \epsilon(Z) = d + \frac{FZ(t)}{C_2^2 + Z(t)^2}$, τ is either zero or nonzero, the model (2) is the same as that in [16]. When $n \geq 2$, the two-patch predator-prey models in [7] without migration or the n-patch ecological model [9] without dispersal are involved in our model (2) as well. It should remark that, currently we only consider the parallel patches, the effect of migration and dispersal will be considered in our future work.

The paper is organized as follows: A single patch with one phytoplankton and one zooplankton model is studied in Section 2, we provide the conditions for the existence of boundary and positive equilibrium points, and further discuss their stabilities. In Section 3, we extend the model and the results in Section 2 to multiple n parallel patches. Numerical simulations are given to compliment the theoretical analysis for such general system. Conclusion is drawn in Section 5.

2. One Phytoplankton and One Zooplankton Model

We start with the simplest case when $n = 1$ which is rewritten as

$$\begin{aligned} \frac{dP}{dt} &= P[\mu(\mathcal{N}) - Zh(P)], \\ \frac{dZ}{dt} &= be^{-d\tau} P(t-\tau) Z(t-\tau) h(P(t-\tau)) - Z\epsilon(Z), \end{aligned} \tag{3}$$

with $\mathcal{N} = T - \alpha P - \beta Z$.

Let $C = C([-\tau, 0), R)$, $C^+ = C([-\tau, 0), R^+)$, first, we have the following positivity and boundedness properties.

Theorem 2.1. *Given the initial condition $P(0), Z(0) \in C^+$, then under the hypothesis (C_1 - C_3), the solutions of (3) are nonnegative in $X = C \times C$. In addition, if $\epsilon(0) > 0$ then all the solutions are ultimately bounded in X.*

Proof. The positivity for $P(t)$ is just followed by standard arguments from the first equation in (3) since it contains a factor of $P(t)$.

For $Z(t)$, on $0 \le t \le \tau$, $Z'(t) \ge -Z\epsilon(Z)$, so $Z(t) \ge Z(0)\ e^{-\int_0^t \epsilon(Z(s))ds} > 0$. By using step-method, we have $Z(t) > 0$ for all $t \in [0, \infty)$. Therefore the solutions of (3) are nonnegative in X.

Under the hypothesis (C_1), $\mu(\mathcal{N})$ is increasing function with respect to $\mathcal{N}$, then

$$P' = P[\mu(\mathcal{N}) - Zh(P)] \le P\mu(\mathcal{N}) \le P\mu(T - \alpha P).$$

Since $\mu(0) = 0$ and $\mu(T - \alpha P)$ is decreasing with respect to P, we know $\lim_{t\to\infty} \sup P(t) \le T/\alpha := \mathcal{T}$, implying $P(t)$ is ultimately bounded.

¿From the hypothesis (C_3), $\epsilon(Z) \ge 0$ for all $Z \ge 0$, multiplying the first equation by $be^{-d\tau}$ and adding it to the second equation in (3), we can obtain

$$\begin{aligned}\left(be^{-d\tau}P(t) + Z(t+\tau)\right)' &= be^{-d\tau}P(t)\mu(\mathcal{N}) - \epsilon(Z(t+\tau))Z(t+\tau)\\ &\le be^{-d\tau}\mu(T)P(t) - \epsilon(0)Z(t+\tau),\end{aligned}$$

which is equivalent to

$$\begin{aligned}&\left(be^{-d\tau}P(t) + Z(t+\tau)\right)'\\ &\le be^{-d\tau}\mu(T)P(t) + be^{-d\tau}\epsilon(0)P(t) - \epsilon(0)\left(be^{-d\tau}P(t) + Z(t+\tau)\right)\\ &\le b\mathcal{T}e^{-d\tau}\left(\mu(T) + \epsilon(0)\right) - \epsilon(0)\left(be^{-d\tau}P(t) + Z(t+\tau)\right).\end{aligned}$$

If $\epsilon(0) > 0$ then $\lim_{t\to\infty} \sup \left(be^{-d\tau}P(t) + Z(t+\tau)\right) \le \frac{b\mathcal{T}e^{-d\tau}(\mu(T)+\epsilon(0))}{\epsilon(0)}$. Therefore, both $P(t)$ and $Z(t)$ are ultimately bounded in X. ■

2.1. *Existence and Stability of Boundary Equilibrium Points*

It is easy to see that in the system (3), there is a trivial equilibrium point $(0, 0)$ and one-species equilibrium point $(\frac{T}{\alpha}, 0) := (\mathcal{T}, 0)$. The stability of these two boundary equilibrium points are given in the following result.

Theorem 2.2.

(i) $(0, 0)$ *is always an unstable saddle point;*

(ii) $(\mathcal{T}, 0)$ *is locally asymptotically stable if* $\epsilon(0) > b\mathcal{T}h(\mathcal{T})e^{-\tau d}$ *and it is unstable if* $\epsilon(0) < b\mathcal{T}h(\mathcal{T})e^{-\tau d}$.

Proof. (i) The linearization of (3) at $(0, 0)$ is

$$\begin{aligned}x'(t) &= \mu(T)x(t)\\ y'(t) &= -\epsilon(0)y(t).\end{aligned}$$

So it is easy to see that $(0, 0)$ is a unstable saddle point since $\mu(T) > 0$ and $\epsilon(0) > 0$.

(ii) The linearization of (3) at $(\mathcal{T},0)$ is

$$x'(t) = -\mathcal{T}[\alpha\mu'(0)x(t) + (\beta\mu'(0) + h(\mathcal{T}))\,y(t)],$$
$$y'(t) = -\epsilon(0)y(t) + be^{-d\tau}\mathcal{T}h(\mathcal{T})y(t-\tau). \tag{4}$$

The characteristic equation of (4) is

$$\Xi(\lambda) = (\lambda + \alpha\mathcal{T}\mu'(0))\left(\lambda + \epsilon(0) - b\mathcal{T}h(\mathcal{T})e^{-\tau(\lambda+d)}\right) = 0.$$

Let $f(\lambda) = \lambda + \epsilon(0) - b\mathcal{T}h(\mathcal{T})e^{-\tau(\lambda+d)}$. Since $f(0) = \epsilon(0) - b\mathcal{T}h(\mathcal{T})e^{-\tau d}$, if $\epsilon(0) < b\mathcal{T}h(\mathcal{T})e^{-\tau d}$, then $f(0) < 0$ and $\lim_{\lambda\to+\infty} f(\lambda) = +\infty$. Therefore there exist $\bar{\lambda} > 0$ such that $f(\bar{\lambda}) = 0$, thus $(\mathcal{T},0)$ is unstable.

When $\epsilon(0) > b\mathcal{T}h(\mathcal{T})e^{-\tau d}$, assume there exists a zero of $f(\lambda)$ with $Re(\lambda) > 0$, then from

$$Re\lambda = Re(b\mathcal{T}h(\mathcal{T})e^{-\tau(\lambda+d)} - \epsilon(0)) \le b\mathcal{T}h(\mathcal{T})e^{-\tau d} - \epsilon(0) < 0,$$

which leads to a contradiction. Therefore all the eigenvalues in (4) have negative real parts, implying $(\mathcal{T},0)$ is locally asymptotically stable. ■

To discuss the global attractivity of the equilibrium point $(\mathcal{T},0)$, we introduce the result given in [8]:

Lemma 2.1. *If $a < b$, then the solution of the equation*

$$u'(t) = au(t-\tau) - bu(t)$$

where $a, b, \tau > 0$, and $u(t) > 0$ for $-\tau \le t \le 0$, satisfies $\lim_{t\to\infty} u(t) = 0$.

Then we have the following:

Theorem 2.3. *With the hypothesis (C_1 - C_3). When $bMe^{-d\tau} < \epsilon(0)$ then the solutions of (3) satisfy $P(t) \longrightarrow \mathcal{T}, Z(t) \longrightarrow 0$ as $t \longrightarrow \infty$.*

Proof. In the proof of Theorem (2.1), we know that $\lim_{t\to\infty}\sup P(t) \le \mathcal{T}$. Therefore there exists $t_\delta > 0$ such that

$$P(t) < \mathcal{T} + \delta, \quad \text{for all } t \ge t_\delta.$$

Following the hypothesis (C_2) and (C_3), we have $\epsilon(Z) \ge \epsilon(0)$ and $P(t-\tau)h(P(t-\tau)) \le M$ for any $t > t_\delta + \tau$ with positive constant M. Thus

$$\begin{aligned} Z' &\le bMe^{-d\tau}Z(t-\tau) - \epsilon(Z)Z(t) \\ &\le bMe^{-d\tau}Z(t-\tau) - \epsilon(0)Z(t). \end{aligned}$$

The comparison theory implies that $Z(t)$ is bounded above by the solution of

$$u'(t) = bMe^{-d\tau}u(t-\tau) - \epsilon(0)u(t), \quad t > t_\delta + \tau$$

such that $u(t) = Z(t)$ for $t \in [t_\delta, t_\delta + \tau]$. The condition $bMe^{-d\tau} < \epsilon(0)$ and the result in *Lemma(2.1)* guarantee $\lim_{t\to\infty} Z(t) = 0$.

Let $\eta \in (0, 1)$ such that $T - \eta(\beta + M) > 0$. Then there exists $t_\eta > 0$ such that for $t \geq t_\eta$, $Z(t) < \eta < \frac{T}{\beta+M}$. By the boundedness of $Ph(P)$, we have $PZh(P) < M\eta$.

Therefore

$$\begin{aligned} P' &= P\mu(\mathcal{N}) - Ph(P)Z > P\mu(\mathcal{N}) - M\eta \\ &= P\mu(T - \alpha P - \beta Z) - M\eta > P\mu(T - \alpha P - \beta\eta) - M\eta. \end{aligned}$$

In general, $T - \alpha P(t) \geq T - \alpha \lim_{t\to\infty} \sup P(t) \geq 0$. Thus

$$P' \geq P\mu(T - \alpha P - \beta\eta) - M\eta - (T - \alpha P - \eta(\beta + M)).$$

That the equation

$$v'(t) = v(t)\mu(T - \alpha v(t) - \beta\eta) - M\eta - (T - \alpha v(t) - \eta(\beta + M))$$

has a solution $v(t) = \frac{T-\beta\eta}{\alpha}$ results $\lim_{t\to\infty} \inf P(t) > \frac{T-\beta\eta}{\alpha}$. Since $\eta \in \left(0, \frac{T}{\beta+M}\right)$ is arbitrary, $\lim_{t\to\infty} \inf P(t) \geq \frac{T}{\alpha} = \mathcal{T}$. Together with $\lim_{t\to\infty} \sup P(t) \leq \mathcal{T}$, we have $\lim_{t\to\infty} P(t) = \mathcal{T}$. ■

Since $Ph(P)$ is bounded by M, we can see the condition for the local stability at $(\mathcal{T}, 0)$ in Theorem (2.2)-(ii) is included in the global stability condition in Theorem (2.3).

2.2. *Existence and Stability of Positive Equilibrium*

When system (3) exists a positive equilibrium point (P^*, Z^*), then P^*, Z^* must satisfy

$$\begin{aligned} \mu(\mathcal{N}^*) &= Z^* h(P^*), \\ be^{-d\tau} P^* h(P^*) &= \epsilon(Z^*), \end{aligned} \tag{5}$$

with $\mathcal{N}^* = T - \alpha P^* - \beta Z^*$. From the second equation in (5), we have

$$\frac{\epsilon(Z^*)e^{d\tau}}{b} = P^* h(P^*) \leq M. \tag{6}$$

Since $\epsilon(0) \leq \epsilon(Z^*)$, (6) implies that τ must be less than $\tau_{\max} = \frac{1}{d} \ln\left(\frac{bM}{\epsilon(0)}\right)$. The condition $\tau < \tau_{max}$ is necessary for the existence of such P^* and Z^* in (5), but is not a sufficient condition in general.

In addition, the existence of such P^*, Z^* and the number of solutions in (5) are uncertain which depends on the choice of the functions $\mu(\mathcal{N}), h(P)$ and $\epsilon(Z)$.

For instance, in Fig. (2), $f(P) = Ph(P)$ and $g(Z) = \frac{\epsilon(Z)e^{d\tau}}{b}$, the second equation of (5) may have infinite solution (Fig. (2a)) or none solution (Fig. (2b)). Let

$$\tilde{\tau} = \sup\{\tau \in (0, \tau_{\max})\,|\text{the second equation of (5) has at least a solution}\}. \quad (7)$$

When $\tau < \tilde{\tau}$, mathematically, we can obtain

$$Z^* = \epsilon^{-1}(be^{-d\tau}P^*h(P^*)),$$

while P^* is determined by

$$\mu(T - \alpha P^* - \beta\epsilon^{-1}(be^{-d\tau}P^*h(P^*))) = h(P^*)\epsilon^{-1}(be^{-d\tau}P^*h(P^*)),$$

which is impossible to find an analytical solution in general. However, when the total higher trophic predation $\epsilon(Z)$ is a constant, then the positive equilibrium point (P^*, Z^*) exists and is unique.

Theorem 2.4. *When $\epsilon(Z) = c$ is a constant and $bMe^{-d\tau} > c$, there exists a unique positive equilibrium point (P^*, Z^*) in (3).*

Proof. When $\epsilon(Z) = c$, $g(Z) = \frac{\epsilon(Z)e^{d\tau}}{b} = \frac{ce^{d\tau}}{b} = \hat{c}$ is a constant with fixed parameters b, c, d and τ. Since $f(P) = Ph(P)$ is a increasing function on the interval $(0, \mathcal{T})$ with $0 < f(P) \leq M$, when $M > \hat{c}$, that is, $bMe^{-d\tau} > c$, there exists a unique $P^* \in (0, \mathcal{T})$ such that $f(P^*) = g(Z^*) = \hat{c}$, i.e $P^* = f^{-1}(\hat{c})$.

In Eq. (5), multiply Z^* in the second and combine with the first equation, we have $be^{-d\tau}P^*\mu(N^*) = Z^*\epsilon(Z^*) = cZ^*$, that is

$$be^{-d\tau}f^{-1}(\hat{c})\mu(T - \alpha f^{-1}(\hat{c}) - \beta Z^*) = cZ^*. \quad (8)$$

In Eq. (8), the left-hand side is decreasing, and the right-hand side is increasing with respect to Z^*, thus there exists a unique Z^* corresponding to $P^* = f^{-1}(\hat{c})$, implying the system (5) has a unique positive equilibrium point (P^*, Z^*). ■

With fixed parameters, feasible values of delay τ and certain functions given in Table (1), Fig. (3) shows the existence of (P^*, Z^*). When $\epsilon(Z)$ is a constant, there is a unique positive equilibrium point shown in Fig.(3a), while when we change ϵ from a constant to a linear function, we can observe three positive equilibrium points, see Fig.(3b).

Table 1.: Functions chosen in Fig. (3). $\alpha = \beta = 1$.

	$h(P)$	$\mu(N)$	$\epsilon(Z)$	T
Fig. (3a)	$\frac{P}{P^2+1}$	$\frac{N}{N+10}$	0.15	5
Fig. (3b)	$\frac{P}{P^2+0.1}$	$\frac{N}{N+15}$	$0.05 + 0.31Z$	9

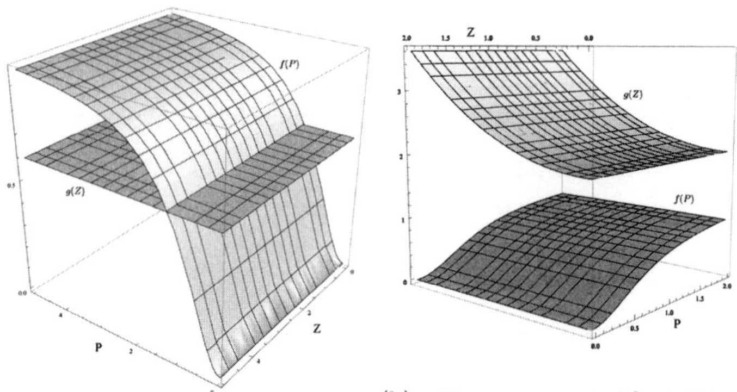

(a) $\epsilon(Z) = 0.15,\ h(P) = \frac{P}{P^2+1}$. (b) $\epsilon(Z) = 0.5 + 0.1Z^2,\ h(P) = \frac{P}{P^2+0.5}$.

Figure 2.: The existence of solution in the second equation of (5). $b = 0.25$, $d = 0.01$, $\tau = 0.2$, and $T = 9$.

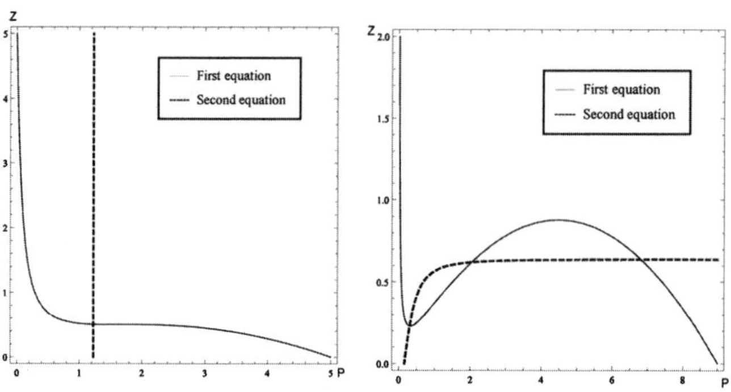

(a) One equilibrium point. (b) Three equilibrium points.

Figure 3.: Existence of (P^*, Z^*) with $b = 0.25, d = 0.01, \alpha = \beta = 1$ and functions in table (1).

When the positive equilibrium point (P^*, Z^*) exists, to study its stability, we linearize (3) at (P^*, Z^*), by setting $x = P - P^*, y = Z - Z^*$, which is,

$$\begin{aligned} \frac{dx}{dt} &= -P^*(\alpha\mu'_* + Z^*h'_*)x(t) - P^*(\beta\mu'_* + h_*)y(t), \qquad (9) \\ \frac{dy}{dt} &= -(\epsilon_* + Z^*\epsilon'_*)y(t) + be^{-d\tau}Z^*(h_* + P^*h'_*)x(t-\tau) + be^{-d\tau}P^*h_*y(t-\tau), \end{aligned}$$

where h_*, ϵ_* denotes the value of the function h, ϵ and $\mu'_*, h'_*, \epsilon'_*$ denotes the value of the derivative of the associated function at (P^*, Z^*) respectively. The characteristic equation of (9) is

$$\Delta(\lambda,\tau) = \lambda^2 + a_1\lambda + a_2 + b_1\lambda e^{-\tau\lambda} + b_2 e^{-\tau\lambda} = 0, \tag{10}$$

with

$$\begin{aligned}
a_1 &= P^*(\alpha\mu'_* + Z^* h'_*) + (\epsilon_* + Z^*\epsilon'_*),\\
a_2 &= P^*(\beta\mu'_* + Z^* h'_*)(\epsilon_* + Z^*\epsilon'_*),\\
b_1 &= -be^{-d\tau}P^* h_* < 0,\\
b_2 &= -be^{-d\tau}(P^*)^2 h_*(\alpha\mu'_* + Z^* h'_*) + be^{-d\tau}Z^*P^*(\beta\mu'_* + h_*)(h_* + P^* h'_*).
\end{aligned}$$

As we know that all the roots of $\Delta(\lambda,\tau)$ with $Re(\lambda) \geq 0$ lie in a bounded domain [15]. When $\tau = 0$. Eq. (10) becomes

$$\Delta(\lambda,0) = \lambda^2 + (a_1 + b_1)\lambda + (a_2 + b_2) = 0,$$

where

$$\begin{aligned}
a_1 + b_1 &= P^*(\alpha\mu'_* + Z^* h'_*) + Z^*\epsilon'_*,\\
a_2 + b_2 &= P^*(\alpha\mu'_* + Z^* h'_*)Z^*\epsilon'_* + be^{-d\tau}P^*Z^*(\beta\mu'_* + h_*)(h_* + P^* h'_*)\\
&= P^*(\alpha\mu'_* + Z^* h'_*)Z^*\epsilon'_* + be^{-d\tau}P^*Z^*(\beta\mu'_* + h_*)\left.\frac{dPh}{dP}\right|_{P=P^*}.
\end{aligned}$$

$a_1 + b_1 > 0$ if and only if $h'_* > -\left(\frac{\alpha\mu'_*}{Z^*} + \frac{\epsilon'_*}{P^*}\right)$ and since $\frac{dPh}{dP} > 0$, we have $a_2 + b_2 > 0$ if and only if $h'_* > \frac{-be^{-d\tau}}{Z^*\epsilon'_*}(\beta\mu'_* + h_*)(\left.\frac{dPh}{dP}\right|_{P=P^*}) - \frac{\alpha\mu'_*}{Z^*}$. Following the Routh-Hurwitz stability criterion we have

Proposition 2.1. *At $\tau = 0$, the equilibrium point (P^*, Z^*) is locally asymptotically stable if and only if*

$$h'_* > -\left(R + \frac{\mu'_*}{Z^*}\right), \tag{$\mathbf{H_0}$}$$

where $R = \min\left\{\frac{\epsilon'_*}{P^*}, \frac{be^{-d\tau}}{Z^*\epsilon'_*}(\beta\mu'_* + h_*)(\left.\frac{dPh}{dP}\right|_{P=P^*})\right\}$.

Let $\tau > 0$ and suppose $\lambda = i\omega$ $(\omega > 0)$ is a purely imaginary root of (10). Substituting it into (10) and separating the real and imaginary parts, we obtain:

$$\begin{aligned}
-\omega^2 + a_2 &= -b_1\omega\sin\omega\tau - b_2\cos\omega\tau\\
a_1\omega &= -b_1\omega\cos\omega\tau + b_2\sin\omega\tau.
\end{aligned} \tag{11}$$

Squaring and adding both equations of (11) lead to

$$F(\omega,\tau) = \omega^4 + q(\tau)\omega^2 + \ell(\tau) = 0,$$

where

$$\begin{aligned}
q(\tau) &= a_1^2 - 2a_2 - b_1^2,\\
\ell(\tau) &= a_2^2 - b_2^2.
\end{aligned}$$

Let $u = \omega^2$. In seek of the positive real roots in $F(u,\tau) = u^2 + q(\tau)u + \ell(\tau) = 0$, at first we need

$$\Theta = q(\tau)^2 - 4\ell(\tau) = \left(a_1^2 - b_1^2\right)\left(a_1^2 - b_1^2 - 4a_2\right) + 4b_2^2 \geq 0. \qquad \textbf{(H}_1\textbf{)}$$

Obviously, $F(u,\tau) = 0$ has no positive roots when $\ell(\tau) \geq 0$ and $q(\tau) \geq 0$ which is equivalent to

$$a_2^2 - b_2^2 \geq 0 \ \ \& \ \ a_1^2 - 2a_2 - b_1^2 \geq 0; \qquad \textbf{(H}_2\textbf{)}$$

and has exactly one positive root if

$$\ell(\tau) = a_2^2 - b_2^2 < 0; \qquad \textbf{(H}_3\textbf{)}$$

or when

$$\ell(\tau) = a_2^2 - b_2^2 = 0 \ \ \& \ \ q(\tau) = a_1^2 - 2a_2 - b_1^2 < 0; \qquad \textbf{(H}_4\textbf{)}$$

or two positive roots when $\ell(\tau) > 0$ and $q(\tau) < 0$ which is equivalent to

$$a_2^2 - b_2^2 > 0 \ \ \& \ \ a_1^2 - 2a_2 - b_1^2 < 0. \qquad \textbf{(H}_5\textbf{)}$$

Furthermore, from (11), we have

$$\sin\omega(\tau)\tau = \frac{a_1 b_2 \omega(\tau) - b_1\omega(\tau)\left(a_2 - \omega(\tau)^2\right)}{b_1^2\omega(\tau)^2 + b_2^2},$$
$$\cos\omega(\tau)\tau = \frac{b_2\left(\omega(\tau)^2 - a_2\right) - a_1 b_1 \omega(\tau)^2}{b_1^2\omega(\tau)^2 + b_2^2}. \qquad (12)$$

Define the function $\theta(\tau) \in [0, 2\pi)$ such that $\sin\theta(\tau)$ and $\cos\theta(\tau)$ are given in (12). Following [2, 13], let

$$S_n(\tau) = \tau - \frac{\theta(\tau) + 2n\pi}{\omega(\tau)}, \quad n \in \mathbb{N}, \quad \tau \in (0, \tau_{\max}), \qquad (13)$$

then $i\omega(\tau^*)$ is a root in (10) if and only if τ^* is a zero of a function S_n for some $n \in \mathbb{N}$.

Proposition 2.2. *If $\tau^* \in (0, \tau_{\max})$ is a positive root of S_n given in (13) for some $n \in \mathbb{N}$, then a pair of purely imaginary roots $i\omega(\tau^*)$ of (10) exist which crosses the imaginary axis from left to right if $S_n'(\tau^*) > 0$ and crosses the imaginary axis from right to left if $S_n'(\tau^*) < 0$, and*

$$Sign\left\{\left.\frac{d\mathrm{Re}(\lambda)}{d\tau}\right|_{\lambda = i\omega(\tau^*)}\right\} = Sign\left\{\left.\frac{dS_n(\tau)}{d\tau}\right|_{\tau=\tau^*}\right\}.$$

Therefore, when the positive equilibrium point (P^*, Z^*) exists, the stability of (P^*, Z^*) is given in the following:

Theorem 2.5.

(i) *When* $(\boldsymbol{H_0})$, $(\boldsymbol{H_1})$ *and* $(\boldsymbol{H_2})$ *hold,* (P^*, Z^*) *is locally asymptotically stable for any feasible time delay* τ*;*

(ii) *When* $(\boldsymbol{H_0})$, $(\boldsymbol{H_1})$ *with either* $(\boldsymbol{H_3})$, $(\boldsymbol{H_4})$ *or* $(\boldsymbol{H_5})$ *hold,* (P^*, Z^*) *is locally asymptotically stable for small time delay* τ*, say* $\tau < \min\{\tau_n^*\}$*;*

(iii) *When* $(\boldsymbol{H_0})$ *does not hold and* $(\boldsymbol{H_1})$ *with either* $(\boldsymbol{H_3})$, $(\boldsymbol{H_4})$ *or* $(\boldsymbol{H_5})$ *hold,* (P^*, Z^*) *is unstable for small time delay* τ*. If there exists a* τ_n^**, such that* $S_n'(\tau_n^*) < 0$*,* (P^*, Z^*) *can become stable with* $\tau > \tau_n^*$*.*

3. n Parallel Phytoplankton Zooplankton Patches

In this section, we discuss the general model (2) with n parallel phytoplankton-zooplankton patches. Rewrite the system (2) as

$$
\begin{aligned}
{P_1}' &= P_1\left[\mu_1(\mathcal{N}) - Z_1 h_1(P_1)\right] \\
{Z_1}' &= b_1 e^{-d_1\tau_1} P_1(t-\tau_1) Z_1(t-\tau_1) h_1(P_1(t-\tau_1)) - Z_1\epsilon_1(\sum_{k=1}^{n} Z_k) \\
&\ \vdots \\
{P_n}' &= P_n[\mu_n(\mathcal{N}) - Z_n h_n(P_n)] \\
{Z_n}' &= b_n e^{-d_n\tau_n} P_n(t-\tau_n) Z_n(t-\tau_n) h_n(P_n(t-\tau_n)) - Z_n\epsilon_n(\sum_{k=1}^{n} Z_k).
\end{aligned}
\tag{14}
$$

Analogue to the positivity and boundedness property given in Theorem (2.1), we have the following result.

Theorem 3.1. *Given the initial condition* $P_i(0), Z_i(0) \in C^+$*, for all* $i = 1, \ldots, n$*, then under the hypothesis* $(C_1$ - $C_3)$*, the solutions of (14) are non-negative in* $\mathcal{X} = C^n \times C^n$*. In addition, if each* $\epsilon_i(0) > 0$*, then the solutions are ultimately bounded in* $\mathcal{X}$*.*

The proof of Theorem (3.1) is similar to that in Theorem (2.1) , we omit here.

System (14) has a trivial equilibrium point $E_0 = \underbrace{(0, 0, \ldots, 0, 0)}_{2n}$ and infinite number of predator-free equilibrium points with the form $\tilde{E} = \underbrace{\left(\alpha_1\tilde{P}_1, 0, \alpha_2\tilde{P}_2, 0, \ldots, \alpha_n\tilde{P}_n, 0\right)}_{2n}$ such that $\sum_{k=1}^{n} \alpha_k\tilde{P}_k = T$, and it is also possible to have the co-existed positive equilibrium point $(P_1^*, Z_1^*, P_2^*, Z_2^* \ldots, P_n^*, Z_n^*)$ under the hypothesis $(C_1$ - $C_3)$ and certain conditions.

To study the stability of any equilibrium point $(x_1, y_1, x_2, y_2, \ldots, x_n, y_n)$, we need to know the general form of the characteristic equation, which is,

$$
\Delta(\lambda) = \det\left(\lambda I - J^0 - \sum_{k=1}^{n} J^k e^{-\lambda\tau_k}\right) = 0
$$

where $J^0 = (a_{ij})$ and $J^k = \left(c_{ij}^k\right)$ are $2n \times 2n$ matrices with

$$a_{ij} = \begin{cases} -x_{\frac{i+1}{2}}\left[\alpha_{\frac{i+1}{2}}\mu'_{\frac{i+1}{2}}(\bar{\mathcal{N}}) + y_{\frac{i+1}{2}}h'_{\frac{i+1}{2}}\left(x_{\frac{i+1}{2}}\right)\right], & i \text{ odd}, j = i; \\ -x_{\frac{i+1}{2}}\left[\beta_{\frac{i+1}{2}}\mu'_{\frac{i+1}{2}}(\bar{\mathcal{N}}) + h_{\frac{i+1}{2}}\left(x_{\frac{i+1}{2}}\right)\right], & i \text{ odd}, \ j = i+1; \\ -\alpha_{\frac{j+1}{2}}x_{\frac{i+1}{2}}\mu'_{\frac{i+1}{2}}(\bar{\mathcal{N}}), & i, j \text{ odd}, i \neq j; \\ -\beta_{\frac{j}{2}}x_{\frac{i+1}{2}}\mu'_{\frac{i+1}{2}}(\bar{\mathcal{N}}), & i \text{ odd}, j \text{ even } j \neq i+1; \\ -\epsilon_{\frac{i}{2}}\left(\sum_{k=1}^{n} y_k\right) - y_{\frac{i}{2}}\epsilon'_{\frac{i}{2}}\left(\sum_{k=1}^{n} y_k\right), & i \text{ even}, j = i; \\ -y_{\frac{i}{2}}\epsilon'_{\frac{i}{2}}\left(\sum_{k=1}^{n} y_k\right), & i, j \text{ even}, i \neq j; \\ 0, & i \text{ even}, j \text{ odd}, \end{cases}$$

where $\bar{\mathcal{N}} = T - \sum_{k=1}^{n} \alpha_k x_k - \sum_{k=1}^{n} \beta_k y_k$ and

$$c_{ij}^k = \begin{cases} 0, & i \text{ odd, any } j; \text{ or } i \text{ even}, j \neq 2k, j \neq 2k-1; \\ b_k e^{-d_k\tau_k} y_k\left(h_k(x_k) + x_k h'_k(x_k)\right), & i = 2k, \ j = 2k-1; \\ b_k e^{-d_k\tau_k} x_k h_k(x_k), & i = 2k, \ j = 2k; \end{cases}$$

Therefore, about the stability of the boundary equilibrium points E_0 and $\tilde{E}$, we have

Theorem 3.2.

(i) E_0 *is always an unstable saddle point;*

(ii) All $\tilde{E}$ *are locally asymptotically stable if* $b_k\tilde{P}_k h_k(\tilde{P}_k)e^{-d_k\tau_k} < \epsilon_k(0)$ *for all* $k = 1, 2, \ldots, n$ *and they are unstable if* $b_k\tilde{P}_k h_k(\tilde{P}_k)e^{-d_k\tau_k} > \epsilon_k(0)$ *for some* $k = 1, 2, \ldots, n$.

Proof (i) At the equilibrium point E_0, $J^k\Big|_{E_0} = [0]$ and

$$J^0 = \left(\begin{array}{cc|cc|c|cc} \mu_1(T) & 0 & 0 & & \cdots & 0 & \\ 0 & -\epsilon_1(0) & & & & & \\ \hline 0 & & \mu_2(T) & 0 & \cdots & 0 & \\ & & 0 & -\epsilon_2(0) & & & \\ \hline \vdots & & \vdots & & \ddots & \vdots & \\ \hline 0 & & 0 & & \cdots & \mu_n(T) & 0 \\ & & & & & 0 & -\epsilon_n(0) \end{array}\right).$$

Obviously E_0 is unstable saddle point since there are positive eigenvalues $\mu_i(T)$ and negative eigenvalues $-\epsilon_i(0)$, $i = 1, 2, \ldots, n$.
(ii) Let $\xi_i = b_i e^{-d_i\tau_i}$. The characteristic equation $\Delta(\lambda) = 0$ at the point $\tilde{E}$ becomes

$$\det\begin{pmatrix} \lambda+\alpha_1\tilde{P}_1\mu_1'(0) & \beta_1\tilde{P}_1\mu_1'(0)+h_1\left(\tilde{P}_1\right) & \alpha_2\tilde{P}_1\mu_1'(0) & \cdots & \beta_n\tilde{P}_1\mu_1'(0) \\ 0 & \begin{matrix}\lambda+\epsilon_1(0)\\ -\xi_1\tilde{P}_1h_1\left(\tilde{P}_1\right)e^{-\lambda\tau_1}\end{matrix} & 0 & \cdots & 0 \\ \vdots & \vdots & \vdots & \ddots & \vdots \\ \alpha_1\tilde{P}_n\mu_n'(0) & \beta_1\tilde{P}_n\mu_n'(0) & \alpha_2\tilde{P}_n\mu_n'(0) & \cdots & \beta_n\tilde{P}_n\mu_n'(0)+h_n\left(\tilde{P}_n\right) \\ 0 & 0 & 0 & \cdots & \begin{matrix}\lambda+\epsilon_n(0)\\ -\xi_n\tilde{P}_nh_n\left(\tilde{P}_n\right)e^{-\lambda\tau_n}\end{matrix} \end{pmatrix}=0$$

which is,

$$\Delta(\lambda)=\left(\prod_{k=1}^{n}\left(\lambda+\epsilon_k(0)-\xi_k\tilde{P}_kh_k\left(\tilde{P}_k\right)e^{-\lambda\tau_k}\right)\right)$$
$$\times\begin{vmatrix} \lambda+\alpha_1\tilde{P}_1\mu_1'(0) & \alpha_2\tilde{P}_1\mu_1'(0) & \cdots & \alpha_n\tilde{P}_1\mu_1'(0) \\ \alpha_1\tilde{P}_2\mu_2'(0) & \lambda+\alpha_2\tilde{P}_2\mu_2'(0) & \cdots & \alpha_n\tilde{P}_2\mu_2'(0) \\ \vdots & \vdots & \ddots & \vdots \\ \alpha_1\tilde{P}_n\mu_n'(0) & \alpha_2\tilde{P}_n\mu_n'(0) & \cdots & \lambda+\alpha_n\tilde{P}_n\mu_n'(0) \end{vmatrix}=0.$$

Since

$$\begin{vmatrix} \lambda+\alpha_1\tilde{P}_1\mu_1'(0) & \alpha_2\tilde{P}_1\mu_1'(0) & \cdots & \alpha_n\tilde{P}_1\mu_1'(0) \\ \alpha_1\tilde{P}_2\mu_2'(0) & \lambda+\alpha_2\tilde{P}_2\mu_2'(0) & \cdots & \alpha_n\tilde{P}_2\mu_2'(0) \\ \vdots & \vdots & \ddots & \vdots \\ \alpha_1\tilde{P}_n\mu_n'(0) & \alpha_2\tilde{P}_n\mu_n'(0) & \cdots & \lambda+\alpha_n\tilde{P}_n\mu_n'(0) \end{vmatrix}=\lambda^{n-1}\left(\lambda+\sum_{i=1}^{n}\left(\alpha_i\tilde{P}_i\mu_i'(0)\right)\right),$$

then the characteristic equation becomes

$$\Delta(\lambda)=\lambda^{n-1}\left(\prod_{k=1}^{n}\left(\lambda+\epsilon_k(0)-\xi_k\tilde{P}_kh_k\left(\tilde{P}_k\right)e^{-\lambda\tau_k}\right)\right)\left(\lambda+\sum_{i=1}^{n}\left(\alpha_i\tilde{P}_i\mu_i'(0)\right)\right)=0.$$

From the result in Theorem (2.2), it is easy to see that if $\xi_k\tilde{P}_kh_k(\tilde{P}_k)<\epsilon_k(0)$ for all $k=1,2,\ldots,n$, then $\tilde{E}$ is locally asymptotically stable and unstable if $b_k\tilde{P}_kh_k(\tilde{P}_k)e^{-d_k\tau_k}>\epsilon_k(0)$ for some $k=1,2,\ldots,n$. ■

Although we cannot show the global attractivity of each equilibrium point $\tilde{E}$, with parallel proof in Theorem (2.3), we have the following "group" property.

Theorem 3.3. *When $b_iM_ie^{-d_i\tau_i}<\epsilon_i(0)$ for all $i=1,\ldots,n$, the solutions of (14) satisfy $\sum_{i=1}^{n}\alpha_iP_i(t)\longrightarrow T$ and $Z_i(t)\longrightarrow 0$ for all $i=1,\ldots,n$ as $t\longrightarrow\infty$.*

As we know from the previous section that, even for the system with one patch ($n=1$), the existence and the number of positive equilibrium point are uncertain which depend on the choices of the functions, parameters and time delay as well. With general functions and $n>1$, this becomes more complicated. Biologically, it is almost impossible to know the environment for the coexistence of each species. However, we know that, if each species can coexist, at least the maturation time delay for each predator cannot be long. In fact, assume there exists a positive equilibrium point $E^* = (P_1^*, Z_1^*, P_2^*, Z_2^*, \ldots, P_n^*, Z_n^*)$, then from

$$\frac{\epsilon_i\left(\sum\limits_{k=1}^{n} Z_k^*\right) e^{d_i\tau_i}}{b_i} = P_i^* h_i\left(P_i^*\right) \leq M_i,$$

and the hypothesis (C_3), we know that τ_i must be less than

$$\tau_{\max} = \min\left\{\frac{1}{d_i}\ln\left(\frac{b_i M_i}{\epsilon_i(0)}\right) : i = 1, 2, \ldots, n\right\}.$$

The condition $\tau < \tau_{\max}$ is necessary for the existence of such E^* in (14).

With some particular choice of the functions, we maybe able to obtain some better result.

Case I: When the total predation terms from higher trophic level from higher trophic level $\epsilon_i\left(\sum\limits_{k=1}^{n} Z_k\right)$ are constants for all $i = 1, 2, \ldots, n$, then the positive equilibrium point E^* exists and is unique under certain condition, which is an extension of Theorem (2.4).

Theorem 3.4. *When* $\epsilon_i\left(\sum\limits_{k=1}^{n} Z_k\right) = c_i$ *are constants and* $b_i M_i e^{-d_i\tau_i} > c_i$ *for* $i = 1, 2, \ldots, n$, *there exist a unique positive equilibrium point* E^* *in (14).*

Case II: When the n parallel food chains are symmetric, that is, the growth functions for all taxa at the same trophic level have the same functional forms and the same parameter values, the maturation delay for each predator is the same, then the system has at least a positive equilibrium point E^* with $P_1^* = P_2^* = \ldots = P_n^* := P^*$, and $Z_1^* = Z_2^* = \ldots = Z_n^* := Z^*$ and the characteristic equation $\Delta(\lambda) = 0$ has the form,

$$\begin{vmatrix}
\lambda+l_1+l_2\alpha & l_3+l_2\beta & l_2\alpha & l_2\beta & l_2\alpha & l_2\beta & \cdots \\
m_1e^{-\tau\lambda} & \begin{matrix}\lambda+m_2e^{-\tau\lambda}\\+m_3+m_4\end{matrix} & 0 & m_4 & 0 & m_4 & \cdots \\
l_2\alpha & l_2\beta & \lambda+l_1+l_2\alpha & l_3+l_2\beta & l_2\alpha & l_2\beta & \cdots \\
0 & m_4 & m_1e^{-\tau\lambda} & \begin{matrix}\lambda+m_2e^{-\tau\lambda}\\+m_3+m_4\end{matrix} & 0 & m_4 & \cdots \\
l_2\alpha & l_2\beta & l_2\alpha & l_2\beta & \lambda+l_1+l_2\alpha & l_3+l_2\beta & \cdots \\
0 & m_4 & 0 & m_4 & m_1e^{-\tau\lambda} & \begin{matrix}\lambda+m_2e^{-\tau\lambda}\\+m_3+m_4\end{matrix} & \cdots \\
\vdots & \vdots & \vdots & \vdots & \vdots & \vdots & \ddots
\end{vmatrix} = 0,$$

which is equal to

$$\Delta(\lambda) = \det\left(A_1 A_2^{n-1}\right) = 0,$$

where

$$A_1 = \begin{pmatrix} \lambda + l_1 + n\alpha l_2 & l_3 + nl_2\beta \\ m_1 e^{-\tau\lambda} & \lambda + m_3 + n\beta m_4 + m_2 e^{-\tau\lambda} \end{pmatrix}$$

and

$$A_2 = \begin{pmatrix} \lambda + l_1 & l_3 \\ m_1 e^{-\tau\lambda} & \lambda + m_3 + m_2 e^{-\tau\lambda} \end{pmatrix},$$

with

$$\begin{aligned} l_1 &= P^* Z^* h'_*, \quad l_2 = P^* \mu'_*, \quad l_3 = P^* h_*, \\ m_1 &= -be^{-d\tau} Z^* (h_* + P^* h'_*), \; m_2 = -be^{-d\tau} l_3, \; m_3 = \epsilon_*, \; m_4 = Z^* \epsilon'_*. \end{aligned}$$

Comparing with the result in [1], without the maturation delay, the effect from A_2 does not affect the stability of the system if the predator functional response is stabilizing. However, the introduction of the delay in the maturation time, the system becomes much more involved, we can expect the system to exhibit rich dynamics, especially when the system owns some symmetry. Certainly it is not trivial, we leave the theoretical analysis for future research, and give some numerical simulations in next section.

4. Numerical Simulations

In this section, we choose some different functions, parameters and time delay in several examples to compliment the theoretical results given previously.

Example 1. $n = 1$. We take the functions and the parameters in system (3) as,

$$\begin{aligned} \mu(\mathcal{N}) &= \frac{\mathcal{N}}{15 + \mathcal{N}}, \; h(P) = \frac{7}{P+1}, \; \epsilon(Z) = 0.05 + 0.8Z \\ d &= 0.13, \; b = 0.05, \; T = 7, \; \alpha = 0.9, \; \beta = 0.4. \end{aligned} \tag{15}$$

First, we can calculate $\tau_{\max} = 14.968$. At $\tau = 0$, the positive equilibrium is $E_1 = (0.393, 0.061)$. By checking the condition $(\mathbf{H_0})$, we know E_1 is unstable. With different delay $\tau > 0$ and the choice in (15), we can obtain $\tilde{\tau} \approx 9.248$ defined in (7), implying the existence of the positive equilibrium point when $\tau \in (0, \tilde{\tau})$. $\tilde{\tau} < \tau_{\max}$ gives an evidence for that $\tau < \tau_{\max}$ is necessary, but not sufficient for the existence of positive steady state.

In Fig. (4), we plot the graphs of the functions $S_0(\tau), S_1(\tau), S_2(\tau)$ and $S_3(\tau)$ for $\tau \in (0, \tau_{\max})$. The curve of $S_1(\tau) = 0$ has one root $\tau^* = 4.347$ and is decreasing from positive to negative, indicating the positive equilibrium is unstable for $\tau \in (0, \tau^*)$ and becomes stable at τ^* before it disappears at $\tilde{\tau}$. Fig. (5) shows the phase portrait of the solutions with different values of τ which tells us that when $0 < \tau < \tau_{\max}$, there exists a stable limit cycle. With increasing of delay τ, the limit cycle is broken, the system approaches to a stable steady state, until τ is sufficient large. Biologically, when the maturation process is short, the concentration of phytoplankton and zooplankton are oscillated regularly; when this process is a little long, all the species move to a certain level with constant concentration; but if the maturation time is too long, the zooplankton cannot survive

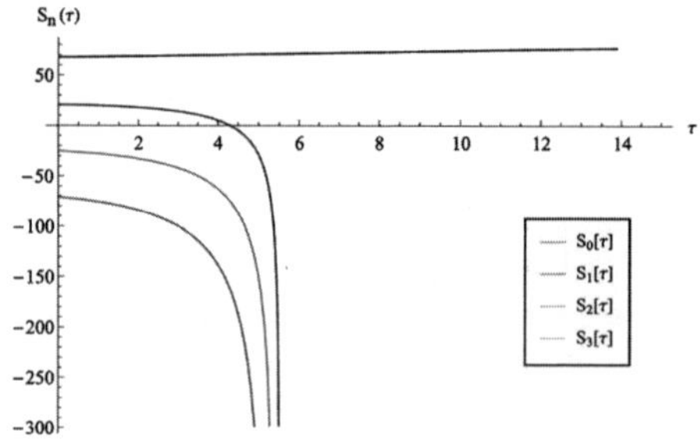

Figure 4.: Functions $S_0(\tau), S_1(\tau), S_2(\tau)$ and $S_3(\tau)$ for $\tau \in (0, 14.968)$ corresponding to the choice in (15).

without enough prey–phytoplankton. Therefore, the maturation time control the dynamical behavior of the system. From this example we can see that, although we have proved the uniqueness of the positive equilibrium point when $\epsilon(Z)$ is constant, the result may hold for other functions such as linear function.

Example 2. $n = 1$. Let

$$\mu(\mathcal{N}) = \frac{\mathcal{N}}{15+\mathcal{N}},\ h(P) = \frac{P}{P^2+0.1},\ \epsilon(Z) = 0.05 + 0.31Z$$
$$\alpha = \beta = 1,\ d = 0.01,\ b = 0.25,\ T = 9 \tag{16}$$

in system (3), then $\tau_{\max} = 160.944$.

There are three positive equilibrium points $E_1 = (P_1^*, Z_1^*)$, $E_2 = (P_2^*, Z_2^*)$ and $E_3 = (P_3^*, Z_3^*)$ with $P_1^* < P_2^* < P_3^*$ for $\tau \in (0, \hat{\tau})$ with $\hat{\tau} = 38.836$, two (P_1^*, Z_1^*), (P_2^*, Z_2^*) at $\tau = \hat{\tau}$ and one (P^*, Z^*) for $\tau \in (\hat{\tau}, \tilde{\tau})$, where $\tilde{\tau} = 160.819$, see Fig. (6). In Fig. (6), the curves $f_1(P^*)$ and $f_2(P^*)$ are obtained from (5) since ϵ is linear. When τ is small, E_1 is stable and E_3 is unstable which can be seen in Figs. (7a) and (7b), while E_2 is a saddle point. With the increasing of τ, E_1 loses the stability at $\tau^* = 1.826$ and a stable limit cycle is bifurcated, E_3 gains the stability at τ^*.

From Fig. (8), we can see that the stability properties of E_1 and E_3 are "local". Since E_2 is a saddle point, there exists a "basin boundary" which divides the phase plane into two parts with different dynamics in each part.

Example 3. When $n = 3$ in the system (2). If we choose all the functional responses and the parameters in each species are the same as,

$$\mu_1(\mathcal{N}) = \mu_2(\mathcal{N}) = \mu_3(\mathcal{N}) = \frac{\mathcal{N}}{10+\mathcal{N}},\ h_1(P) = h_2(P) = h_3(P) = \frac{P}{P^2+1},$$
$$\epsilon_1(Z) = \epsilon_2(Z) = \epsilon_3(Z) = 0.15,\ d_1 = d_2 = d_3 = 0.01,\ b_1 = b_2 = b_3 = 0.25,$$
$$T = 7,\ \alpha_1 = \alpha_2 = \alpha_3 = 0.8,\ \beta_1 = \beta_2 = \beta_3 = 0.5. \tag{17}$$

then $\tau_{\max} = 51.083$ and $\tilde{\tau} = 28.122$. We can check that the condition given in Theorem (3.4) is satisfied for small value of delay $\tau_1 = \tau_2 = \tau_3 := \tau$, thus there

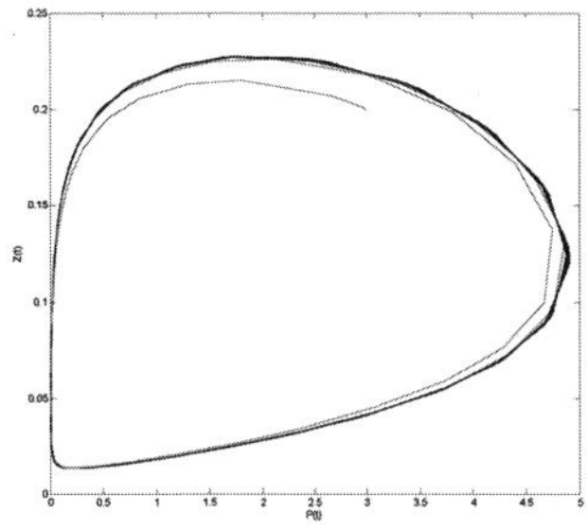

(a) $\tau = 0.1$.

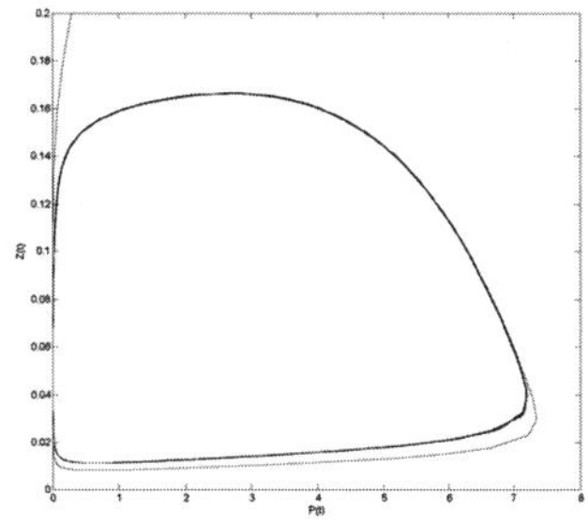

(b) $\tau = 2.9$.

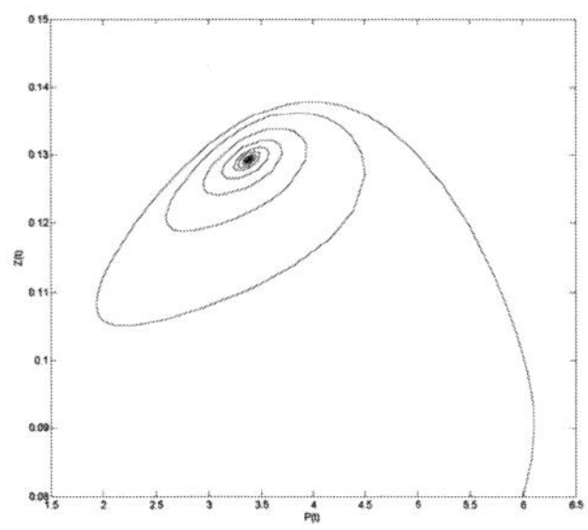

(c) $\tau = 4.5$.

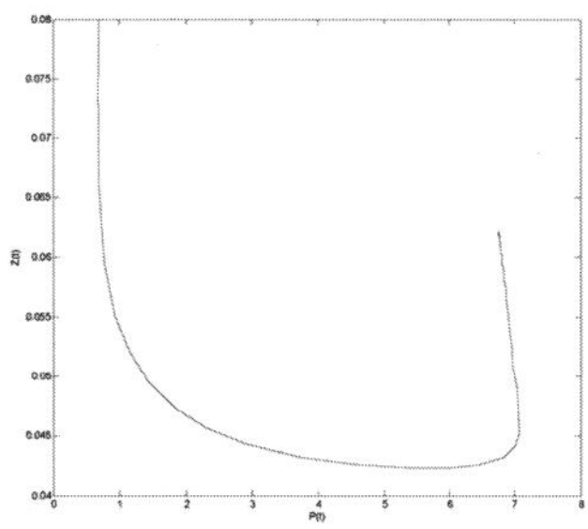

(d) $\tau = 8.6$.

Figure 5.: Phase portrait of (3) with different values of time delay τ and the functions and parameters given in (15).

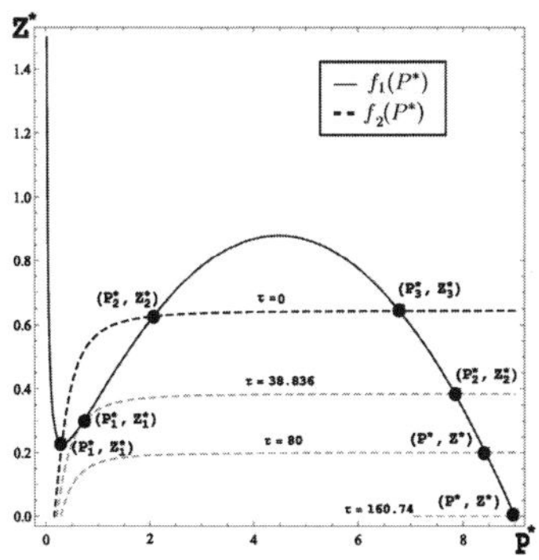

Figure 6.: Existence of different number of positive equilibrium points with different τ.

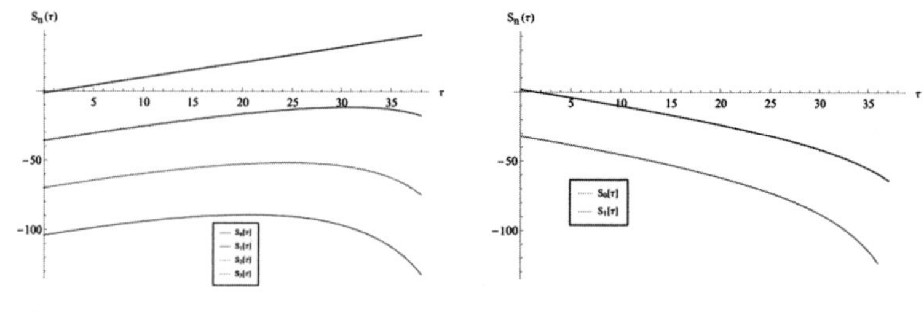

(a) at P_1^*.   (b) at P_3^*.

Figure 7.: Graphs of the functions $S_0(\tau)$ and $S_1(\tau)$ for $\tau \in (0, \hat{\tau})$ with the functions and parameters chosen in (16).

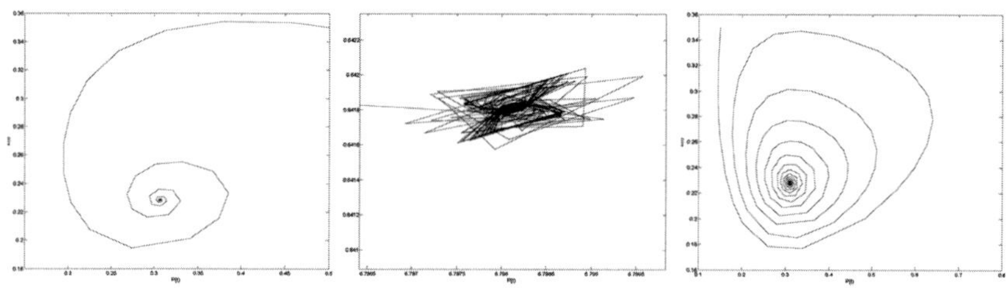

(a) $\tau = 0.2$, P_1^*.   (b) $\tau = 0.2$, P_3^*.   (c) $\tau = 1.5$, P_1^*.

(d) $\tau = 1.5$, P_3^*.   (e) $\tau = 2.1$, P_1^*.   (f) $\tau = 2.1$, P_3^*.

Figure 8.: Phase portrait of the system (3) with functions and parameters given in (16).

exists a unique positive equilibrium point. Since all the functional responses are the same, it is expected that all the species will be synchronized, i.e., with different initial condition, after a transaction period, each phytoplankton and each zooplankton approach to the same level by group which is confirmed in Fig. (9a,9b). With the increasing of the maturation time, the steady state loses the stability and the system becomes oscillatory. See Fig. (9c,9d).

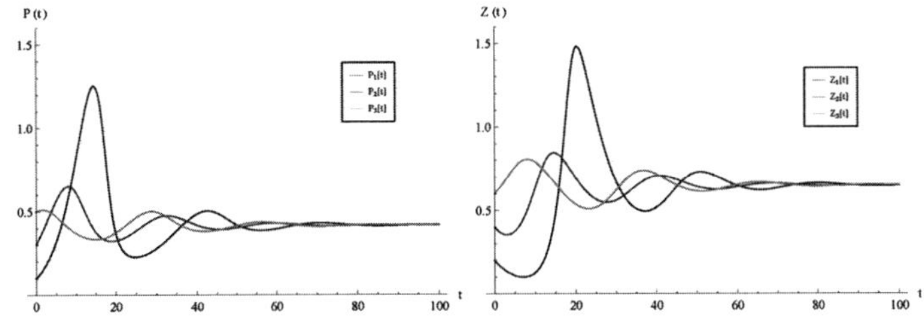

(a) Time series $P_i(t), (i = 1, 2, 3)$. (b) Time series $Z_i(t), (i = 1, 2, 3)$.

$\tau = 0.8$, $P_1(0) = 0.1, Z_1(0) = 0.2, P_2(0) = 0.3, Z_2(0) = 0.4, P_3(0) = 0.5$ and $Z_3(0) = 0.6$.

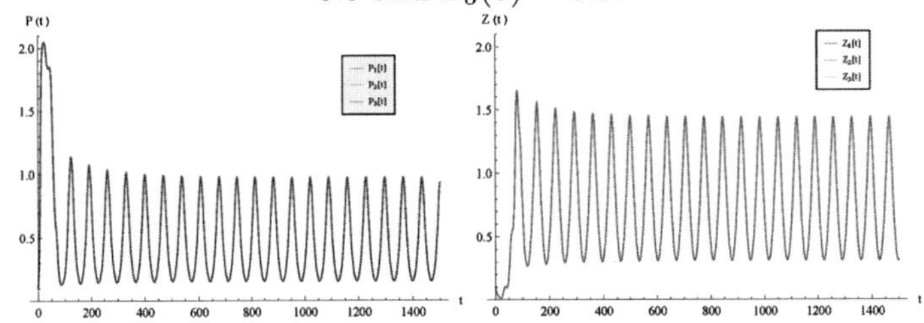

(c) Time series $P_i(t), (i = 1, 2, 3)$. (d) Time series $Z_i(t), (i = 1, 2, 3)$.

$\tau = 20$, $P_1(0) = P_2(0) = P_3(0) = Z_1(0) = Z_2(0) = Z_3(0) = 0.1$.

Figure 9.: Time series $P_i(t)$ and $Z_i(t)(i = 1, 2, 3)$ of system (2) with functions and parameters given in (17).

Example 4. When $n = 3$ in the system (2). Based on the choice in (17), we take different functional responses and some parameters, such as,

$$\begin{aligned}
&\mu_1(\mathcal{N}) = \frac{\mathcal{N}}{15 + \mathcal{N}}, \mu_2(\mathcal{N}) = \mu_3(\mathcal{N}) = \frac{4\mathcal{N}}{10 + \mathcal{N}}, \ h_1(P) = \frac{P}{P^2 + 0.1}, \\
&h_2(P) = h_3(P) = \frac{1}{P + 1}, \ \epsilon_1(Z) = 0.05 + 0.31Z, \ \epsilon_2(Z) = \epsilon_3(Z) = 0.15, \\
&T = 9, d_1 = d_2 = d_3 = 0.01, \ b_1 = b_2 = b_3 = 0.25, \\
&\alpha_1 = \alpha_2 = \alpha_3 = 1, \ \beta_1 = \beta_2 = \beta_3 = 1. \qquad (18)
\end{aligned}$$

We can observe interesting dynamical behavior, see Fig. (10). With the same initial condition, when the maturation time is very small, a doubly periodic solution exists, implying a parallel condition given in $(\mathbf{H_5})$ may be satisfied for the feasible value of τ and the choice in (18). While when we just change the initial condition a little bit, say, keeping all are the same as that in Fig. (10) except changing $Z_2(0)$ from 0.1 to 0.2, then the doubly periodic orbit is disappeared, although a periodic solution still exists and the second and the third patches have almost the same behavior, see Fig. (11). Therefore the dynamical behavior is sensitive to the initial condition. Is it possible to have chaotic motion and/or

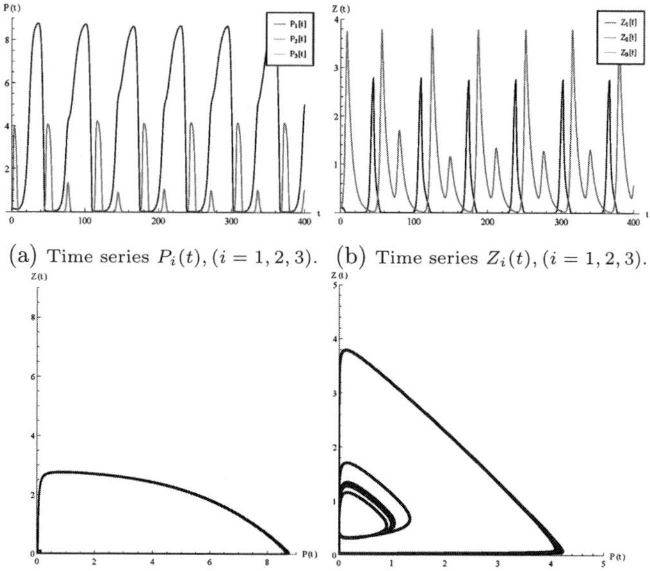

(a) Time series $P_i(t), (i = 1, 2, 3)$. (b) Time series $Z_i(t), (i = 1, 2, 3)$.

(c) $P_1(t)$ and $Z_1(t)$. (d) $P_i(t)$,$Z_i(t)$, $i = 2, 3$.

$\tau = 0.1,\ P_1(0) = P_2(0) = P_3(0) = Z_1(0) = Z_2(0) = Z_3(0) = 0.1.$

Figure 10.: (a),(b): Time series $P_i(t)$, $Z_i(t), (i = 1, 2, 3)$; (c), (d): Phase portrait of the system (2) with functions and parameters given in (18).

does this due to the partial symmetry of the system since we choose two of the three patches have same response? We leave these as our future pursuit. As the maturation time is relative large, even with the same initial condition, the double periods coincide, see Fig. (12).

5. Conclusion

In this paper, we propose a general model with n parallel food chains through the stage structured maturation time delay. We have carried out mathematical analysis to discuss the existence of the steady states and their stabilities. When $n = 1$, we provide the explicit conditions for the local stability of the one-species or the co-existed species equilibrium points and the global stability of the predator-free equilibrium point. Further, we show that the time delay can, not only destroy the existence, but also destabilize the positive equilibrium even it exists. Later, we extend some of the results for the single patch to the model with $n > 1$ multiple parallel patches. To complement the analytical results, we illustrate some numerical simulation to show the rich dynamics in the system.

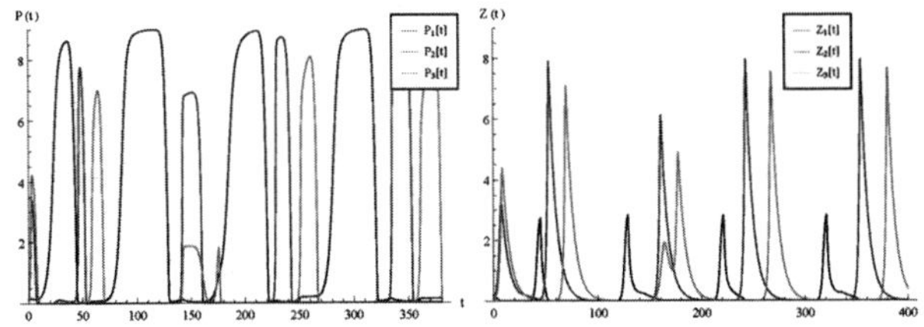

(a) Time series $P_i(t), (i = 1, 2, 3)$. (b) Time series $Z_i(t), (i = 1, 2, 3)$.

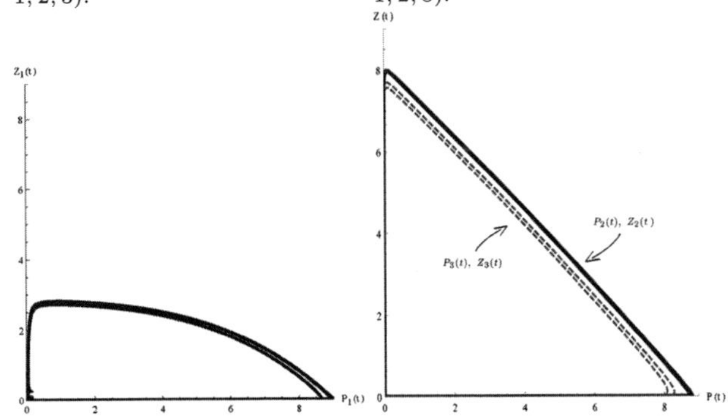

(c) $P_1(t)$ and $Z_1(t)$. (d) $P_i(t)$,$Z_i(t)$, $i = 2, 3$.

$\tau = 0.1,\ P_1(0) = Z_1(0) = 0.1, P_2(0) = 1,\ Z_2(0) = 0.2,\ P_3(0) = Z_3(0) = 0.1.$

Figure 11.: (a),(b): Time series $P_i(t)$, $Z_i(t), (i = 1, 2, 3)$; (c), (d): Phase portrait of (2) with functions and parameters given in (18).

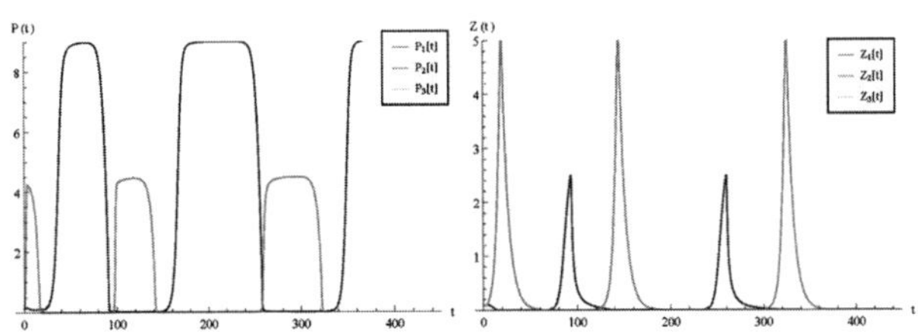

(a) Time series $P_i(t), (i = 1, 2, 3)$. (b) Time series $Z_i(t), (i = 1, 2, 3)$.

$\tau = 2.6,\ P_1(0) = P_2(0) = P_3(0) = Z_1(0) = Z_2(0) = Z_3(0) = 0.1.$

Figure 12.: Time series $P_i(t)$ and $Z_i(t)(i = 1, 2, 3)$ of system (2) with functions and parameters given in (18).

References

1. R.A. Armstrong, *Stable model structures for representing biogeochemical diversity and size spectra in plankton communities*, Journal of Plankton research, 21(3) (1999), 445–464.
2. E. Beretta, Y. Kuang, *Geometric stability switch criteria in delay differential systems with delay dependent parameters*, SIAM J Math Anal., 33 (2002), 1144-1165.
3. C. D. Collins, *Formulation and validation of a mathematical model of phytoplankton growth*, Ecology 61 (1980), 639–649.
4. A. M. Edwards and J. Brindley, *Oscillatory behavior in a three-component plankton population model*, Dynamics and Stability of Systems, 11(4) (1996), 347–370.
5. A. M. Edwards and A. Yool, *The role of higher predation in plankton population models*, J. Plankton Research, 22 (2000), 1085–1112.
6. H. Cheng and T. Zhang. *A new predator-prey model with a profitless delay of digestion and impulsive perturbation on the prey*, Applied Mathematics and Computation, 217 (2011), 9198–9208.
7. W. Feng, B. Rock and J. Hinson, *On a new model of two-patch predator-prey system with migration of both species*, J. Applied Analysis and Computation, 1 (2011), 193–203.
8. S. A. Gourley and Y. Kuang, *A stage structured predator-prey model and its dependence on maturation delay and death rate*, J. Math. Biol., 49 (2004), 188–200.
9. M.D. Holland and A. Hastings, *Strong effect of dispersal network structure on ecological dynamics*, Nature, 456(2008), 792–795.
10. W. Ko and K. Ryu, *A qualitative study on general Gause-type predator-prey models with constant diffusion rates*, J. Mathematical Analysis and Applications, 344 (2008), 217 – 230.
11. T. Kolokolnikov, C. Ou and Y. Yuan, *Phytoplankton depth profiles and their transitions near the critical sinking velocity*, J. Math. Biol., 59 (2009), 105–122.
12. E. Litchman and C.A. Klausmeier,*Trait-based community ecology of phytoplankton*, Annu. Rev. Ecol.Evol.Syst., 39 (2008), 615–639.
13. M.Y. Li and H. Shu,*Impact of intracellular delays and target-cell dynamics on in vivo viral infections*, SIAM J. Appl. Math., 70(7) (2010), 2434–2448.
14. A. Medvinsky, S. Petrovskii, I. Tikhonova, H. Malchow and B. Li, *Spatiotemporal Complexity of plankton and fish dynamics*, SIAM Review, 44 (2002), 311–370.
15. F. Wang, Y. Kuang, C. Ding, S. Zhang, *Stability and bifurcation of a stage-structured predatorprey model with both discrete and distributed delays*, Chaos, Solitons & Fractals, 46 (2013), 1927.
16. Yuan, *A coupled plankton system with instantaneous and delayed predation*, J. Biological Dynamics, 6 (2012), 148–165.

COMPLEX DATA CLUSTERING: FROM NEURAL NETWORK ARCHITECTURE TO THEORY AND APPLICATIONS OF NONLINEAR DYNAMICS OF PATTERN RECOGNITION *

GUOJUN GAN

Research and Development, Global Variable Annuity Hedging,
Manulife Financial, Canada
Email: Guojun_Gan@manulife.com

JIALUN YIN

National University of Defense Technology
Changsha, Hunan, China
E-mail: genial@126.com

YULIA WANG AND JIANHONG WU†

Laboratory for Industrial and Applied Mathematics, York University,
Toronto, Ontario, M3J 1P3, Canada
Centre for Disease Modelling, York Institute of Health Research,
Toronto, Ontario, M3J 1P3, Canada
E-mail: wujh@mathstat.yorku.ca, terry_312@hotmail.com

*This work at the Laboratory for Industrial and Applied Mathematics on the theoretical foundation and applications of projected clustering of high dimensional and big data has been supported by a number of programs and funding agencies including the Canada Research Chairs program, the Natural Sciences and Engineering Research Council of Canada (discovery grant, collaborative research development program), the Mitacs globalink program, the Mitacs accelerate program, the Fields-Mitacs summer research program, the Canada Foundation for Innovation and the Ontario Innovation Trust. A few industrial partners have also been involved, these partners include the Generation 5 Mathematical Technologies Inc. and the InferSystems Corporation.

†Corresponding author.

We present some progress in high dimensional data clustering, made at the Laboratory for Industrial and Applied Mathematics over the last ten years. The focus is on the role of information processing delay as an adaptive mechanism for pattern recognition in subspaces of high dimensional data. Our objective is to develop both mathematical foundation and effective techniques/tools for projective clustering. We also present some applications to gene filtering, cancer diagnosis, neural spike trains pattern recognition, text mining, stock associations, and online social network news aggregation.

1. Introduction

The purpose of this survey is to organize a few PhD theses and MSc dissertations, research publications, and projects conducted at the York University's Laboratory for Industrial and Applied Mathematics (LIAM) in a coherent framework about information processing delay, high dimension data clustering, and nonlinear neural dynamics.

The objective of this decade long effort at LIAM is to develop both mathematical foundation and effective techniques/tools for pattern recognition in high dimensional data. We refer to the monograph[21] for our collection of existing clustering algorithms, and the survey paper[44] for a heuristic description of our philosophy that the nonlinear dynamic systems theory may provide some theoretical foundation and principles based on recent biological evidences for novel neural network inspired clustering architectures.

In the papers[9,10] and in the thesis by Cao[8], we developed a novel neural network architecture and algorithm to detect low dimensional patterns in a high dimensional data set. These patterns are associated with the projective clusters introduced by Aggarwal and his co-workers from the IBM Watson Centre[2]. The developed projective adaptive resonance theory (PART) has received much attention by data clustering researcher community and industry, and formed the core of a Collaborative Research Development project funded by the Natural Science and Engineering Research Council of Canada (NSERC) in collaboration with Generation 5 Mathematical Technologies Inc. The PART algorithm has since been used in a number of applications. For example, it was used to develop a powerful gene filtering and cancer diagnosis method in [39], which shows that PART was superior for gene screening. As will be documented in later sections, the PART was also used for clustering neural spiking trains, ontology construction, stock associations, and online social network news aggregation. The PART algorithm was further extended to deal with categorical data in the thesis [20], and with supervised clustering in the dissertations [36,28].

The PART architecture is based on the well known ART developed by Carpenter and Grossberg, with a selective output signalling (SOS) mechanism to deal with the inherent sparsity in the full space of the data points in order to focus on dimensions where information can be found. The key feature of the PART network is a hidden layer of neurons which incorporates SOS to calculate the dissimilarity between the output of a given input neuron with the corresponding component of the template (statistical mean) of a candidate cluster neuron and to allow the signal to be transmitted to the cluster neuron only when the similarity measure is sufficiently large. Recently discovered physiological properties of the nervous system, the adaptability of transmission time delays and the signal losses that necessarily arises in the presence of transmission delay, enabled us to interpret SOS as a plausible mechanism from the self-organized adaptation of transmission delays driven by the aforementioned dissimilarity. The result is a novel clustering network, termed PART-D, with physiological evidence from living neural network and rigorous mathematical proof of exceptional computational performance. This clustering network was developed in [45].

Such an adaptation can be regarded as a consequence of the Hebbian learning law, and the dynamic adaptation can be modelled by a nonlinear differential equation. As a result, we obtained a new class of multi-scale systems of delay differential equations with adaptive delay. A key issue then is how to analytically formulate the delay adaptation. This links to another PhD thesis by Beamish[3], which proposed an alternative neural network formulation of the Fitts' law for the speed-accuracy trade-off of information processing. A number of publications have been resulted from this thesis work, including [5,6,4,7]. It remains an open problem how to use this alternative neurodynamical formulation to obtain a precise delay adaption rule of the PART-D neural network architecture for projective clustering.

When the delay adaption rates are in certain ranges, we anticipate nonlinear oscillatory behaviors of the PART-D neural network as the signal processing delay has been recognized as a major mechanism for nonlinear oscillation in the form of Hopf bifurcations, and this oscillation slows down the convergence of the clustering algorithm. How to detect the birth and to describe the global persistence of these nonlinear oscillations is the central subject of the thesis by Hu[29] and the subsequent publications[32,31,30].

In summary, there have been increasing physiological evidences to support the idea of projective clustering using neural networks with delay adaption, there has been some preliminary theoretical analysis to show why such a network architecture works well for high dimensional data, and there have been sufficient applications to illustrate our PART network based clustering algorithm is efficient. An interdisciplinary approach for high dimensional data clustering clearly shows the potential to develop a dynamical system framework for pattern recognition in high dimensional data.

2. Clustering and clustering neural networks

Data clustering, a common cognitive task effectively performed by our central nervous system routinely, becomes increasingly important and challenge in today's "big data" reality. It aims to finding certain homogeneous patterns in data sets containing many heterogeneous structures. The goal of data clustering is to reorganize subsets of data points into groups, called clusters, so that the data points within the same group share some common features while points in different clusters are distinguished by some of these common features. In unsupervised clustering, these features have to be identified during the process of clustering.

The approach taken by the Laboratory for Industrial and Applied Mathematics (LIAM) towards data clustering problem is to consider the clustering process as an *inverse process* of *pattern formation* of complex dynamical systems. In this approach, the goal for clustering a data is to *construct a dynamical system* to *automatically and adaptively* identify patterns hidden in the given data set. Namely, for a given data set D in R^m, we try to construct a dynamical system with data-specified local attractors (such as equilibria or periodic orbits) $CR_1, \cdots, CR_n$ so that each CR_i represents a cluster (for example, the centre of a cluster) and its domains of attraction gives the cluster criterion that distinguishes this cluster from others.

There have been a few effective projected clustering algorithms developed such as CLIQUE and PROCLUS, see [44,21] and references therein. Here we describe a neural dynamics inspired architecture with which a dynamical system is constructed from adaptively processing a high dimension data. A key issue is what constitutes of the minimal size and structure of a nonlinear dynamical system required to identify the clusters hidden in arbitrary unknown subspaces of the given data set.

We consider a given data set D of n points in the m dimensional Euclidean space. If we try to mimic the clustering functioning of our central nervous system to construct a network of neurons to identify the hidden patterns of the data set, we will need:

- a layer of m neurons to process inputs (the input layer);
- a layer of neurons to represent clusters (the clustering layer), with the number of clustering neurons unspecified;
- synaptic connections (bottom-up weights) between the input layer and the clustering layer to weight appropriately the output (activation) of each input neuron so that every clustering neuron can calculate the sum of weighted activations generated by a given input vector for the purpose of selecting a candidate cluster;
- synaptic connections among clustering neurons so that this layer of neurons can automatically select a winner as the candidate cluster;
- a mechanism to update the feature (statistical mean) of the selected cluster neuron and to store the updated feature at the synaptic connections (top-down weights, or templates);
- a mechanism and learning rule with which the top-down weights and bottom-up weights are updated to learn the experience.

Specific principles for the connection topology (competitive network) of the clustering layer and for the learning rules to update top-down and bottom-up weights have led to the renowned ART (Adaptive Resonance Theory) neural networks, which have been shown to be very effective in self-organized clustering in *full dimensional spaces*. ART was first introduced by Grossberg in 1976[26,27] in order to analyze how brain networks can learn in real time about a changing world in a rapid but stable fashion, based on which Capenter and Grossberg[14,13,16] developed two classes of ART neural network architectures ART1 and ART2, whose computational performance (dynamics) is described by systems of differential equations. ART1 self-organizes recognition categories for arbitrary sequences of binary input patterns, while ART2 does the same for either binary or continuous inputs. Some other classes of ART neural network architectures such as Fuzzy ART[12], ARTMAP[11], Fuzzy ARTMAP[17], and Gaussian ARTMAP[43] were then developed with increasingly powerful learning and patten recognition capabilities in either an unsupervised or a supervised mode.

Examples have been provided in Cao-Wu [9] to show that the ART neural network needs additional structure in order to perform the task of subspace clustering in high dimensional data sets since ART focuses on similarity of patterns in the *full* dimensional space. The first paper [9] of a series of studies introduces a new mechanism to deal with the identification of subspaces where clusters are formed, this is the so-called *selective output signaling* (SOS in short) and the corresponding ART is termed PART. This SOS mechanism selectively selects the signal from an input neuron only when the signal is similar to the top-down weight (template) between the input neuron and the targeted clustering neuron, hence PART focuses on only those dimensions where information is relevant for a particular cluster. We refer to [9] for a schematic illustration of the PART architecture.

3. Projected ART with Adaptive Delay

Cao and Wu [9,10] implemented PART and demonstrated that PART networks outperform ART networks for pattern recognition in high dimensional spaces. The key feature of a PART network is a hidden layer which incorporates the SOS mechanism to calculate the similarity between the output of a given input neuron with the corresponding component of the template of a candidate cluster neuron and allows the signal to be selectively transmitted to the cluster neuron only when the similarity measure is sufficiently large. So, in PART the output signal of an input neuron will be completely prohibited to be transmitted to its target cluster neuron if the similarity measure is small although in practice, this output signal may still play a (relatively minor) role in the final clustering result. This issue has been successfully addressed by the novel clustering network, termed PART–D, which interprets the SOS mechanism in terms of two recently emphasized properties of the nervous system, namely the adaptability of transmission time delays and the signal losses that necessarily arises in the presence of transmission delay. In PART-D, the SOS mechanism is shown to arise because the self-organized adaptation of transmission delays is driven by the dissimilarity between the input pattern and the stored pattern (represented by the template of a cluster neuron). Such an adaptation can be regarded as a consequence of the Hebbian learning law, and the dynamic adaptation can be modeled by a nonlinear differential equation. As a result, we obtain a new class of systems of delay differential equations with adaptive delay as follows:

$$\epsilon_p \frac{dx_i(t)}{dt} = -x_i(t) + I_i(t), \tag{1}$$

$$\epsilon_c \frac{dy_j(t)}{dt} = -y_j(t) + [1 - Ay_j(t)][f_c(y_j(t)) + T_j(t)] \tag{2}$$

$$- [B + Cy_j(t)] \sum_{k\neq j, k\in\Lambda_2} f_c(y_k(t)), \tag{3}$$

$$T_j(t) = D \sum_{1\leq i\leq m} z_{ij}(t) f_p(x_i(t - \tau_{ij}(t)) e^{-\alpha\tau_{ij}(t)}, \tag{4}$$

$$\beta\frac{d\tau_{ij}(t)}{dt} = -\tau_{ij}(t) + E[1 - h_{ij}(t)], \tag{5}$$

$$h_{ij}(t) = h_\sigma(d(f_p(x_i(t)), w_{ji}(t)))l_\theta(z_{ij}(t)), \tag{6}$$

$$\delta\frac{dz_{ij}(t)}{dt} = f_c(y_j(t))[(1 - z_{ij}(t))Lf_p(x_i(t - \tau_{ij}(t))e^{-\alpha\tau_{ij}(t)}, \tag{7}$$

$$- z_{ij}(t) \sum_{k\neq i, k\in\Lambda_1} f_p(x_k(t - \tau_{kj}(t)e^{-\alpha\tau_{kj}(t)}], \tag{8}$$

$$\gamma\frac{dw_{ji}(t)}{dt} = f_c(y_j(t))[-w_{ji}(t) + f_p(x_i(t - \tau_{ij}(t))e^{-\alpha\tau_{ij}(t)})]. \tag{9}$$

In the above model, the activation of the i-th inout neuron is denoted by x_i, the activation of the j-th clustering neuron is denoted by y_j; the bottom-up weight between the i-th input neuron and the j-th clustering neuron is denoted by z_{ij}, while the top-down weight is denoted by w_{ji}.

In the Short Term Memory trace equations for input neurons, $0 < \epsilon_p \ll 1$, I_i is the constant input imposed on the i-th neuron. This is based on the assumption that for an isolated neuron, the dynamics is the balance of the internal decay and the external input excitation. For the change of the Short Term Memory trace equations for clustering neurons, we assume that the activation of the cluster neuron depends on the internal decay, the excitation from self-feedback, the inhibition from other cluster neurons and the excitation by the bottom-up filter inputs from input neurons. In the equations, $0 < \epsilon_c \ll 1$, $f_c : R \to R$ is a signal function, A, B, and C are non-negative constants. In the the bottom-up filter input T_j calculation, D is a scaling constant, and $f_p : R \to R$ is the signal function of the input layer. It is assumed the signal transmissions between two layers are not instantaneous and the signal decays exponentially at a rate $1/\alpha > 0$.

The term τ_{ij} is the signal transmission delay between the ith input neuron and the j-th clustering neuron. We assume this delay is driven by the dissimilarity in the sense that the signal processing from the input neuron to the cluster neuron is faster when the output is similar to the corresponding component of w_{ji} of the feature vector $w_j = (w_{ji})_{1\leq i\leq m}$ of the cluster neuron. In the equation for the delay adaptation, $\beta > 0$, $E \in (0, 1)$ are constants and $h_{ij}(t) = S(d(f_p(x_i(t)), w_{ji}(t)), z_{ij}(t))$ is the similarity measure between the output signal $f_p(x_i(t))$ and the corresponding component $w_{ji}(t)$ of the feature vector of the cluster neuron, with respect to the significance factor of the bottom-up synaptic weight $z_{ij}(t)$, here d is the usual distance function in the one dimensional Euclidean space and $S : R^+ \times [0, 1] \to [0, 1]$ is a given function, non-increasing with respect to the first argument and non-decreasing with respect to the second argument. Moreover, $S(0,1) = 1$ (The similarity measure is 1 with complete similarity and maximal synaptic bottom-up weight) and $S(+\infty, z) = S(x, 0) = 0$ for all $z \in [0,1]$ and $x \in R^+ := [0, \infty)$ (The similarity measure is 0 with complete dissimilarity or minimal bottom-up synaptic weight). In the above formulation, we used some special function of S where $h_{ij}(t)$ is determined by the distance between the output signal $f_p(x_i(t))$ and the corresponding component $w_{ji}(t)$ of the feature vector of the cluster neuron, multiplied by the significance factor of the bottom-up synaptic weight $z_{ij}(t)$, with a threshold parameter $\theta > 0$.

The equation governing the change of the weights follows from the synaptic conservation rule of [41] and only connections to activated neurons are modified. The top-down weights are modified so that the template will point to the direction of the delayed and exponentially decayed outputs from the input layer (with the exponential decay rate $\gamma > 0$). The bottom-up weights are changed according to the competitive learning law and Weber Law Rule that says that LTM (Long Term Memory) size should vary inversely with input pattern scale to present a clustering neuron that has learned a particular pattern from also coding every superset pattern (see [15]). In the equation, $0 < \delta \ll \gamma = O(1)$ and $L > 0$ is a given constant.

We refer to [9] for the equations of the LTM equations for non-committed candidate neurons and the discussion of a reset mechanism. In particular, a candidate (active) node will be reset if at any given time $t \geq 0$, the degree of match is less than a prescribed vigilance. Namely, reset occurs if and only if $\sum_{1 \leq i \leq m} h_{ij}(t) < \rho$, here $\rho \in \{1, 2, \ldots, m\}$ is a vigilance parameter.

The following theorem describes the computational dynamics during a trial.

Theorem 3.1. *We can choose small ϵ_p, ϵ_c, and δ so that:*

(i) *(Inhibition of Non-Candidate Neurons): For $j \neq J$ and $t \geq 0$, $y_j(t) < \eta_c$ and $f_c(y_j(t)) = 0$;*
(ii) *(Sustained Excitation of the Candidate Neuron): There exists $\Gamma > 0$ such that $y_J(t) < \eta_c$ and $f_c(y_J(t)) = 0$ when $t < \Gamma$, and $y_J(t) \geq \eta_c$ and $f_c(y_J(t)) = 1$ when $t \geq \Gamma$;*
(iii) *(Invariance of Similarity): For any $t \geq 0$, $h_{ij}(t) = h_{ij}(0)$;*
(iv) *(Learning at Infinity): For any $j \in \Lambda_c$ with $j \neq J$, $z_{ij}(t)$ and $w_{ji}(t)$ remain unchanged for all $t \geq 0$. But $\lim_{t \to \infty} w_{Ji}(t) = f_p(I_i) e^{-\alpha \tau_{iJ}^*}$*

and

$$\lim_{t\to\infty} z_{iJ}(t) = \begin{cases} 0 & \textit{if } h_{iJ}(0) = 0, \\ \frac{L}{L+l_i} & \textit{if } h_{iJ}(0) = 1, \end{cases}$$

where $l_i = \#\{k \neq i; h_{kJ}(0) = 1\}$.

Learning may take place in a much faster pace, the following theorem describes the transit computation performance and it also shows an amazing choice of a discrete Lyapunov function that was used to prove both theorems.

Theorem 3.2. *We can choose small* ϵ_p, ϵ_c, *and* δ *so that:*

(v) (Fast Excitation): $\Gamma \in (0,1)$;

(vi) (Fast Learning): Write $z_{ij}^{\epsilon_p,\epsilon_c,\delta}$ *and* $w_{ji}^{\epsilon_p,\epsilon_c,\delta}$ *to indicate explicitly the dependence on* $(\epsilon_p, \epsilon_c, \delta)$. *Then we have (with* $q = 1 - e^{-1/\gamma}$*)*

$$\lim_{\delta\to 0} z_{iJ}^{\epsilon_p,\epsilon_c,\delta}(1) = \begin{cases} 0 & \textit{if } h_{iJ}(0) = 0, \\ \frac{L}{L+l_i} & \textit{if } h_{iJ}(0) = 1, \end{cases}$$

$$\lim_{\epsilon_p\to 0,\beta\to 0} w_{Ji}^{\epsilon_p,\epsilon_c,\delta}(1) = (1-q)w_{Ji}(0) + q f_p(I_i) e^{-\alpha\tau_{iJ}^*};$$

(vii) (Convergence of Projective Subspace) Let $D_j(t) = \{i; l_\theta(z_{ij}(t)) = 1\}$. *Then, as* $\epsilon_p, \epsilon_c, \delta \to 0$, *we have*

$$\begin{aligned} &D_j(t) = D_j(0) \textit{ for any } j \neq J; \\ &D_J(t_2) \subseteq D_J(t_1) \textit{ if } t_2 \geq t_1 \geq 0; \\ &D_J(t) = D_J(1) \textit{ for all } t \geq 1. \end{aligned}$$

These theorems describe the computational performance of PART during a trial, either in terms of long-term behaviours or transit behaviours. It confirms that the network does provide the winner-take-all paradigm: all clustering neurons with $j \neq J$ are always inactive, but the J-th clustering neuron will be activated after some finite time. This activated clustering neuron receives the the largest bottom-up filter input (T_J), and (v) shows that the identification of a clustering neuron can take place very fast. The above theorems also give the learning formulae (either long-term or fast learning). Note also that $D_j(t)$ is the set of dimensions of projected subspace associated with cluster representing by the j-th neuron at time t, and above results indicate that the set of dimensions is non-increasing during the learning. This non-increasing property of dimensions contributes to stabilizing learning in response to arbitrary sequences of input patterns.

4. Algorithms and applications

An effective algorithm based on the above results, specially the fast learning rules, has been developed in [9,10] (for PART) and then in [45] for PART-D. These algorithms consist of the following major steps:

- Input Processing and Select Output Signals from Input Layer;
- Activation, Inhibition, and Identification of a Potential Cluster;
- Confirmation, Vigilance and Reset;
- Fast Learning;
- Identification of Subspaces;
- Outliers collection.

The time cost of these algorithms is $O(mnNM)$, where m is the number of dimensions of data space, n is the number of clustering neurons, m is the number of all data points and M is the number of iterations.

4.1. *Experiments on synthetic data*

Extensive simulations on high dimensional synthetic data showed that the clustering layer becomes stable after only a few iterations. Here we describe one example on a high dimensional synthetic data generated via the method introduced by Aggarwal et al [1] in 1999. The input data has 20,000 data points in a 100-dimensional space, which has 6 clusters generated in 20, 24, 17, 13, 16 and 28-dimensional subspaces respectively. The data points are presented in random order, and the clustering results can be reported as number of clusters found, dimensions found, centers of clusters found, and the contingency table of input clusters (original clusters) and output clusters (clusters found).

Output\Input	1	2	3	4	5	6	Sums
1	5144	0	0	0	0	0	5144
2	0	1878	0	0	0	0	1878
3	0	0	4412	0	0	0	4412
4	0	0	0	2716	0	0	2716
5	0	0	0	0	2608	0	2608
6	0	0	1	0	0	1185	1186
Outliers	106	66	239	290	68	287	2056
Sums	5250	1944	4652	3006	2676	1472	20000

The above table shows the simulation results with $\rho = 10$. Note that in the reported results we have treated as outliers the data points in the clustering neurons with very small sizes (less than 1.2% of total data points). The simulation results show that the PART algorithm succeeds in finding the exact number of original clusters and in finding almost exact centers of all original clusters. Note that the dimensions found in different clusters are different, for example, cluster 1 is formed with respect to dimensions $10, 12, 17, 37, 46, 58, 61, 79, 81, 99$, while cluster 2 is formed with respect to dimensions $5, 8, 13, 15, 16, 18, 30, 70, 85, 92$. Also note that the dimensions found are not identical to those of the original clusters (for example, the dimensions of the original cluster 1 include $1, 6, 10, 12, 15, 17, 31, 36, 37, 38, 45, 46, 52, 54, 58, 61, 67, 79, 81, 99$), but these found dimensions are contained as subsets of the associated dimensions of original clusters. These subsets are sufficiently large so that, after a further reassignment procedure, we are able to reproduce the original clusters from the found cluster centers, the found number of clusters and the found dimensions.

4.2. *Application to neural spiking trains clustering*

In [33], PART was used as an effective tool for clustering neural spiking trains via transient behaviors. It was noted that "the detection of non-stationarities in neural spike trains recorded from chronically implanted multielectrode grids, such as transient synchronizations in a neural subpopulation, becomes increasingly difficult as the number of electrodes increases". This calls for unsupervised learning algorithms that can be used to "group, or cluster, spike trains based on the presence of local, shared features". The feature of PART that allows comparisons be made between inputs and learned patterns using a subset of the total number of spikes available enables the network to learn the characteristics that defines each cluster making as few assumptions about the statistical properties of the spike trains as possible.

> "The result is an extremely powerful tool for clustering neural spike trains that is computationally inexpensive. The fact that projective clustering dramatically increases the ability of an artificial neural network to discover patterns in its sensory inputs raises the question of whether analogous mechanisms operate in the nervous system. Thus we anticipate that PART neural networks will not only have increasing applications for data analysis, but also have the

potential to provide insights into the computational activities of the nervous system."

Spike train inputs for the PART neural network have the general form

$$\left(\overbrace{s_{11}, s_{12}, \cdots, s_{1p}}^{bin_1}, \overbrace{s_{21}, s_{22}, \cdots, s_{2p}}^{bin_2}, \cdots, \overbrace{s_{k1}, s_{k2}, \cdots, s_{kp}}^{bin_k}\right),$$

where the dimension, m, is equal to the number of bins, bin_k, of size Δt times the number of statistical features of interest, and the notation s_{kp} denotes the p–th statistical feature evaluated for the k–th bin. The number of input neurons is m and the number of clustering neurons is much greater than the expected number of clusters. At onset all of the clustering neurons are non–committed. The few round of trails generates a committed neuron to represent a cluster. Once the committed clustering neuron has been determined, the next spike train is presented. All spike trains that belong to the same committed neuron belong to the same cluster.

The number of input patterns that can be learned by a PART neural network is limited only by the finiteness of the number and length of spike trains that can be presented to it. There are a number of consequences for the practical application of PART neural networks:

- it is better to cluster data sets with respect to a few, e.g. one, statistical features at a time;
- the order of presentation of spike trains may have an influence of the clustering results;
- the number of clustering neurons must be larger than the number of suspected clusters;
- there will always be a small number of spike trains which do not cluster well: following [9] we placed all such data into an *outlier node*.

The PART clustering algorithm was validated on populations of neural spike trains constructed using two types of model neurons: 1) the leaky integrate and fire (LIF) mode, and 2) a reduced Hodgkin–Huxley model. The goal in constructing these data sets was to pose a difficult clustering problem consistent with the known physiological responses of neurons. Validation using this procedure is facilitated by the fact that the natures and numbers of the true clusters are known.

4.3. *Experiments on online social network news aggregation*

We have recently considered [19,46] the issue of news aggregation in online social networks with a pilot project on the social news website Digg.com, a content discovery and sharing application launched in 2004. According to the traffic statistics by Alexa.com in 2010, Digg is the 117th most popular website globally and 52nd in the USA[40]. Digg allowed people to vote web content up or down, called digging and burying, respectively. Users in Digg can share the content with other users who are connected to them by voting for or against the news. We have used the dataset from K. German (http://www.isi.edu/lerman/downloads/digg2009.html), who collected the information of stories on the Digg's front page over a period of a month in 2009. 3553 popular stories are voted for 3,018,197 times by 139,409 distinct users and on average, each story received about 850 votes.

All of the stories were provided the voter ID as well as the exact time of when voted. We are able to obtain the time series curve of each story. Apparently, large amount of superficial information can be found, such as how the popularity was, when the curve started and how the voting rate was going. When thousands of news gather together, they show some similarities to the tendency of the curves. We formatted the votes data over a period of 50 hours, as is used in [42]. Most of the stories were almost fixed and experienced little change on the vote number at the end of 50 hours. For each story, at the end of each hour, we obtained the cumulative number of voted users and used the value as a measurement of voting density. Here we can get a 3553×50 matrix indicating the increasing vote trend for all the stories.

Obviously, clustering these data in the 50 dimensional spaces is meaningless since every news distinguished itself from all others. Indeed, when we try to cluster these news in the full space, we found a large number of clusters with every cluster containing very small number of pieces of news. Projective clustering in relatively lower dimensional subspaces does generate some meaningful clusters, for example, Figure 1 gives a cluster, when we specify the PART algorithm to find clusters in subspaces of at least 30 dimension. Effectively, these pieces of news are grouped in one single cluster as they all reached the equilibrium states (final size of the news outbreak) after 20 hours from the source, and the final accumulated numbers of votes are close to each other. A better way is to look at the number of new votes each hour, and this give a curve of "influence votes" for each news, very much similar to the typical epidemic curve of an outbreaking infectious disease. Then we can define certain features for each news, such as the initial time t_i when the total number of votes reach a pre-assigned number (say 50 in Figure 2), the beginning and ending times (t_b and t_e) of the so-called "viral period"–when the "epidemics" starts and ends, the turning point t_{tu} when the growth rate of the number of influenced users changes from being positive to negative. In this way, each time series of a given news is characterized by these features. Figure 2 gives a projective cluster of news with respect to the subspaces (t_i, t_b, t_{tu}, t_e).

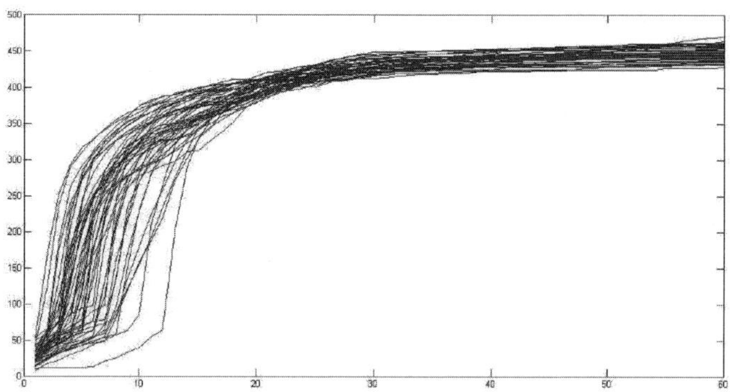

Figure 1. An example of projective clustering of the time series for the accumulated votes in the Digg networks. The cluster is formed based on the final size, and for those news reaching the equilibrium state within 20 ours since their release from the sources.

4.4. *Application to gene filtering and cancer diagnosis*

In [37], PART was used as a gene filtering method for the construction of robust prognostic predictors. Subspace clustering is essential for establishing prognostic predictors of various diseases using DNA microarray analysis technology, since it is desired to selectively find significant genes for constructing the prognostic model and also necessary to eliminate nonspecific genes or genes with error before constructing the model. According to the authors,

> "Genes selected by PART were subjected to our FNN-SWEEP modeling method for construction of a cancer class prediction model. The model performance was evaluated through comparison with a conventional screening signal-to-noise (S2N) method

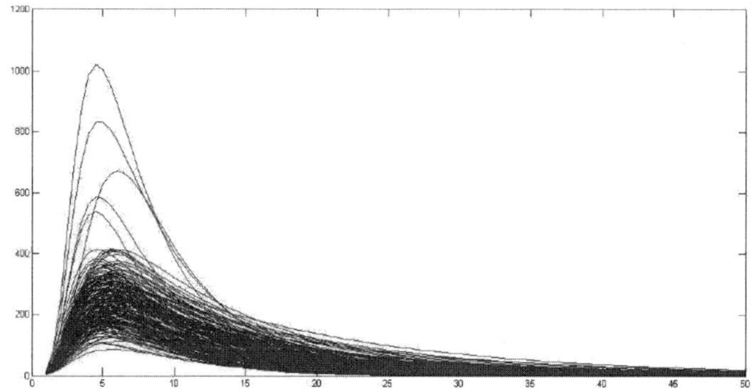

Figure 2. An example of projective clustering of the news in the Digg network, using features that are selected to reflect some "epidemic" nature of the number of new "influence votes."

> or nearest shrunken centroids (NSC) method. The FNN-SWEEP predictor with PART screening could discriminate classes of acute leukemia in blinded data with 97.1% accuracy and classes of lung cancer with 90.0% accuracy, while the predictor with S2N was only 85.3% and 70.0% or the predictor with NSC was 88.2% and 90.0%, respectively. The results have proven that PART was superior for gene screening."

The PART network was also used in [39] as a gene filtering method for cancer diagnosis marker extraction for soft tissue sarcomas. The authors noted that

> "In a previous study, we developed the PART filtering method by modifying PART, and reported that PART exhibited a higher performance than conventional methods, such as S2N and NSC. The combination method of PART and BFCS (PART-BFCS) was developed and applied to gene expression data, such as lymphoma and esophageal cancer. In the present study, we applied the various filtering methods to the gene expression profile data for the STS subtypes and constructed SVM models using the filtered genes. The results showed that the accuracy of the model based on the genes filtered by PART was the highest."

Further, the same group of authors combined PART with boosted fuzzy classifier and the SWEEP operator method effectively for gene selection. They concluded that [38]

> "In the present study, we investigated combinations of various filter and wrapper approaches, and found that combination method of PART and BFCS (a kind of boosting) is significantly superior to other methods with regard to high prediction accuracy for construction of class predictor from gene expression data. This method could select some marker genes related to cancer outcome. In addition, we proposed improved RIBFCS of PART-BFCS. Based on this new index, the discriminated group with over 90% prediction accuracy was separated from the others. It is necessary that there are about 90% or more prediction accuracy in the practical diagnosis application. These results suggest that the PART-BFCS method has a high potential to function as a new method of marker gene selection for the diagnosis of patients, using high dimensional data such as DNA microarray, mass spectrometry (MS), and two-dimensional polyacrylamide gel electrophoresis (2D-PAGE)."

4.5. *Application to text mining*

In [34], PART was used as an effective tool for reducing multidimensional text document space and also the text document clustering. It was shown that PART overcomes some lacks of computational complexity in traditional clustering algorithms in multidimensional space. They noted that with appropriate parameter settings of distance, the PART neural network achieved very good results on the clustering of multidimensional text documents and sorts precisely selected documents to corresponding supposed clusters. In addition to the correct classification of the text documents, PART was able to distinguish the projective dimension centers in each cluster and group noisy documents included in the outlier cluster. They stated that

> "Clustering algorithm via PART can correctly collect input documents to corresponding clusters, when the number of dimensions in a text document dataset increases, distance measure does not become increasingly meaningless and effort of system do not go down. The PART with appropriate input parameters enables to find the correct number of clusters, the correct centers for each

cluster and sufficiently large subset of dimensions in which clusters are formed. Results of our approach show, that application of PART for clustering of text documents can easy discover intrinsic clusters and also discover noisy patterns in datasets of text documents. Thanks to using average similarity degree function and exact settings of input parameters has this modification of PART very good computational efficiency in process of C."

4.6. *Application to stock associations*

In [35], the PART algorithm was improved with buffer management known as BPART to overcome the disadvantage of PART depending on accurate parameters and orders of input data sets. The authors noted that although "Projective Adaptive Resonance Theory, based on the ART and PROCLUS, is very good at recognizing self-organizing patterns in arbitrary sequences," the clustering accuracy may be degraded if an incorrect value ρ is chosen. They proposed an improvement-buffer management, which can neglect the noise data sets and achieve a parameter-free algorithm.

> "We find that there are 4 out of 100 stocks which have concurrence associations. In detail, 00001 Cheung Kong, 00004 Wharf (Hldgs), 00012 Henderson Land and 00293 Cathay Pac Air are related (or concurrence) in 90 days out of 481 transaction days, and partly related in 105 days out of 481. Compared with PART, our algorithm initializes the important parameter ρ to 2, which is easily estimated and applied. And from this result above, any value more than 4 fails to find the concurrence of four stocks. Therefore, our algorithm over PART can obtain the good result without an accurate parameter ρ."

4.7. *Application to ontology construction*

In [18], the PART algorithm was used along with Bayesian network probability theory to construct an ontology in the system. In details, it was an efficacious tool for clustering the web pages based on the frequency of the term. It was shown that the PART tree can provide critical information about the hierarchical relation of the projective clusters. The PART neural network does a good job because it not only considers the data points but also the dimensions. What's more, it can deal with the lack of flexibility in the cluster.

"PART shows better results when the quantity of data is large. In this experiment, we attempt to demonstrate that PART is better than ART in web page clustering. We used the ART neural network to cluster all the web pages (1523 web pages) and compared the result with the results of clustering by PART. In order to emphasize the equity of comparison, the parameter settings of ART were identical to those of PART .... Afterward, we used the method described in Chen et al. (2008) to generate the pattern for ART. It is clear that the concept precision (C_P) and concept location precision (C_L_P) of PART are both better than those of ART."

4.8. *Comments on challenges and future directions*

Seeking a neural-network inspired dynamical system architecture that automatically identifies projected clusters in high dimensional data leads us very naturally to the extension of the celebrated Adaptive Resonance Theory by incorporating the delay adaptation in neural computation. This extension generated an effective Projective Adaptive Resonance Theory neural network, whose global dynamical behaviour is governed by a large scale system of delay differential equations with adaptive delay. This adaptive delay, a special case of the so-called state-dependent delay, has been the focus of recent and intensive study in the field of functional differential equations and infinite dimensional dynamical systems. Consequently, we hope further development, in terms of neural physiological evidence, the qualitative and numerical theory, and applications, of this neural network architecture should provide inspiration for the development of a comprehensive theory for systems of state-dependent delay differential equations. Specifically, we have mentioned the PhD thesis of Qingwen Hu and the subsequent publications about the nonlinear oscillations in the form of Hopf bifurcation and global continuation. There is some evidence in the paper of [45] that PART-D may exhibit some oscillatory behaviours in certain parameter ranges, and how the general theory can be applied to exclude or confirm this oscillatory behaviours of a clustering algorithm remains to be a subject for future study.

We have shown that delay in neural networks may play a very useful role in regulating the speed with which different set of information is processed in order to identify hidden patterns in subspaces. This is based on the assumption that delay decays naturally without learning and external stimuli, and that adaptive delay is increased if the input pattern is different from the stored pattern of a potential cluster. Our PART-D model formulates this as a simple linear equation with forcing and we had some very preliminary justification of this delay decay and adaptation law using the cable equation in the appendix of [45]. We would like to derive this delay decay and adaptation law from some first principles in neural dynamics. It would be interesting to see how this is linked to the thesis of Beamish[3] and its subsequent publications [5,6,4,7]. In particular, we would like to know how to use this alternative neurodynamical formulation to obtain a precise delay adaption rule of the PART-D neural network architecture for projective clustering.

PART has been successfully extended to deal with category data and fuzzy clustering in the thesis of Gan [20] and in its subsequent publications [22,24,23,25] . How to extend PART-D along this direction remains open. There have been two MSc theses [36,28] at LIAM dedicated to extending PART for supervised clustering to deal with the useful annotation information of some data sets, further development would need good indices to evaluate the effectiveness of a good clustering algorithm and clustering results, as functions of algorithm-relevant parameters.

Finally, we note that high-dimensional data clustering poses significant challenges for traditional clustering algorithms when correlations among features appear as a result of increasing number of dimensions. These local arbitrarily oriented correlations are the interesting hidden patterns in many applications. In [47], we developed a new correlation clustering algorithm by designing an ART-type neural network architecture. Our new iterative clustering algorithm PART-D-MCA incorporates minor component analysis to a delay-driven winner-take-all architecture. The resulting method shows very promising properties, and demonstrates the potential of extending this theory for clustering data sets in nonlinear submanifolds.

References

1. C.C. Aggarwal, J.L. Wolf, P.S. Yu, C. Procopiuc, and J.S.Park. Fast algorithms for projected clustering. In *Proceedings of the 1999 ACM SIGMOD international conference on Management of data*, pages 61–72. ACM Press, 1999.
2. R. Agrawal, J. Gehrke, D. Gunopulos, and P. Raghavan. Automatic subspace clustering of high dimensional data for data mining applications. In *SIGMOD Record ACM Special Interest Group on Management of Data*, pages 94–105, New York, NY, USA, 1998. ACM Press.

3. D. Beamish. *50 years later: a neurodynamic explanation of Fitts' law*. PhD thesis, Department of Mathematics and Statistics, York University, 2004.
4. D. Beamish, M. Bhatti, S. MacKenzie, and J. Wu. Fifty years later: a neurodynamic explanation of fitts' law. *Journal of the Royal Society Interface*, 3(10):649–654, 2006.
5. D. Beamish, S.A. Bhatti, C.S. Chubbs, I.S. MacKenzie, J. Wu, and Z. Jing. Estimation of psychomotor delay from the fitts' law coefficients. *Biological Cybernetics*, 101(4):279–296, 2009.
6. D. Beamish, S.A. Bhatti, J. Wu, and Z. Jing. Performance limitation from delay in human and mechanical motor control. *Biological Cybernetics*, 99(1):43–61, 2008.
7. D. Beamish, S. MacKenize, and J. Wu. Speed-accuracy trade-off in planned arm movements with delayed feedback. *Neural Networks*, 19(5):582–599, 2006.
8. Y. Cao. *Neural Networks for Clustering: Theory, Architecture, Algorithms and NeuralDynamics*. PhD thesis, Department of Mathematics and Statistics, York University, Toronto, ON, Canada, October 2002.
9. Y. Cao and J. Wu. Projective ART for clustering data sets in high dimensional spaces. *Neural Networks*, 15(1):105–120, January 2002.
10. Y. Cao and J. Wu. Dynamics of projective adaptive resonance theory model: the foundation of PART algorithm. *IEEE Transactions on Neural Networks*, 15(2):245–260, 2004.
11. G. A. Carpenter, S. Grossberg, and J. H. Reynolds. ARTMAP: Supervised real-time learning and classification of nonstationary data by a self-organizing neural network. *Neural Networks*, 4:565–588, 1991.
12. G. A. Carpenter, S. Grossberg, and D. B. Rosen. Fuzzy ART: Fast stable learning and categorization of analog patterns by an adaptive resonance system. *Neural Networks*, 4:759–771, 1991.
13. G.A. Carpenter and S. Grossberg. ART2: Self-organization of stable category recognition codes for analog input patterns. *Applied Optics*, 26:4919–4930, 1987.
14. G.A. Carpenter and S. Grossberg. A massively parallel architecture for a self-organizing neural pattern recognition machine. *Computer Vision, Graphics and Image Processing*, 37:54–115, 1987.
15. G.A. Carpenter and S. Grossberg. Neural dynamics of category learning and recognition: Attention, memory consolidation, and amnesia. In S. Grossberg, editor, *The Adaptive Brain I Cognition, Learning, Reinforcement, and Rhythm*, volume 42 of *Advances in Psychology*, pages 239 – 286. North-Holland, 1987.
16. G.A. Carpenter and S. Grossberg. ART3: Hierarchical search using chemical transmitters in self-organizingpattern recognition architectures. *Neural Networks*, 3:129–152, 1990.
17. G.A. Carpenter, S. Grossberg, N. Markuzon, J.H. Reynolds, and D.B. Rosen. Fuzzy ARTMAP: A neural network architecture for incremental supervised learning of analog multidimensional maps. *IEEE Transactions on Neural Networks*, 3:698–713, 1992.

18. R.-C. Chen and C.-H. Chuang. Automating construction of a domain ontology using a projective adaptive resonance theory neural network and bayesian network. *Expert Systems*, 25(4):414–430, 2008.
19. M. Freeman, J. McVittie, I. Sivak, and J. Wu. An epidemiological approach to information propagation in the digg online social network. *submited*, 2013.
20. G. Gan. Subspace clustering for high dimensional categorical data. Master's thesis, Department of Mathematics and Statistics, York University, Toronto, Canada, October 2003.
21. G. Gan, C. Ma, and J. Wu. *Data Clustering: Theory, Algorithms, and Applications*, volume 20 of *ASA-SIAM Series on Statistics and Applied Probability.* SIAM Press, SIAM, Philadelphia, ASA, Alexandria, VA, USA, 2007.
22. G. Gan and J. Wu. A convergence theorem for the fuzzy subspace clustering (fsc) algorithm. *Pattern Recognition*, 41(6):1939–1947, 2008.
23. G. Gan, J. Wu, and Z. Yang. A fuzzy subspace algorithm for clustering high dimensional data. In X. Li, O.R. Zaiane, and Z. Li, editors, *Lecture Notes in Artificial Intelligence*, volume 4093, pages 271–278. Springer, August 2006.
24. G. Gan, J. Wu, and Z. Yang. A genetic fuzzy k-modes algorithm for clustering categorical data. *Expert Systems with Applications*, 36(2):1615–1620, 2009.
25. G. Gan, Z. Yang, and J. Wu. A genetic k-modes algorithm for clustering categorical data. In X. Li, S. Wang, and Z.Y. Dong, editors, *Proceedings on Advanced Data Mining and Applications: First InternationalConference, ADMA 2005, Wuhan, China*, volume 3584 of *Lecture Notes in Artificial Intelligence*, pages 195–202. Springer-Verlag GmbH, July 2005.
26. S. Grossberg. Adaptive pattern classification and universal recoding, i: parallel development and coding of neural feature detectors. *Biological Cybernetics*, 23:121–134, 1976.
27. S. Grossberg. Adaptive pattern classification and universal recoding, ii: feedback, expectation, olfaction, and illusions. *Biological Cybernetics*, 23:187–202, 1976.
28. H. Habibkhani. Supervised projective adaptive resonance theory with top scoring pair (part-tsp). *MSc Dieertation, York University*, 2012.
29. Q. Hu. *Differential equations with state-dependent delay: global Hopf bifurcation and smoothness dependence on parameters*. PhD thesis, Department of Mathematics and Statistics, York University, 2008.
30. Q. Hu and J. Wu. Global continua of rapidly oscillating periodic solutions of state-dependent delay differential equations. *Journal of Dynamics and Differential Equations*, 22(2):253–284, 2010.
31. Q. Hu and J. Wu. Global hopf bifurcation for differential equations with state-dependent delay. *Journal of Differential Equations*, 248(12):2801–2840, 2010.
32. Q. Hu, J. Wu, and X. Zou. Estimates of periods and global continua of periodic solutions for state-dependent delay equations. *SIAM Journal on Mathematical Analysis*, 44(4):2401–2427, 2012.
33. J. Hunter, J. Wu, and J. Milton. Clustering neural spike trains with transient responses. In *Decision and Control, 2008. CDC 2008. 47th IEEE Conference on*, 2008.

34. R. Krakovsky, S. Ruzomberok, and I. Mokris. Clustering of text documents by projective dimension of subspaces using PART neural network. In *Applied Computational Intelligence and Informatics (SACI), 2012 7th IEEE International Symposium on*, pages 203–207, 2012.
35. L. Liu, L. Huang, M. Lai, and C. Ma. Projective ART with buffers for the high dimensional space clustering and an application to discover stock associations. *Neurocomputing*, 72(4-6):1283–1295, 2009.
36. W. Liu. Supervised projective adaptive resonance theory. *MSc Dieertation, York University*, 2007.
37. H. Takahashi, T. Kobayashi, and H. Honda. Construction of robust prognostic predictors by using projective adaptive resonance theory as a gene filtering method. *Bioinformatics*, 21:179–186, 2005.
38. H. Takahashi, Y. Murase, T. Kobayashi, and H. Honda. New cancer diagnosis modeling using boosting and projective adaptive resonance theory with improved reliable index. *Biochemical Engineering Journal*, 33:100–109, 2007.
39. H. Takahashi, T. Nemoto, T. Yoshida, H. Honda, and T. Hasegawa. Cancer diagnosis marker extraction for soft tissue sarcomas based on gene expression profiling data by using projective adaptive resonance theory (PART) filtering method. *BMC Bioinformatics*, 7:1–11, 2006.
40. S. Tang, , N. Blenn, C. Doerr, and P. Van Mieghem. Digging in the digg social news website. *IEEE Transactions on Multimedia*, 13(5):1163–1175, 2011.
41. C. von der Malsburg. Self-organization of orientation sensitive cells in the striate cortex. *Kybernetik*, 14:85–100, 1973.
42. F. Wang, H. Wang, K. Xu, J. Wu, and X. Jia. Characterizing information diffusion in online social networks with linear diffusive model. In *2013 IEEE 33rd International Conference on Distributed Computing Systems*, 2013.
43. J.R. Williamson. Gaussian ARTMAP: a neural network for fast incremental learning of noisy multidimensional maps. *Neural Networks*, 9:881–897, 1996.
44. J. Wu. High dimensional data clustering from a dynamical systems point of view. In W. Nagata and N. Sri. Namachchivaya, editors, *Fields Institute Communications*, volume 49 of *Bifurcation Theory and Spatio-Temporal pattern Formation*, pages 117–150, American Mathematical Society 2005.
45. J. Wu, H. ZivariPiran, J. Hunter, and J. Milton. Projective clustering using neural networks with adaptive delay and signal transmission loss. *Neural Computation*, 23(6):1568–1604, 2011.
46. J. Yin, J. McVittie, and J. Wu. Dynamic modelling assisted feature selection for online social network news aggregation. *submitted*, 2013.
47. Hossein ZivariPiran and J. Wu. Similarity-driven delay adaptation for clustering skewed high-dimensional data. *submitted*, 2013.

A TWO-STEP KINETIC MODEL OF INSULIN AGGREGATION WITH A COMPETITIVE INHIBITOR *

MARK WHIDDEN

Department of Molecular & Integrative Physiology,
University of Michigan Medical School,
Ann Arbor, MI 48109-5622, USA
E-mail: mwhidden@umich.edu

ALLISON HO

Department of Molecular & Integrative Physiology,
University of Michigan Medical School,
Ann Arbor, MI 48109-5622, USA
E-mail: allho@umich.edu

SANTIAGO SCHNELL

Department of Molecular & Integrative Physiology,
Department for Computational Medicine & Bioinformatics,
Brehm Center for Diabetes Research,
1000 Wall Street, Brehm Tower 5132,
Ann Arbor, MI 48105-1912, USA
E-mail: schnells@umich.edu

We present a novel model of the inhibition of insulin aggregation using a two-step autocatalytic reaction mechanism with an explicitly competitive inhibition pathway. The kinetic rate constants are estimated using nonlinear regression to analyze previously obtained experimental data for inhibition by synthetic hybrid peptides. This framework provides a method for uniquely characterizing the inhibitory capacity of compounds on protein aggregation.

1. Introduction

Insulin is a globular protein hormone produced in pancreatic β-cells that is central to regulating carbohydrate and fat metabolism in organisms. The

*This work is supported by the James S. McDonnell Foundation (Grant No. 220020223) under the 21st Century Science Initiative Studying Complex Systems

mature form of the protein is composed of two chains: A (21 residues) and B (30 residues).[1,2] During biosynthesis, insulin naturally aggregates into dimers in the endoplasmic reticulum[3] and Zn^{2+} mediated hexamers in the Golgi apparatus.[4] Upon exocytosis of insulin from the β-cells in response to elevated levels of blood glucose, aggregates dissociate into the functional monomeric form of insulin and are transferred to the blood stream.[5] If insulin aggregates in the blood stream, it could induce an immune response with unfavorable biological consequences.[6] Biosynthetic human insulin is manufactured for widespread clinical treatment of diabetes, and its aggregation in the blood stream makes it pharmacologically ineffective.[1,7,8]

Evidence suggests that Zn^{2+} mediated insulin aggregation confers protection from insulin misfolding and aberrant aggregation during storage.[9] The formation of amyloid fibrils has been observed during the manufacture, purification, storage, and infusion of insulin.[10,11] The insulin monomer is susceptible to aberrant aggregation by forming fibrils at millimolar concentration and body temperature.[12,13] Partial unfolding of insulin can also occur in environments with acidic pH and elevated temperature, under aggressive agitation, or contact with hydrophobic surfaces, leading to aggregation and subsequent amyloid fibril formation.[1] Experimentally, insulin monomers self-assemble into dimers, tetramers, and hexamers, depending on the concentration and the solvent conditions.[9,14] For example, insulin tends to form hexamers in higher concentrations.[8] Understanding the mechanisms of insulin aggregation is important in the pharmacological industry to develop effective methods for preserving insulin in its monomeric and functional form in solution and in the blood stream.[15]

Insulin aggregation is inhibited by compounds that either prevent unfolding of the native insulin or sequester partially folded aggregation-prone intermediates.[1] Among these compounds we find lecithins, cyclodextrins, and polymeric surfactants,[16–19] carbohydrates and glycerols,[20–22] low molecular weight compounds such as betaine, trehalose, and citrulline,[23] and small hybrid peptides.[1] The chemical kinetic characterization of inhibitors is a valuable tool for investigating the mechanisms of aggregation, but it is also of practical importance in the search and design of specific protein aggregation inhibitors.

The two-step kinetic model of insulin aggregation by Gibson and Murphy[1] is a classical example of an autocatalytic reaction, where aggregation-prone species form more aggregation-prone species in the presence of native protein. This model of insulin aggregation has been employed to characterize the difference between the measured lag time of insulin aggregation in the absence and presence of aggregation inhibitors.[1] Unfortunately, this approach to characterize aggregation inhibitors neglects the fundamental competitive nature of inhibitors in chemical kinetics, which is of increasing relevance in the design of highly specific pharmacological agents.[24,25] In this paper, we introduce a competitive inhibition pathway to the two-step reaction mechanism of insulin aggregation originally developed by Gibson and Murphy.[1]

2. A two-step kinetic model of insulin aggregation

Insulin aggregation requires association between specific domains of the B-chain of two insulin molecules. In Pro human insulin (DKP-insulin), the histidine in position 10 of the B-chain (His^{B10}) is considered essential for the formation of hexamers, while the proline in position 28 of the B-chain (Pro^{B28}) and lysine in position 29 of the B-chain (Lys^{B29}) are essential for insulin dimerization.[26,27] Gibson and Murphy[1] identified short insulin fragments that bind to full-length insulin by mixing insulin fragments produced by limited proteolysis and incubated with immobilized insulin. Utilizing reverse-phase high-performance liquid chromatography to detect peptides that bind to the full insulin molecule, their analysis suggested that the fragment CGSHLVEAL bound to full-length insulin. Interestingly, this fragment overlaps with an insulin domain, LVEALYLV, which has been previously proposed to be involved in insulin misfolding and aggregation.[16,28] Gibson and Murphy postulated that part or all of the B-chain segment B7–B18 of insulin, or CGSHLVEALYLV, could serve as an effective recognition or binding domain for an insulin hybrid dimer,[1] which is depicted in Figure 1.

To determine a set of conditions leading to amyloidogenesis, 0.52 mM insulin in 1 M acetic acid (pH 2.0) was incubated at 37°C, and aliquots were taken at various time points.[1] The presence of insulin aggregation-prone species was quantified with thioflavin T (ThT) fluorescence intensity assays, assuming that the fluorescence of insulin aggregation-prone species is proportional to the total mass of amyloid insulin fibrils in the experiment. They reported a lag phase of 45-55 h, after which a rapid increase in ThT fluorescence was detected, indicating the growth of insulin aggregates.[1]

Based on the above observations, Gibson and Murphy[1] introduced a two-step reaction model of insulin aggregation. In the first step, native insulin N is converted with a slow first-order rate into the aggregation-prone insulin A. The second step is a fast second-order autocatalytic conversion

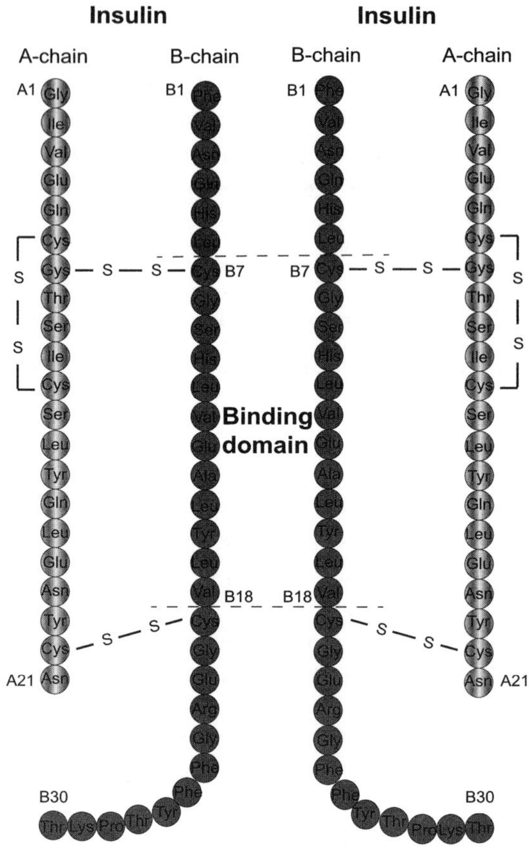

Figure 1. An illustration of insulin dimerization. Gibson and Murphy[1] postulate that the B7–B18 region of the amino acid sequence serves as the effective recognition or binding domain of two insulin monomers.

of the N into A by using an existing A as a template. It is assumed that N is irreversibly converted into A. The reaction mechanism for the two-step model is:

$$\begin{aligned} N &\xrightarrow{k_s} A \\ N + A &\xrightarrow{k_f} 2A \end{aligned} \tag{1}$$

where k_s and k_f are rate constants. Reaction mechanism (1) has been used before to investigate amyloid β, α-synuclein, polyglutamine, prions, and human calcitonin aggregation.[29–31]

Applying the law of mass action to reaction mechanism (1), the governing differential equations for the two-step kinetic model are:

$$\begin{aligned}\frac{dN}{dt} &= -k_s N - k_f N A \\ \frac{dA}{dt} &= k_s N + k_f N A.\end{aligned} \tag{2}$$

Using the mass conservation law for the reaction mechanism,

$$\frac{dN}{dt} + \frac{dA}{dt} = 0, \tag{3}$$

the system (2) can be analytically solved to obtain a closed-form expression for the aggregate-prone insulin concentration as a function of time:

$$A = \frac{N_0(e^{k_s t(1+KN_0)} - 1)}{e^{k_s t(1+KN_0)} + KN_0}, \tag{4}$$

where N_0 is the initial amount of native insulin, and $K = k_f/k_s$. In this solution, it is assumed that aggregates are initially absent from the reaction. Gibson and Murphy[1] used Eq. (4) to fit experimental data for the insulin aggregation-prone species A and characterize the fibrillogenesis of insulin.

We assume that conformational changes in native insulin (occurring with rate k_s) and recruitment by amyloidogenic species (taking place with rate k_f) are biophysical properties of differing conformational states of insulin in the absence of inhibitor. Furthermore, although the number of experimental points given by Gibson and Murphy[1] is small and the measured experimental errors are relatively large, the kinetic rate constants k_s and k_f act on disparate timescales. Therefore, we elect to simultaneously estimate these parameter values using nonlinear regression, which are listed in Table 1. The optimization routine `lsqnonlin` from Matlab R2012b with the trust-region-reflective algorithm was utilized to produce the nonlinear fits to the experimental data for insulin fibrillogenesis. The results of our fitting are shown in Figure 2. As $t \to \infty$, the nonamyloidogenic species N is completely converted into the amyloidogenic species A.

Table 1. The parameter estimates obtained by the nonlinear fitting of Eq. (4) using the experimental data of Gibson and Murphy,[1] as shown in Figure 2. 0.52 mM insulin in 1 M acetic acid (pH 2.0) was incubated at 37°C, and aliquots were taken at various time points.[1]

Rate Constant	Value	Units
k_s	$(0.561 \pm 2.597) \times 10^{-7}$	h^{-1}
k_f	$(2.415 \pm 0.665) \times 10^{-4}$	(fluorescence units)$^{-1}$ (h)$^{-1}$

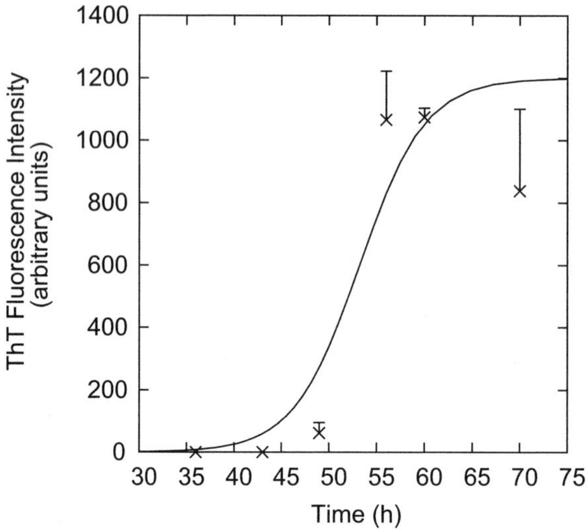

Figure 2. Time course of the two-step kinetic model given by reaction mechanism (1). 0.52 mM insulin in 1 M acetic acid (pH 2.0) was incubated at 37°C, and aliquots were taken at various time points.[1] Parameters k_s and k_f were fit simultaneously using non-linear regression. The parameter estimates used for the fitting are given in Table 1.

3. Inhibition of insulin aggregation in the presence of small hybrid peptides

As we mentioned in the introduction, insulin aggregation (or fibrillation of insulin) can be inhibited by compounds that either prevent the unfolding of a native protein or sequester partially folded aggregation-prone intermediates.[1] Some of the most commonly used aggregation inhibitors are insulin co-solutes[20–22] and low molecular weight compounds such as betaine, arginine, or trehalose.[23] These solutes inhibit aggregation through their effect on the protein-solvent interface. However, they are non-specific and only effective at very high concentrations. For instance, a significant increase in lag time of aggregation of 0.5 mM insulin required arginine concentrations of at least 100 mM.[1]

To obtain more effective insulin aggregation inhibitors, Gibson and Murphy[1] synthesized hybrid peptides using overlapping sequence fragments of B7–B18 as a binding domain. The hybrid peptides consist of an effective binding domain (with sequence GSHLV, GSHLVEAL, VEALYL, or ALYLV) and a C- or N-terminal placement of the arginine hexamer as the disrupting domain. Both domains were necessary for inhibitory activity of the hybrid peptides.

To test for inhibition, insulin was dissolved at 0.52 mM in 1 M acetic acid (pH 2.0), mixed with hybrid peptides at 2 mM, and kept at a constant temperature of 37°C.[1] In Figure 3, we illustrate the inhibition of insulin dimerization by the RRRRRRLVEALYLV hybrid peptide, which was estimated by Gibson and Murphy[1] to be the most effective inhibitor at low concentrations.

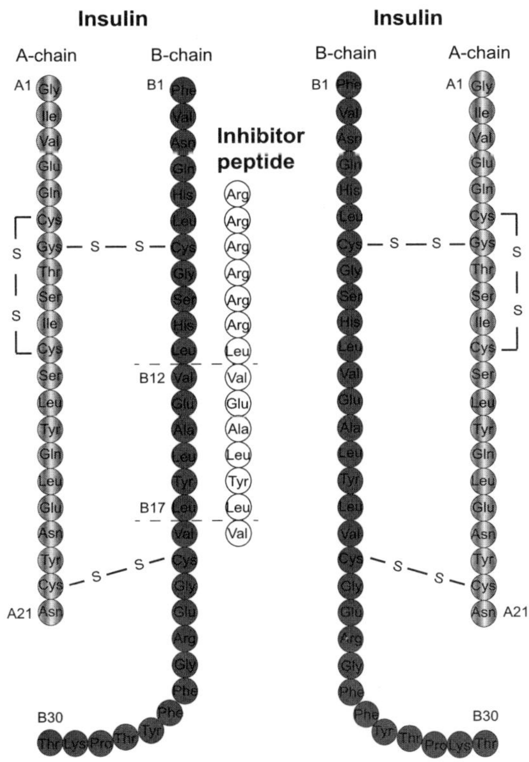

Figure 3. An illustration of the association between insulin and the most effective hybrid peptide RRRRRRLVEALYLV. The recognition domain binds to the region B12-B17, and the disrupting domain, hexameric arginine, is attached at the N-terminus.

4. A two-step kinetic model of insulin aggregation with a competitive inhibitor

Based on the above observations, we introduce a competitive inhibitor in the two-step model of insulin aggregation. The inhibitor I reversibly associates with a native protein to form a complex C, thereby sequestering insulin and preventing aggregation. In this case, the reaction mechanism is given by

$$\begin{aligned} N &\xrightarrow{k_s} A \\ N + A &\xrightarrow{k_f} 2A \\ N + I &\underset{k_d}{\overset{k_b}{\rightleftharpoons}} C \end{aligned} \tag{5}$$

where k_b and k_d are rate constants for the association and dissociation of the complex, respectively. This yields the following system of differential equations:

$$\frac{dN}{dt} = -k_s N - k_f NA - k_b NI + k_d C \tag{6}$$

$$\frac{dA}{dt} = k_s N + k_f NA \tag{7}$$

$$\frac{dI}{dt} = -k_b NI + k_d C \tag{8}$$

$$\frac{dC}{dt} = k_b NI - k_d C. \tag{9}$$

We assume that the concentrations of aggregates and complexes are initially absent, and N_0 and I_0 are the initial concentrations (in fluorescence units) of native proteins and inhibitors, respectively.

The above model has two conserved quantities, which we exploit to simplify the dimensionality of the system:

$$\frac{dI}{dt} + \frac{dC}{dt} = 0 \tag{10}$$

$$\frac{dN}{dt} + \frac{dA}{dt} + \frac{dC}{dt} = 0 \tag{11}$$

which implies that

$$I = I_0 - C \tag{12}$$

$$N = N_0 - A - C. \tag{13}$$

Substituting Eqs. (12) and (13) into Eqs. (7) and (9) yields

$$\frac{dA}{dt} = k_s(N_0 - A - C) + k_f(N_0 - A - C)A \tag{14}$$

$$\frac{dC}{dt} = k_b(N_0 - A - C)(I_0 - C) - k_d C. \tag{15}$$

Assuming that the binding of native insulin and competitive inhibitor is at equilibrium, we take $\frac{dC}{dt} = 0$, for which Eq. (15) can be written as

$$0 = C^2 - \left(I_0 + N_0 - A + \frac{k_d}{k_b}\right) C + (N_0 - A)I_0 \tag{16}$$

whose solutions are given by

$$C = \frac{1}{2}\left[\left(I_0 + N_0 - A + \frac{k_d}{k_b}\right) \pm \sqrt{\left(I_0 + N_0 - A + \frac{k_d}{k_b}\right)^2 - 4(N_0 - A)I_0}\,\right]. \tag{17}$$

We rewrite Eq. (17) for C as

$$C = \frac{1}{2}\left(I_0 + N_0 - A + \frac{k_d}{k_b}\right)\left[1 \pm \sqrt{1 - \frac{4(N_0 - A)I_0}{\left(I_0 + N_0 - A + \frac{k_d}{k_b}\right)^2}}\,\right]. \tag{18}$$

Following the work of Schnell and Hanson,[32] we consider the quantity

$$\epsilon = \frac{4(N_0 - A)I_0}{\left(I_0 + N_0 - A + \frac{k_d}{k_b}\right)^2} \ll 1. \tag{19}$$

Then expanding Eq. (18) approximates the two solution branches as

$$C^+ = \frac{1}{2}\left(I_0 + N_0 - A + \frac{k_d}{k_b}\right)\left[2 - \frac{1}{2}\epsilon + \mathcal{O}(\epsilon^2)\right] \tag{20}$$

$$C^- = \frac{1}{2}\left(I_0 + N_0 - A + \frac{k_d}{k_b}\right)\left[\frac{1}{2}\epsilon + \mathcal{O}(\epsilon^2)\right]. \tag{21}$$

Note that if the inhibitor is initially absent ($I_0 = 0$), Eq. (20) does not recover the simplified model given by Eq. (2). Therefore, this branch is considered unphysical, and we assume C is given by Eq. (21). Substitution of Eq. (21) into Eq. (14) reveals

$$\frac{dA}{dt} = \frac{(N_0 - A)(k_s + k_f A)}{1 + \dfrac{I_0}{N_0 - A + K_I}}. \tag{22}$$

where $K_I = k_d/k_b$. Here, if $I_0 = 0$ in Eq. (22), then the exact solution reverts back to Eq. (4).

Using the previously determined values of k_s and k_f (see Table 1), the parameter K_I is estimated using the time course data for insulin aggregation inhibited by each hybrid peptide (see Section 3 for more details).

These estimated values are listed in Table 2. With the exception of the peptide RRRRRR, the estimates yield a standard error that is at least an order of magnitude lower, and give excellent fits to the inhibitory data, which are shown in Figure 4.

Table 2. The parameter values and standard errors of K_I corresponding to each inhibitory hybrid peptide. Estimates were obtained by the nonlinear fitting of Eq. (22) using the experimental data of Gibson and Murphy,[1]. Insulin was dissolved at 0.52 mM in 1 M acetic acid (pH 2.0), and mixed with 2 mM hybrid peptide at 37°C.

Hybrid Peptide	K_I (fluorescence units) $\times 10^{-5}$
RRRRRR	1.844 ± 1.563
ALYLVRRRRRR	1.202 ± 0.567
GSHLVEALRRRRRR	0.175 ± 0.011
RRRRRRGSHLVEAL	0.333 ± 0.097
LVEALYLVRRRRRR	0.084 ± 0.002
RRRRRRLVEALYLV	0.050 ± 0.001

Since K_I is the ratio of the dissociation rate k_d to the binding rate k_b, higher values of K_I correspond to weaker inhibitory compounds. This corresponds well to the experimental data, and the magnitude of this parameter represents a quantification of the relative strength of a compound's inhibitory capacity. In Figure 4, we show the nonlinear fits to the experimental data using the values of K_I listed in Table 2 and the values of k_s and k_f in Table 1.

5. Comparison with the analysis of Gibson and Murphy

Gibson and Murphy[1] had shown that hybrid peptides can significantly reduce the rate of insulin aggregation by analyzing the two-step model of aggregation kinetics (1). In their analysis, they quantify the strength of individual inhibitors by estimating k_s, which is listed in Table 3. Note that Gibson and Murphy's relative strength of inhibition among distinct peptides is comparable to our results given in Table 2. On the other hand, the estimates of Gibson and Murphy are subject to significantly higher standard error, which originate from analyzing their data using a reaction mechanism without an explicitly competitive inhibition pathway.

Gibson and Murphy[1] also estimate the change in lag time for individual inhibitors relative to insulin alone. In protein folding kinetics, the aggregration lag time is a common description of influential magnitude on protein

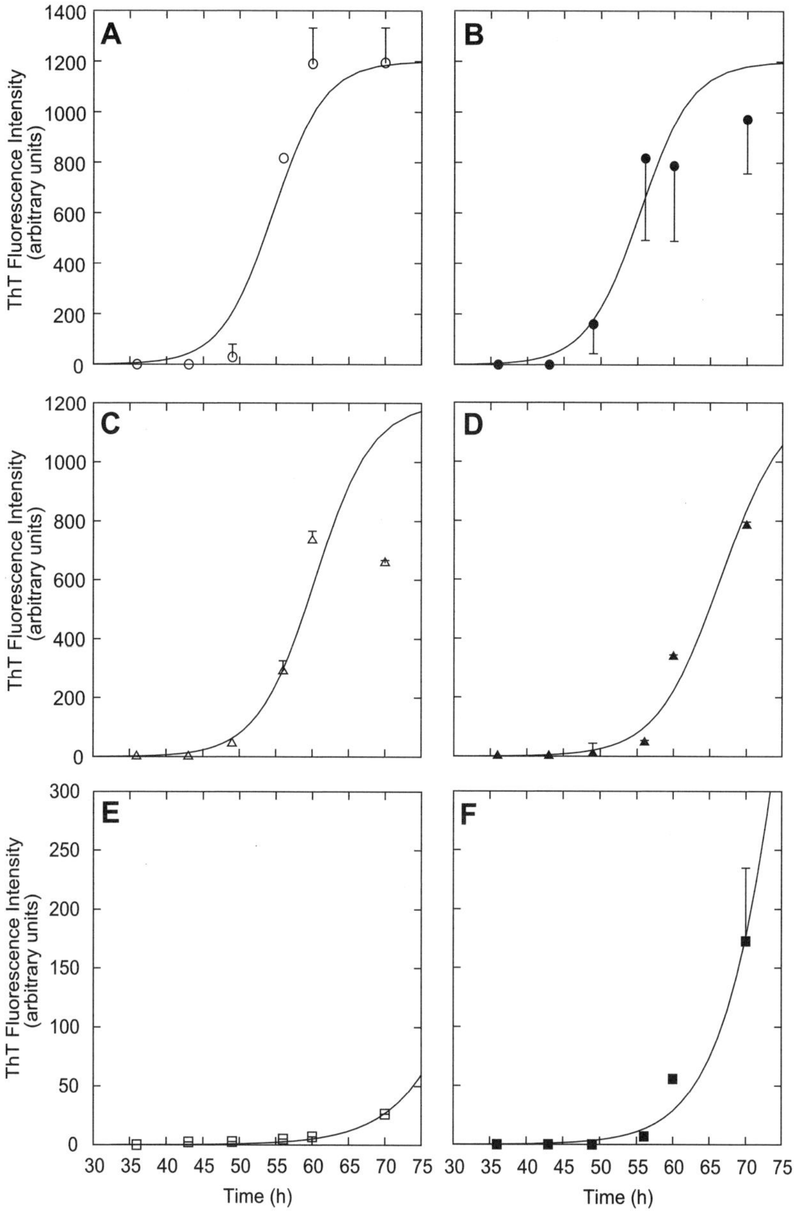

Figure 4. Time course of the two-step aggregation model with competitive inhibition given by reaction mechanism (5). The inhibitors analyzed were the hybrid peptides (A) RRRRRR, (B) ALYLVRRRRRR, (C) RRRRRRGSHVEAL, (D) GSHVEALRRRRRR, (E) RRRRRRLVEALYLV, and (F) LVEALYLVRRRRRR, as described in Section 3.

aggregation. These empirical calculations contend that the hybrid peptides ALYLVRRRRRR and RRRRRRGSHVEAL are comparable in their delay of the onset of aggregation, and RRRRRR and GSHVEALRRRRRR are similarly comparable. However, this is a poor measure of the inhibitory capacity of these synthetic compounds, as it is unable to uniquely delineate each hybrid peptide. In our analysis, the explicit inclusion of competitive inhibition affords a higher resolution in characterizing each inhibitor by the parameter K_I.

Table 3. The parameter values of k_s and corresponding lag times for each hybrid peptide, reported in Gibson and Murphy.[1] Insulin was dissolved at 0.52 mM in 1 M acetic acid (pH 2.0), and mixed with 2 mM hybrid peptide at 37°C.

Hybrid Peptide	k_s (h^{-1}) $\times 10^6$	Change in lag time (h)
RRRRRR	1.1 ± 1.1	6
ALYLVRRRRRR	1.6 ± 0.9	0
GSHLVEALRRRRRR	0.19 ± 0.07	0
RRRRRRGSHLVEAL	0.17 ± 0.17	6
LVEALYLVRRRRRR	0.015 ± 0.004	12
RRRRRRLVEALYLV	0.002 ± 0.001	58

6. Discussion

In this work, we introduce a novel model of protein inhibition aggregation to characterize the effects of hybrid peptides in inhibiting insulin aggregation. Using the foundation of a two-step reaction mechanism of insulin aggregation by Gibson and Murphy,[1] we have expanded this model to include a competitive inhibitor in the reaction to investigate inhibitor kinetics. The inhibitor reversibly binds with nonamyloidogenic species of insulin by forming a complex, which temporarily prevents insulin aggregation.

The nonlinear fitting of the model to the experimental data of Gibson and Murphy[1] shows that our model represents an adequate description of the insulin aggregation time course and provides an effective approach for quantitatively estimating the kinetics of competitive inhibition of hybrid peptides in insulin aggregation. Our analysis gives estimates of the inhibitor effects with lower standard errors than those previously reported by Gibson and Murphy[1]. Nonetheless, our model can be improved from the biophysical chemistry point of view. For example, the formation of an intermediate complex in the recruitment of native insulin by amyloidogenic species to become aggregation-prone could provide a more biochemically realistic representation of the reaction steps. At the same time, this more detailed chemical model could potentially alter the qualitative dynamics of insulin aggregation and provide a better fitting to the experimental data.

Although the time course data gives important dynamic information about insulin aggregation, further insights about the reaction mechanism can be obtained through initial rate experiments and using Eq. (22). By testing various initial insulin concentrations in the presence of differing inhibitor concentrations, the initial rate of aggregate formation can be used to estimate the kinetic parameters and distinguish different mechanisms of inhibition. This will be explored elsewhere in a future study.

Acknowledgments

The authors would like to thank Roberto Miguez for his critical review of this manuscript prior to its publication. This work is supported by the James S. McDonnell Foundation (Grant No. 220020223) under the 21st Century Science Initiative Studying Complex Systems.

References

1. T. J. Gibson and R. M. Murphy, *Protein Sci.*, 2006, **15**, 1133–1141.
2. M. A. Weiss, *FEBS Lett.*, 2013, **587**, 1942–1950.
3. G. Dodson and D. Steiner, *Curr. Opin. Struc. Biol.*, 1998, **8**, 189–194.
4. X. F. Huang and P. Arvan, *J. Biol. Chem.*, 1995, **270**, 20417–20423.
5. M. F. Dunn, *Biometals*, 2005, **18**, 295–303.
6. S. E. Fineberg, T. T. Kawabata, D. Finco-Kent, R. J. Fountaine, G. L. Finch and A. S. Krasner, *Endocr. Rev.*, 2007, **28**, 625–652.
7. V. Sluzky, A. M. Klibanov and R. Langer, *Biotechnol. Bioeng.*, 1992, **40**, 895–903.
8. V. Sluzky, J. A. Tamada, A. M. Klibanov and R. Langer, *P. Natl. Acad. Sci. USA*, 1991, **88**, 9377–9381.
9. Q.-x. Hua and M. A. Weiss, *J. Biol. Chem.*, 2004, **279**, 21449–21460.
10. L. Nielsen, S. Frokjaer, J. Brange, V. N. Uversky and A. L. Fink, *Biochemistry-US*, 2001, **40**, 8397–8409.
11. L. Nielsen, S. Frokjaer, J. F. Carpenter and J. Brange, *J. Pharm. Sci.*, 2001, **90**, 29–37.
12. A. Ahmad, I. S. Millett, S. Doniach, V. N. Uversky and A. L. Fink, *Biochemistry-US*, 2003, **42**, 11404–11416.
13. J. Sharp, J. Forrest and R. Jones, *Biochemistry-US*, 2002, **41**, 15810–15819.

14. E. J. Nettleton, P. Tito, M. Sunde, M. Bouchard, C. M. Dobson and C. V. Robinson, *Biophys. J.*, 2000, **79**, 1053–1065.
15. S. Frokjaer and D. E. Otzen, *Nat. Rev. Drug Discov.*, 2005, **4**, 298–306.
16. J. Brange, L. Andersen, E. D. Laursen, G. Meyn and E. Rasmussen, *J. Pharm. Sci.*, 1997, **86**, 517–525.
17. M. E. Brewster, M. S. Hora, J. W. Simpkins and N. Bodor, *Pharm. Res.*, 1991, **8**, 792–795.
18. U. Grau and C. D. Saudek, *Diabetes*, 1987, **36**, 1453–1459.
19. H. Thurow and K. Geisen, *Diabetologia*, 1984, **27**, 212–218.
20. P. J. Blackshear, T. D. Rohde, J. L. Palmer, B. D. Wigness, W. M. Rupp and H. Buchwald, *Diabetes Care*, 1983, **6**, 387–392.
21. J. Brange and S. Havelund, *Acta Med. Scand.*, 1983, **213**, 135–138.
22. M. Katakam and A. K. Banga, *J. Pharm. Pharmacol.*, 1995, **47**, 103–107.
23. A. Arora, C. Ha and C. B. Park, *FEBS Lett.*, 2004, **564**, 121–125.
24. S. Schnell and C. Mendoza, *J. Math. Chem.*, 2000, **27**, 155–170.
25. S. Schnell and C. Mendoza, *B. Math. Biol.*, 2000, **62**, 321–336.
26. D. N. Brems, L. A. Alter, M. J. Beckage, R. E. Chance, R. D. DiMarchi, L. K. Green, H. B. Long, A. H. Pekar, J. E. Shields and B. H. Frank, *Protein Eng.*, 1992, **5**, 527–533.
27. S. E. Shoelson, Z. X. Lu, L. Parlautan, C. S. Lynch and M. A. Weiss, *Biochemistry-US*, 1992, **31**, 1757–1767.
28. J. Brange, G. Dodson, D. Edwards, P. Holden and J. Whittingham, *Proteins*, 1997, **27**, 507–516.
29. M. Kamihira, A. Naito, S. Tuzi, A. Y. Nosaka and H. Saito, *Protein Sci.*, 2000, **9**, 867–877.
30. A. M. Morris, M. A. Watzky and R. G. Finke, *BBA-Proteins Proteom.*, 2009, **1794**, 375–397.
31. M. A. Watzky and R. G. Finke, *J. Am. Chem. Soc.*, 1997, **119**, 10382–10400.
32. S. Schnell and S. M. Hanson, *Biophys. Chem.*, 2007, **125**, 269–274.

OPTIMAL CONTROL OF RESOURCE COEFFICIENT IN A PARABOLIC POPULATION MODEL

J. BINTZ, H. FINOTTI, AND S. LENHART*
University of Tennessee
Department of Mathematics
Knoxville, TN 37996-1320, USA
E-mail: bintz@math.utk.edu, heather@math.utk.edu, lenhart@math.utk.edu

We study the problem of allocating resources to maximize the net benefit in the conservation of a single species while the cost of the resource allocation is minimized. The population model is a parabolic differential equation with density dependent growth and spatial-temporal resource control coefficient. The existence of an optimal control and the uniqueness and characterization of the optimal control are established. Numerical simulations illustrate several cases with Dirichlet and Neumann boundary conditions.

1. Introduction

The effect of resource allocation on the dynamics of a population is an important conservation issue. One aspect is how the arrangement of resources can effect the survival of a population.[13,4,5,11] This issue of arranging favorable parts of the habitat for long-term survival was first addressed by Cantrell and Cosner[2] in partial differention equations (PDEs) of the form:

$$u_t = \mu \Delta u + m(x)u - u^2 \qquad \text{in } \Omega,$$

subject to Dirichlet, Robin, or Neumann boundary conditions. The solution $u(x,t)$ is the density of the population, and the coefficient $m(x)$ represents the intrinsic growth rate of the population and is a proxy for the availability of resources. Their fundamental and other following related work studied positive attracting steady states and showed in some cases that a single favorable area in the habitat would be best given limited resources.[3,6,12,19]

*Work partially supported by the National Institute for Mathematical and Biological Synthesis through the National Science Foundation grant EF-0832858

A related question is: How does resource allocation effect population size? Population size is clearly a reasonable goal in conservation effort. Optimal control of PDEs [17,18] can give some input on such questions. Ding et al.[8] investigated this question, where the control was the resource function $m(x)$ in an elliptic PDE

$$\mu\Delta u + m(x)u - u^2 = 0 \qquad \text{in } \Omega,$$

with Neumann boundary conditions. The objective was to choose $m(x)$ to maximize the population size while minimizing a quadratic cost of implementing the control. They also considered an integral constraint on m (fixed amount of resource). The optimal control depended on the diffusion coefficient μ and the term for the cost of the control.

In this paper, we consider a time-varying problem, using a parabolic PDE with a resource coefficient depending on space and time. We seek to balance the two features of maximizing the population and minimizing the cost of the control, representing the resources. Using a finite time interval, results from Dirichlet and Neumann boundary conditions are compared.

The next section formulates our PDE with control and the objective functional. In the third section, *a priori* estimates for the solutions are given, and the existence of an optimal control is presented in the fourth section. Characterization of an optimal control using the adjoint problem and the uniqueness of the optimal control are presented in sections 5 and 6. Numerical results for several illustrative cases are shown in section 7. Last, we present some conclusions and plans for related future work.

2. Problem Formulation

Let Ω be an open, bounded, C^1 subset of $\mathbb{R}^n$ and let $Q = \Omega \times (0,T)$ where $T \in (0,\infty)$. Let $u = u(x,t)$ be the population density of a species at $(x,t) \in Q$. We consider the problem:

$$\begin{cases} u_t - \mu\Delta u = u(m-u), & Q, \\ u(\cdot,0) = u_0, & \Omega \times \{t=0\}. \end{cases} \tag{1}$$

Here $\mu > 0$ is the diffusion coefficient which is fixed. The function, $m = m(x,t)$, is the "resource" function which is assumed to be bounded. The initial function u_0 is given and assumed to be non-negative bounded and is in the Sobolev space $H^1(\Omega)$. We will emphasize Neumann boundary conditions

$$\frac{\partial u}{\partial \nu} = 0 \text{ on } \partial\Omega \times (0,T) \tag{2}$$

where ν is the outward normal vector on $\partial\Omega$, but will also compare results with Neumann boundary conditions to those with Dirichlet boundary conditions,

$$u = 0 \text{ on } \partial\Omega \times (0, T). \tag{3}$$

Biologically, Neumann boundary conditions represent a closed habitat for the species under management. For comparison, we also consider Dirichlet boundary conditions which means the population is 0 at the boundary due to a hostile area outside the boundary.

Resources impact a population through influencing the intrinsic growth rate of the species. We consider, therefore, m to be our control. The goal is to maximize the overall population size while minimizing the costs associated with providing resources. We formulate the following control problem. With a control set U given by

$$U = \{m \in L^\infty(Q) : 0 \leq m \leq M\}, \tag{4}$$

we seek m^* such that

$$J(m^*) = \sup_{m \in U} J(m)$$

where the objective functional, J, is defined by

$$J(m) = \int_Q \left[u(x,t) - Bm(x,t)^2\right] dxdt, \tag{5}$$

where B is a positive weight constant on the cost term.

Weak Formulation for Neumann Boundary Conditions The function $u \in L^2((0,T), H^1(\Omega))$ with $u_t \in L^2((0,T), H^1(\Omega)^*)$ and $u(x,0) = u_0(x)$ is said to be a weak solution of (1) if and only if for a.e. $t \in (0,T)$

$$\int_\Omega u_t \phi dx + \int_\Omega (\mu \nabla u) \cdot \nabla \phi dx = \int_\Omega u(m-u)\phi \, dx, \quad \forall \phi \in H^1(\Omega). \tag{6}$$

For Dirichlet boundary conditions, the solution space is

$$u \in L^2((0,T), H_0^1(\Omega))$$

with $u_t \in L^2((0,T), H^{-1}(\Omega))$.

3. A priori Estimates and Existence of Solutions

In this section, we state some preliminary results needed for proving the existence of and for characterizing our optimal solutions. These results hold for both Dirichlet and Neumann boundary conditions. The first lemma states that all solutions u of (1) must be positive and the proof of this first lemma can be found in Finotti et al.[10]

Lemma 3.1. *Assume that $m \in U$ and $u_0 \in L^\infty(\Omega)$ and $u_0 \geq 0$. Then, any solution u must be non-negative on Q.*

The next lemma will be used to show our objective functional J is bounded above.

Lemma 3.2. *Under above assumptions, for each $m \in U$, any solution u of (1) satisfies*

$$\int_\Omega u(x,t)dx \leq e^{tC} \int_\Omega u_0(x)dx, \ \forall\, t \in [0,T).$$

In particular, there exists a positive constant C such that

$$J(m) \leq \frac{[e^{CT}-1]}{C}||u_0||_{L^\infty}, \qquad \textit{for all} \ \ m \in U.$$

Proof. From Lemma 3.1, we see that $u \geq 0$. Then, the lemma follows directly by using the weak formulation with $\phi = 1$ and using Gronwall's inequality. □

The rest of the section is devoted to prove the existence and *a priori* estimates of the solutions u of the equation (1). Our first result is an energy estimate.

Lemma 3.3. *For every m in U, there exists a constant $C > 0$ depending on $\mu, |\Omega|$ and T such that the following estimate holds for any solution u of (1)*

$$\sup_{t\in[0,T]} \int_\Omega |u(x,t)|^2 dx + \int_Q |\nabla u(x,t)|^2 dxdt \leq C||u_0||^2_{L^2(\Omega)}.$$

Proof. It follows from Lemma 3.1 that $u \geq 0$. Thus, multiplying the first equation of (1) with u and applying integration by parts, we get

$$\begin{aligned} \frac{1}{2}\frac{d}{dt}\int_\Omega u^2 dx + \mu \int_\Omega |\nabla u|^2 dx &= \int_\Omega [mu^2 - u^3] dx \\ &\leq M \int_\Omega u^2 dx, \end{aligned} \tag{7}$$

where we used the upper bound on m and the non-negativity of u. This implies that there is a constant C_1 depending on the upper bound of our controls such that

$$\frac{d}{dt}\int_\Omega u(x,t)^2 dx + \mu \int_\Omega |\nabla u(x,t)|^2 dx \leq C_1 \int_\Omega u^2 dx. \tag{8}$$

Gronwall's inequality implies

$$\int_\Omega u(x,t)^2 dx \leq e^{C_1 T} \int_\Omega u_0(x) dx, \quad \forall\, 0 \leq t < T. \tag{9}$$

Integrating inequalities (8) and (9) over $(0,T)$ leads to our estimate. □

Using the maximum principle and standard existence and uniqueness results,[9] one obtains this result. The result below is also a special case of Theorem 3.1 in Finotti et al.[10]

Theorem 3.1. *Let $0 < T < \infty$ and u_0 be non-negative, bounded and in $H^1(\Omega)$. Then, for each $m \in U$, there is a unique weak solution $u = u(m)$ of (1). Moreover, there is a finite constant $C > 0$ depending only on $|\Omega|, \mu, T, and ||u_0||_{L^\infty}$ such that*

$$0 \leq u(m) \leq C, \quad \forall\ (x,t) \in Q. \tag{10}$$

4. Existence of an Optimal Control

To investigate the maximum of our objective functional, we first show the existence of an optimal control.

Theorem 4.1. *Assume that $0 < T < \infty$, $u_0 \in L^\infty(\Omega) \cap H^1(\Omega)$ and u_0 is non-negative. There exists an optimal control $m^* \in U$ maximizing the objective functional $J(m)$.*

Proof. By Lemma 3.2, $J(m) \leq C$ where C is a constant depending on $||m||_{L^\infty(Q)}$, $||u_0||_{L^\infty(Q)}$ and T only. Therefore, $\sup_{m\in U} J(m)$ exists. Let $m^n \in U$ be maximizing sequence, i.e.

$$\lim_{n\to\infty} J(m^n) = \sup_{m\in U} J(m). \tag{11}$$

Let $u^n = u(m^n)$ be the corresponding state solution when the control is m^n. We have shown *a priori* estimates for the states in $L^2((0,T), H^1(\Omega))$ and in $L^\infty(Q)$.

By passing to a subsequence and using that $L^2(Q)$ is weakly compact, there exists u^* and m^* such that

$$u^n \rightharpoonup u^* \text{ in } L^2(0,T,H^1(\Omega)), \tag{12}$$

$$m^n \rightharpoonup m^* \text{ in } L^2(Q).$$

For each n and each $\phi \in L^2(0,T,H^1(\Omega))$, the weak form of the solution u^n is

$$\int_Q u_t^n \phi \, dxdt = -\mu \int_Q \nabla u^n \cdot \nabla \phi \, dxdt + \int_Q m^n u^n \phi \, dxdt - \int_Q (u^n)^2 \phi \, dxdt.$$

Tne following estimate,

$$\left| \int_Q u_t^n \phi \, dxdt \right| \leq C ||\phi||_{L^2(0,T,H^1(\Omega))}, \quad \forall \, n \in \mathbb{N},$$

comes from our bounds on the states in $L^2(0,T,H^1(\Omega))$ and the weak form of the PDE for u^n. Then we have

$$\|u_t^n\|_{L^2(0,T,H^1(\Omega)^*)} \leq C, \quad \forall n \in \mathbb{N}. \tag{13}$$

with the constant C depending only on $\mu, |\Omega|, T, ||u_0||_{L^\infty}$ and $||m||_{L^\infty}$. Using Brezis[1] and by passing to a subsequence, we have

$$\begin{gathered} u^n \to u^* \quad \text{in} \quad L^2(Q), \quad \nabla u^n \rightharpoonup \nabla u^* \quad \text{in} \quad L^2(Q), \quad \text{and} \\ u_t^n \rightharpoonup u_t^* \text{ in } L^2((0,T),H^1(\Omega)^*). \end{gathered} \tag{14}$$

We need to show that $u^* = u(m^*)$, meaning that u^* is the state corresponding to m^*, and that m^* is an optimal control.

From the weak convergences, it follows

$$\begin{aligned} \lim_{n\to\infty} \int_Q u_t^n v \, dxdt &= \int_Q u_t^* v \, dxdt, \\ \lim_{n\to\infty} \int_Q (u^n)^2 \phi \, dxdt &= \int_Q (u^*)^2 \phi \, dxdt. \\ \lim_{n\to\infty} \int_Q \nabla u^n \cdot \nabla v \, dxdt &= \int_Q \nabla u^* \cdot \nabla v \, dxdt. \end{aligned} \tag{15}$$

From the strong convergence of the sequence $\{u^n\}$ and the weak convergence of the sequence $\{m^n\}$, we obtain

$$\begin{aligned}
&\Big| \int_Q (m^n u^n \phi dxdt - \int_Q m^* u^* \phi \, dxdt \Big| \\
&= \Big| \int_Q m^n \phi (u^n - u^*) \, dxdt - \int_Q (m^n - m^*) \, \phi u^* \, dxdt \Big| \\
&\leq C_1 \|\phi\|_{L^2(Q)} \|u^n - u^*\|_{L^2(Q)} + \Big| \int_Q [m^n - m^*] \phi u^* \, dxdt \Big| \\
&\to 0, \quad \text{as} \quad n \to \infty.
\end{aligned}$$

Collecting the four convergence limits above, we pass to the limit in the four terms of our weak formulation of the PDEs for the u^n sequence and obtain $u^* = u(m^*)$.

Using the strong convergence in $L^2(Q)$ of the sequence u^n and the fact that the function $m \mapsto \int_Q |m|^2 dxdt$ is weakly lower semi-continuous in $L^2(Q)$ (see Brezis[1]), we obtain

$$\sup_{minU} J(m) = \lim_{n \to \infty} J(m^n) = \int_Q [u^* - B|m^*|^2] dxdt = J(m^*).$$

We conclude that $m^* \in U$ is an optimal control. □

5. Necessary Conditions

In order to characterize the optimal control, we need to differentiate the map $m \mapsto J(m)$ with respect to the control m. We denote by $u = u(m)$ the unique, positive solution of (1). Since $u = u(m)$ appears in $J(m)$, we first prove the appropriate differentiability of the map $m \mapsto u(m)$ whose derivative is called the *sensitivity*.

Sensitivity 5.1. The mapping $m \in U \mapsto u(m)$ is differentiable in the following sense: for each m, l in U such that $m + \epsilon l \in U$ for all ϵ sufficiently small, then there is a uniform constant $C > 0$ such that

$$\psi^\epsilon = \frac{u(m + \epsilon l) - u(m)}{\epsilon}$$

satisfies

$$||\psi^\epsilon||_{L^2((0,T);H^1(\Omega))} \leq C.$$

Moreover, there exists $\psi = \psi(m, l) \in L^2((0,T), H^1(\Omega))$, such that

$$\psi^\epsilon \rightharpoonup \psi \text{ weakly in } L^2((0,T), H^1(\Omega)) \text{ as } \epsilon \to 0,$$

and the sensitivity ψ satisfies

$$\begin{cases} \psi_t - \mu\Delta\psi - (m-2u)\psi = ul & \text{in } Q, \\ \frac{\partial\psi}{\partial\nu} = 0 & \text{on } \partial\Omega \times (0,T), \\ \psi(x,0) = 0 & \text{on } \Omega \times \{t=0\}. \end{cases} \tag{16}$$

Proof. *A priori* estimates on the quotients ψ^ϵ can be shown similar to the estimates on the states and one can pass to the limit in the ψ^ϵ PDE weak formulation to show that the sensitivity ψ satisfies the desired PDE. □

Next, we characterize our optimal control solution m^* by differentiating the map $m \mapsto J(m)$. We use the sensitivity equation to find our adjoint equation and our characterization.

Theorem 5.2. *Given an optimal control m^* and corresponding state u^*, there exists a solution p in $L^2(0,T,H^1(\Omega))$ which satisfies $p_t \in L^2((0,T),H^1(\Omega)^*)$ and*

$$\begin{cases} -p_t - \mu\Delta p - (m^*-2u^*)p = 1, & \text{in } Q, \\ \frac{\partial p}{\partial\nu} = 0, & \text{in } \partial\Omega \times (0,T), \\ p = 0 & \text{in } \Omega \times \{t=T\}. \end{cases} \tag{17}$$

Furthermore, m^* is characterized by

$$m^* = \min\left\{1, \max\left\{\frac{u^*p}{2B}, 0\right\}\right\}. \tag{18}$$

Proof. Suppose m^* is an optimal control. Let $l \in U$ such that $m^* + \epsilon l \in U$ for sufficiently small $\epsilon > 0$. Let $u^\epsilon = u(m^* + \epsilon l)$ denote the unique solution of (1) when the control term is $m^* + \epsilon l$.

The operator in the adjoint equation is the formal analysis "adjoint" of the operator in the sensitivity equation (16) at m^*. The nonhomogeneous term, 1, comes from differentiating the integrand of the objective functional with respect to the state. The final time condition on the adjoint function is the transversality condition.

The equation (17) is linear in p and its coefficients are measurable and bounded. By the change of variable $t \to T-t$, the existence and uniqueness of the weak solution p of (17) follows by Galerkin's method (e.g. see [9]).

Now, observe that the directional derivative of J with respect to the control at m^* in the direction of l satisfies

$$
\begin{aligned}
0 &\geq \lim_{\epsilon \to 0^+} \frac{J(m^* + \epsilon l) - J(m^*)}{\epsilon} \\
&= \lim_{\epsilon \to 0^+} \frac{1}{\epsilon} \int_Q [u^\epsilon - B(m^* + \epsilon l)^2 - (u^* - Bm^{*2})]\, dxdt \\
&= \lim_{\epsilon \to 0^+} \int_Q [\frac{u^\epsilon - u^*}{\epsilon} - B(2m^* l + \epsilon l^2)]\, dxdt \\
&= \int_Q [\psi - 2Bm^* l]\, dxdt.
\end{aligned}
$$

Using the weak solution formulation for the adjoint problem with test function ψ, we obtain

$$
\begin{aligned}
0 &\geq \int_Q \psi - 2Bm^* l \, dxdt \\
&= \int_Q [\psi(-p_t - (m^* - 2u^*)p) + \mu \nabla p \cdot \nabla \psi - 2Bm^* l]\, dxdt \qquad (19) \\
&= \int_Q u^* lp - 2Bm^* l \, dxdt = \int_Q l(u^* p - 2Bm^*)\, dxdt.
\end{aligned}
$$

Taking cases on the sets where $0 \leq m^* \leq M$, $m^* = M$, and $m^* = 0$, one can obtain the desired control characterization. □

6. Uniqueness Result

The equations (1), (17) with boundary conditions and the control characterization form the optimality system. In this section, we show the uniqueness of solutions to this optimality system under a small condition on T, which gives the uniqueness of the optimal control. Such a condition on T is quite common in optimal control with time-varying dynamics.

Theorem 6.1. *There exists a positive number T_0 such that if $0 < T \leq T_0$, then there is a unique optimal control.*

Proof. From the result in section 4, there exists an optimal control, and corresponding adjoints and states satisfying the optimality system. Thus, we only need to prove the uniqueness of the solutions of the optimality system. Then for $i = 1, 2$, we us denote u_i, p_i, m_i to be the state solution and adjoint solution and control solving the optimality system.

Let $u_i = e^{\lambda t} w_i$ and $p_i = e^{-\lambda t}\pi_i$ for λ a positive constant to be chosen below. Then

$$\lambda w_i + (w_i)_t = \mu \Delta w_i + w_i(m_i^* - e^{\lambda t} w_i)$$

and

$$\lambda \pi_i - (\pi_i)_t = \mu \Delta \pi_i + \pi_i(m_i^* - 2e^{\lambda t} w_i) \ .$$

We also have

$$m_i^* = \min(\max(\frac{u_i p_i}{2B}, 0), 1) \ .$$

Now if we let $W = w_1 - w_2$, $\Pi = \pi_1 - \pi_2$, and $M = m_1^* - m_2^*$, subtracting we get

$$\lambda W + W_t = \mu \Delta W + m_1^* W + w_2 M - e^{\lambda t}(w_1 + w_2)W$$

and

$$\lambda \Pi - \Pi_t = \mu \Delta \Pi + m_1^* \Pi + \pi_2 M - 2e^{\lambda t}(\pi_1 W + w_2 \Pi) \ .$$

Multiplying by W and Π in the weak formulations respectively, and integrating over Q, we obtain

$$\int_\Omega \frac{1}{2} W^2(x,T)\, dx + \lambda \int_Q W^2 \, dxdt + \mu \int_Q (\nabla W)^2 \, dxdt$$

$$= \int_Q (m_1^* - e^{\lambda t}(w_1 + w_2))W^2 + w_2 M W \, dxdt$$

and

$$\int_\Omega \frac{1}{2} \Pi^2(x,0)\, dx + \lambda \int_Q \Pi^2 \, dxdt + \mu \int_Q (\nabla \Pi)^2 \, dxdt$$

$$= \int_Q (m_1^* - 2e^{\lambda t} w_2)\Pi^2 + \pi_2 M \Pi - 2e^{\lambda t}\pi_1 W \Pi \, dxdt \ .$$

Now, we estimate M:

$$|M| = |m_1^* - m_2^*| \leq |\frac{u_1 p_1 - u_2 p_2}{2B}| = \frac{1}{2B}|w_1 \pi_1 - w_2 \pi_2| \tag{20}$$

$$\leq \frac{1}{2B}|w_1 \Pi + \pi_2 W| = \frac{1}{2B}|\pi_1 W + w_2 \Pi| \tag{21}$$

The spatial gradient terms from the right hand sides and the $e^{\lambda t}(w_1 + w_2)W^2$ and $2e^{\lambda t} w_2 \Pi^2$ terms from the left hand sides can be eliminated due to their signs. Then also using the estimate on $|M|$ and the upper bound on m_1^*, we have

$$\int_\Omega \frac{1}{2}W^2(x,T)\,dx + \lambda\int_Q W^2\,dxdt \le \int_Q [W^2 + \frac{|w_2|}{2B}(|\pi_2|W^2 + |w_1\Pi W|)]\,dxdt$$

and

$$\int_\Omega \frac{1}{2}\Pi^2(x,0)\,dx + \lambda\int_Q \Pi^2\,dxdt$$

$$\le \int_Q [\Pi^2 + \frac{|\pi_2|}{2B}(|w_2|\Pi^2 + |\pi_1 W\Pi|) - 2e^{\lambda t}\pi_1 W\Pi]\,dxdt\ .$$

Summing these two inequalities gives

$$\begin{aligned}\int_\Omega \frac{1}{2}W^2(x,T)\,dx &+ \lambda\int_Q [W^2+\Pi^2]\,dxdt + \int_\Omega \frac{1}{2}\Pi^2(x,0)\,dx \\ &\le \int_Q (1+\frac{|w_2||\pi_2|}{2B})W^2 + \int_Q (1+\frac{|\pi_2||w_2|}{2B})\Pi^2 dxdt \\ &+ \int_Q (\frac{|\pi_2|}{2B}|\pi_1| + 2e^{\lambda T}|\pi_1| + \frac{|w_2||w_1|}{2B})W\Pi\,dxdt\end{aligned}$$

Using L^∞ bounds on π_i and w_i gives

$$\begin{aligned}\int_\Omega \frac{1}{2}W^2(x,T)\,dx &+ \int_Q (\lambda - (C_3 + C_1 e^{\lambda T} + \frac{C_2}{B})[W^2+\Pi^2]\,dxdt \\ &+ \int_\Omega \frac{1}{2}\Pi^2(x,0)\,dx \le 0\end{aligned}$$

Thus, taking λ large enough and then T small enough, we get that $W = \Pi = 0$ must be true, so that there is a unique optimal control. □

7. Numerical Results

Using numerical simulations, we illustrate various scenarios with both Neumann and Dirichlet boundary conditions varying the final time T and the initial conditions, as well as the maximum control value. Since we have an initial state condition and a final time adjoint condition, a forward-backward sweep iterative method is used to solve the optimality system.[16] Finite differences are used to approximate derivatives with forward difference in time and centered difference in space.

We begin by contrasting Neumann and Dirichlet boundary conditions. For this we use an initial population with one peak in the center of the domain. Our final time T is 1, diffusion coefficient μ is 0.1, and the maximum control value is 1. We can see (Figure 1) that more control is applied for Dirichlet boundary conditions. This makes sense since we expect our control to be more effective for a population not subject to an inhospitable boundary.

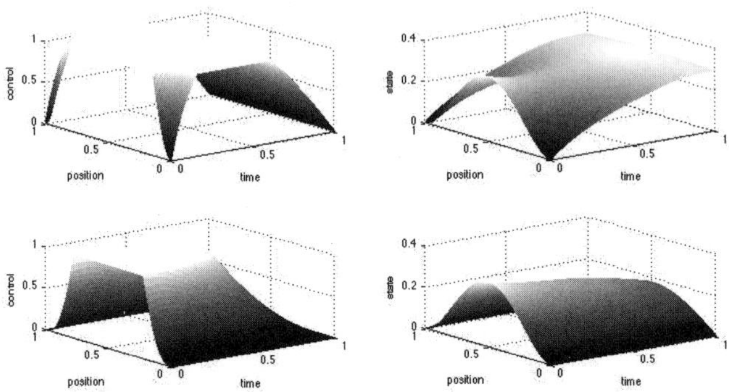

Figure 1. Optimal control and corresponding states for Neumann (top) and Dirichlet (bottom) boundary conditions.

Figure 2 compares the case of Neumann boundary conditions with varying values for the final time, $T = 1$ and $T = 0.2$ respectively. We see that in the case with more time, more control is applied. Ecologically, it is worthwhile to invest more in resources if there is more time to allocate them.

In Figure 3 we consider an initial population distribution with two peaks of the same size: one in the middle of the domain and another to one side of the domain. We again contrast the case with Neumann boundary conditions to that with Dirichlet boundary conditions. In each case the maximum control value is $M = 2$ and $T = 0.2$. In the case of Neumann boundary conditions, there is no distinction between the control levels applied to the two population concentrations. This is not the case, however, when we consider Dirichlet boundary conditions. In the figure it is clear that more control is applied to the population in the center. From an ecological perspective, it makes sense to invest more in the population that is further away from the inhospitable boundary.

Finally, in Figure 4 we consider the case of a single peak initial population located near one boundary of the domain. The interesting case

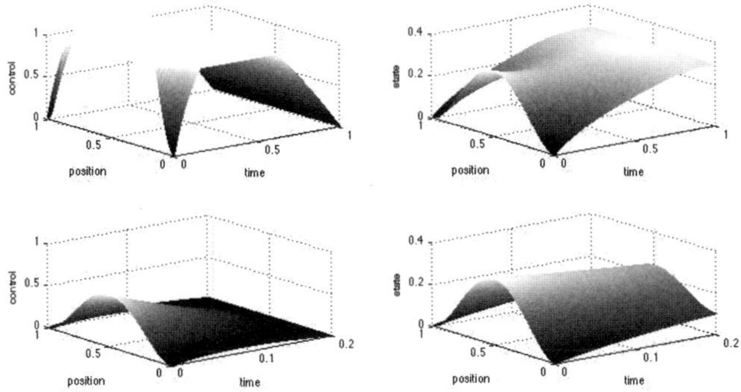

Figure 2. Optimal control and corresponding states for Neumann boundary conditions varying final time with T=1 (top) and T=0.2 (bottom).

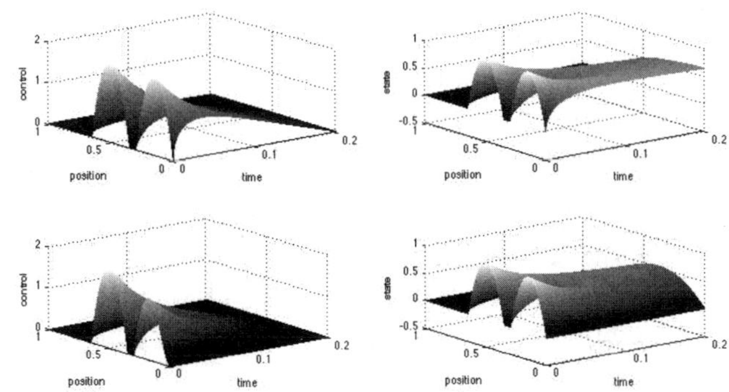

Figure 3. Optimal control and corresponding states for Neumann (top) and Dirichlet (bottom) boundary conditions with two peak initial condition.

is that of an inhospitable boundary (Dirichlet boundary conditions). The maximum control value is $M = 2$, and the diffusion coefficient is $\mu = 0.1$. One can see how an optimal control will "drive" the population toward the center of the domain and away from the hostile boundary.

8. Conclusions

In this preliminary investigation, we have considered the problem of optimal resource allocation for a diffusive population with logistic growth. Optimal resource allocation will maximize the total population over time and space while minimizing the associated costs for control. We have shown the ex-

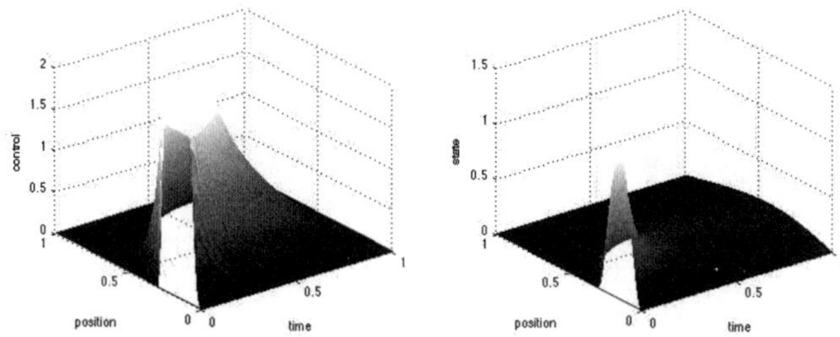

Figure 4. Optimal control and corresponding state for Dirichlet boundary conditions with one peak initial condition.

istence of an optimal control and derived a characterization for the same. Further, for sufficiently small values of final time T we have demonstrated uniqueness of the optimal control. A few numerical simulations have been given to demonstrate our preliminary investigation. We show results for a one dimensional domain that compare Neumann and Dirichlet boundary conditions with various initial conditions as well as various final times and maximum control values.

There are several questions we are currently investigating that extend these results. In particular we consider an additional constraint on our control reflecting the availability of only a fixed amount of resource. Moreover, we are considering two dimensional habitats and ecologically more realistic initial conditions for numerical simulations. Plans for future investigation include considering other objectives such as maximizing the final time population size.

Another interesting feature to be considered involves a population choosing the direction of movement. Along this line, recent work investigated effects of an advection coefficient when the resource coeffient is known.[7,10,14,15]

References

1. H. Brezis, Functional Analysis, Sobolev Spaces and Partial Differential Equations, Universitext. Springer, New York, 2011.
2. R. S. Cantrell and C. Cosner, *Diffusive logistic equations with indefinite weights: population models in a disrupted environments*, Proc. Roy. Soc. Edinburgh A **112** (1989) 293–318.
3. R. S. Cantrell and C. Cosner, *The effects of spatial heterogeneity in population dynamics*, J. Math. Biol. **29** (1991) 315–338.

4. R. S. Cantrell and C. Cosner, Spatial Ecology in Reaction-Diffusion Equations, Wiley Press (2003).
5. R. S. Cantrell, C. Cosner and S. Ruan, Spatial Ecology, CRC Press (2010).
6. X. F. Chen and Y. Lou, *Principal eigenvalue and eigenfunction of elliptic operator with large convection and its application to a competition model*, Indiana Univ. Math. J. **57** (2008), 627–658.
7. C. Cosner and Y. Lou, *Does movement toward better environments always benefit a population*? J. Math. Anal. Appl. **277** (2003), no. 2, 489503.
8. W. Ding, H. Finotti, S. Lenhart, Y. Lou and Q. Ye *Optimal control of growth coefficient on a steady-state population model*, Nonlinear Anal. Real World Appl. **11** (2010), no. 2, 688–704.
9. L. C. Evans, Partial differential equations, 2nd edition, Graduate Studies in Mathematics, 19. American Mathematical Society, Providence, RI, 2010.
10. H. Finotti, S. Lenhart and T. V. Phan, Optimal control of advection direction on reaction-diffusion population models, Evolution Equations and Control Theory 1 (2012), 81-107.
11. E. E. Holmes, M. A. Lewis, J. E. Banks and R. R. Veit, *Partial differential equations in ecology: Spatial interactions and population dynamics*, Ecology **75** (1994), 17 - 29.
12. C. Y. Kao, Y. Lou, and E. Yanagida, Principal eigenvalue for an elliptic problem with indefinite weight on cylindrical domains, Math. Biosc. Eng. 5 (2008) 315-335.
13. P. Kareiva, *Population dynamics in spatially complex environments: Theory and data*, Phil. Trans. Riy. Soc. London Ser. B, **330** (1987), 175 - 190.
14. K. Y. Lam, *Concentration phenomena of a semilinear elliptic equation with large advection in an ecological model*, J. Diff. Eqns. **250** (2011), 161–181.
15. K. Y. Lam and W.-M. Ni, *Limiting profiles of semilinear elliptic equations with large advection in population dynamics*, Discrete Contin. Dyn. Syst. Series A, **28** (2010), 1051–1067.
16. S. Lenhart and T. J. Workman, Optimal control applied to biological models, Chapman & Hall/CRC Mathematical and Computational Biology Series. Chapman & Hall/CRC, Boca Raton, FL, 2007.
17. X. J. Li and J. M. Yong, Optimal Control Theory for Infinite-dimensional Systems, Birkhäuser Boston, Inc., Boston, MA, 1995.
18. J. L. Lions, Optimal Control Systems Governed by Partial Differential Equations, Springer-Verlag, New York, 1971.
19. Y. Lou and E. Yanigda, Principal eigenvalue for an elliptic boundary value problem with indefinite weight and application to population dynamics, Japan. J. Indust. Appl. Math. 23 (2006) 275-292.

OPTIMIZATION OF COSTS FOR COMBATING *AEDES AEGYPTI* IN OPTIMAL TIME-WINDOWS*

W. O. DIAS, G. A. XAVIER, D. A. P. LIMA,
E. F. WANNER AND R. T. N. CARDOSO

Centro Federal de Educação Tecnológica de Minas Gerais
Av. Amazonas, 7675 - Nova Gameleira
Belo Horizonte, MG, 30.510-000, Brasil
E-mail: wodbyte@gmail.com, gisxavier@gmail.com, srdiogolima@gmail.com, efwanner@dppg.cefetmg.br, rodrigoc@des.cefetmg.br

Dengue is one of the most important viral disease worldwide, whereas it is the most important one in Brazil. To avoid a Dengue epidemy, it is necessary to control the transmission vector *Aedes aegypti*. This paper uses a dynamic model of the mosquitoes population to analyze the social and economic cost of combined policies using chemical control (via application of insecticides) and biological control (via insertion of sterile males produced by irradiation). A mono-objective optimization problem is analysed and three different approaches are implemented: constant control, simultaneous step-size control and alternated step-size control. Each optimization problem is solved using a Real-Coded Genetic Algorithm. The results show that the alternated step-size control is preferable over the constant control since each control strategy is used during a smaller time interval while obtaining a smaller cost function value and similar effects in the environment. Furthermore, it is possible to find an optimal number of days in which each control strategy must be applied and the respective rate of insecticide and sterile males that should be used. In this way, the best policy to avoid a Dengue epidemy consists in applying insecticide only at the beginning of the time-window during a small time interval and, then, keeping an acceptable level of natural mosquitoes into environment using a small quantity of biological control.

1. Introduction

Dengue is a disease transmitted by the bite of an *Aedes* mosquito, mainly the *Aedes aegypti*, infected with one of the four Dengue viruses (DENV1, DENV2, DENV3, DENV4). Dengue is a severe, flu-like illness that affects infant, young children and adults. According to World Health Organization

*This work is supported by the Brazilian agencies CAPES, FAPEMIG, CNPQ and CEFET-MG.

(WHO), in 2009, 40% of world population lives in a risk area - which represents approximately 2.5 billion of human beings - showing that Dengue is one of the major public health problems in tropical and subtropical regions around the world. Furthermore, it is estimated that there are 50 to 100 million new infections per year. In Brazil, Dengue is a very important epidemic disease and nowadays it is considered the most important viral disease.

The life cycle of the mosquito has two phases: immature (water) phase consisting of the eggs, larvae and puppae and adult (winged) phase consisting of the adult mosquitoes. The female mosquitoes lay eggs in containers of standing water. In this way, sanitization is an important factor in the battle against Dengue disease, since the vector *Aedes aegypti* can develop in water with high level of pollution. A poor infrastructure and a high level of solid waste allow the water storage which is essential to the development of the vector egg. As reported by National Survey of Brazilian Sanitation 2008, almost half of the states (44.8%) does not have sewage and water distribution system which imply a higher number of breeder to vector reproduction. Nowadays, there are several procedures used by SUCEN (Superintendence for Endemic Disease Control) in Brazil to inhibit the proliferation of the vector such as mechanical control, chemical control and, recently, biological control.

In a type of mechanical control, public health officials are responsible to visit households and to remove all kind of objects that can be used by the vector as a place to deposit eggs. It can inhibit the immature phase of the mosquito but it is a very expensive control. The chemical control uses pesticide in the streets and aims to eliminate the adult mosquitoes. It is very effective and cheap, however it can kill different mosquitoes and can cause health problem in population. In addition to those control types, a biological control is now being used: to release sterile males into the natural environment. The main idea behind this control is that the sterile males will mate with the female mosquitoes and sterile eggs will be produced. These eggs do not hatch and the number of mosquitoes in the environment will be reduced. Seeing this control as an important method, Brazilian government has just launched a factory facility in the city of Juazeiro, Bahia, to produce the sterilized mosquitoes. Since 1958, the sterile insect procedure has been used successfully around the world, for example in USA, Mexico and Canada. This type of control is effective, safe but very expensive. An alternative is to combine pesticide control with sterile male insertion.

Thome *et al.*[1] proposed a mathematical dynamic system to describe the dynamics of the mosquito population when sterile male mosquitoes are introduced, as a biological control, together with the application of insecticide. However, this optimal strategy is hard to be implemented in practice. Barsante *et al.*[2] used the model proposed in Ref. 1 to study the optimization of the effect of the simultaneous insertion of insecticide and sterile males into the environment via a mono-objective optimization approach using a Genetic Algorithm. In this work, the authors used constant values for both control variables. Despite the good results, it was a preliminary work.

In the first place, this work performs a comparative study considering the results obtained by the optimal control strategy[1] with the constant control mono-objective optimization problem using the same the objective function and scenarios, allowing a search within a closed interval for each control variable.

This preliminary work leads to two questions:

i Do the two control actions (application of pesticide and insertion of sterile male in the environment) need to be applied during all the time interval?

ii Do the control actions need to be applied simultaneously?

With the goal of trying to answer these questions, the same mathematical model is used and two different approaches are proposed: to find optimal policies using simultaneous step-size control and using alternated step-size control. As expected, the results show that the alternated step-size control is preferable over the constant control since each control strategy is used during a smaller time interval while obtaining a smaller cost function value and similar effects in the nature. Furthermore, it is possible to find an optimal number of days in which each control strategy must be applied and the respective rate of insecticide and sterile males that should be used.

This article is organized as follow: Section 2 presents the mathematical model for the life cycle of the *Aedes aegypti* mosquitoes and the proposed control models; Section 3 introduces the Genetic Algorithm used as the optimization machinery; Section 4 shows the simulation results; and Section 5 presents final discussions and the conclusions of the work.

2. The Problem Model

The mathematical model, proposed in Ref. 3, describes a biological dynamic control considering that the vector life cycle is divided into six compartments:

- $A(t)$: immature phase of the insect at time t.
- Adult form: $I(t)$: females before mating (singles) at time t; $F(t)$: mating fertilized females at time t; $U(t)$: mating unfertilized females at time t; $M(t)$: non-sterile male insects at time t; $S(t)$: sterile male insects at time t.

The dynamic system, given by Equation (1), is used in Refs. 1 and 2, and it intents to evaluate how the vector population evolves.

$$\begin{cases} \dfrac{dA}{dt} & = \phi(1 - \frac{A}{C})F - (\gamma + \mu_A)A \\ \dfrac{dI}{dt} & = r\gamma A - \beta\frac{M}{M+S}I - \beta_S\frac{S}{M+S}I - \mu_I I \\ \dfrac{dF}{dt} & = \beta\frac{M}{M+S}I - \mu_F F \\ \dfrac{dM}{dt} & = (1-r)\gamma A - \mu_M M \\ \dfrac{dS}{dt} & = \alpha - \mu_S S \end{cases} \tag{1}$$

in which μ is the natural mortality rate of the vectors, ϕ determines the intrinsic rate of oviposition of fertilized females; C is the ability of the environment related to the number of nutrients available; γ refers to the rate at which the vectors are evolving from water phase to winged phase, where r a proportion of females and are $(1-r)$ are males; β represents the mating rate between natural males and immature females; β_s refers to the mating rate between sterile males and immature females; α is the rate at which sterile males are sprayed.

The system (6) has a trivial equilibrium point $P_2 = \left(0, \;\; 0, \;\; 0, \;\; 0, \frac{\alpha}{\mu_T}\right)$. The equilibrium point corresponds to the state where there is only a constant population of sterile male mosquitoes. The non-trivial steady state is $\left(\bar{A}, \;\; \bar{I}, \;\; \bar{F}, \;\; \bar{M}, \;\; \frac{\alpha}{\mu_T}\right)$, in which $\bar{I}$, $\bar{F}$ and $\bar{M}$ are given by Equation (2) and $\bar{A}$ is a solution of the equation $aA^2 + bA + c = 0$, with coefficients shown in Equation (3).

$$\begin{cases} \bar{I} &= \dfrac{r\gamma\bar{A}(\bar{M}+\alpha/\mu_S)}{(\mu_I+\beta)\bar{M}+(\mu_I+\beta_S)\alpha/\mu_S} \\ \bar{F} &= \dfrac{(\gamma+\mu_A)C\bar{A}}{\phi(C-\bar{A})} \\ \bar{M} &= \dfrac{(1-r)\gamma\bar{A}}{\mu_M} \\ \bar{A} &= \dfrac{(1-r)\gamma\bar{A}}{\mu_M} \end{cases} \tag{2}$$

$$\begin{cases} a = \dfrac{1}{C}\dfrac{\phi r\gamma\beta}{(\gamma+\mu_A)(\beta+\mu_I)\mu_F} \\ b = 1 - \dfrac{\phi r\gamma\beta}{(\gamma+\mu_A)(\beta+\mu_I)\mu_F} \\ c = \dfrac{(\beta_S+\mu_I)\mu_M\alpha}{(\beta+\mu_I)(1-r)\gamma\mu_S} \end{cases} \tag{3}$$

The value R_0, the average number of secondary female insects produced by a single female insect, and R^*, the ratio of mated but not fertilized female insects with respect to the fertilized ones, are shown in Equation (4). According to Ref. 4, the equilibrium point P_2 of System (1) is always stable, and when $R_0 > 1$ and $(R-1)^2 > 4RS$, there are two non-trivial equilibria, P_{3-} and P_{3+}, both feasible. In this case P_{3-} is always unstable and P_{3+}, stable.

$$R_0 = \frac{\phi r\gamma\beta}{(\mu_A+\gamma)(\beta+\mu_I)\mu_F} \text{ and } R^* = \frac{(\beta_S+\mu_I)\mu_M\alpha}{(\beta+\mu_I)(1-r)\gamma C\mu_S} \tag{4}$$

2.1. *Optimization Approach*

In the work Ref. 1, the social and economic cost of chemical and biological controls to combat Dengue is analyzed. To perform the control, the authors considers two optimization variables: u_1, related to the pesticide rate that should be applied; and u_2, related to the rate of sterile male to be released in the environment. This policies are to be applied in a 120 day interval. The time interval corresponds to the average length (in days) of the Brazilian summer, the rainy season, at which an effective control policy must be applied. The dynamic system characterizes a mixed control policy, because it evolves two different strategies.

The objective function to be minimized is shown in Equation (5). It is a quadratic function, composed by the sum of four parcels: the cost of pesticides, the cost of sterile males to be released, the social cost of the infected people (measured according to the number of fertilized females), and the environment preservation cost (measured according to the opposite of the total number of sterile males).

$$J(u) = \frac{1}{2}\int_0^T (c_1 u_1^2 + c_2 u_2^2 + c_3 F^2 - c_4 S^2)dt \tag{5}$$

in which c_1, c_2, c_3, c_4 represents the relative importance of the pesticide cost, of the sterile male production, of social cost and of sterile male preservation in the environment, respectively.

The objective function (Equation 5) is to be minimized subject to the dynamic system shown in Equation 6, in which the influence of insecticide and sterile males are considered: a parcel proportional to u_1 is added in variables I, F and M, and the constant α is replaced by u_2. The autonomous stable nontrivial equilibrium is used as an initial condition of the Initial Value Problem.

$$\begin{cases} \dfrac{dA}{dt} &= \phi(1 - \frac{A}{C})F - (\gamma + \mu_A)A \\ \\ \dfrac{dI}{dt} &= r\gamma A - \beta\frac{M}{M+S}I - \beta_S\frac{S}{M+S}I - \mu_I I - u_1 I \\ \\ \dfrac{dF}{dt} &= \beta\frac{M}{M+S}I - \mu_F F - u_1 F \\ \\ \dfrac{dM}{dt} &= (1-r)\gamma A - \mu_M M - u_1 M \\ \\ \dfrac{dS}{dt} &= u_2 - \mu_S S - u_1 S \end{cases} \tag{6}$$

In Ref. 1, the problem is solved solved using mathematical analytical methods based on the Pontryagin Maximum's Principle finding, for some scenarios, the optimal functions $u_1^*(t)$ and $u_2^*(t)$. It is worthwhile to notice that this strategy is hard to be implemented in practice since it is difficult to build a mechanism in which the level of pesticide and the rate of insertion of sterile mosquitoes, varying continuously in time, are released into the environment. However, the results show that the control variables exhibit, respectively, chair (bottom of the chair is almost a constant plateau) and bell (top of bell is almost constant plateau) shapes. These shapes indicate that constant functions for $u_1(t)$ and $u_2(t)$ can be used. For purposes of comparison, the obtained plateau values are used in this work, and will be referred as obtained by the optimal control problem (OCP).

Similarly to Ref. 2, this work considers, in first instance, the study of the effect of the insertion of both insecticide (u_1) and sterile males (u_2) into the environment during all the time horizon. The control variables, $u_1(t) = \bar{u_1}$ and $u_2(t) = \bar{u_2}$, are used as constant ones. It is clear that this policy is easier to be implemented in practice. This first optimization problem will be referred as constant control problem (CCP). Having in mind the comparison with the plateaus obtained via OCP, closed intervals for each control variables are pre-determined in CCP and the optimization problem consists in a search for constant optimal values within those intervals. In this way, constant-variable control policies are considered, using $u_1 \in [0;\ u_1^F]$ and $u_2 \in [0;\ u_2^F]$, in which the extremal values of the intervals are obtained from the respective extremes found in Ref. 1.

Furthermore, this paper proposes two different approaches: simultaneous and alternated step-size control. The purpose is to realize if it is necessary to apply the control during all the time interval and if the application of pesticide and insertion of sterile male need to be applied simultaneously. It is necesary to consider two more decision variables, t_1 and t_2: the numbers of days in which each control strategy will be applied. Mathematically, $t_1 \in [0;\ T_f]$ and $t_2 \in [0;\ T_f]$), with $T_f = 120$. It is important to highlight that the decision variables space becomes a discontinuous one. The constants, c_i, $i = 1, \cdots, 4$ have the same meaning.

One of the control policies proposed in this work consists in applying insecticide and sterile males simultaneously until they reach its optimum intervals. This optimization problem will be referred as **simultaneous step-size control problem (SSP)**. The other control policy consists in applying the insecticide during its optimal interval and, after that, to insert the sterile males also during its optimal interval. This optimization problem will be referred as **alternated step-size control (ASP)**. It can be highlighted that all cases can be seem as sub-cases of the full optimal control problem.

On the one hand, the optimization problems are constrained to a nonlinear dynamic system, making it difficult to solve using classical analytical or exact methods. On the other one, evolutionary algorithms have become an important computational tool for solving problems with objective functions which can be discontinuous, nonconvex and multimodal. These algorithms are heuristics, being designed to find feasible solutions but there is no guarantee for how close those solutions are to the true optimal solution. This paper uses a parallel mono-objective version of Genetic Algorithm to solve the optimization problem.

3. Genetic Algorithms

Genetic algorithms (GA), introduced by John Holland in the 1960s[5,6], are computational methods of search based on the mechanisms of genetics and natural evolution. In these algorithms, a set of candidate solutions (a population) evolves according to probabilistic rules inspired by biological metaphors: in each generation, the individuals combine to produce (on average) improved descendants. This class of algorithms has reached importance in many fields of optimization. In general, GAs have the following characteristics[7]: GAs use a set of candidate solutions (population); GAs operate on a codified search space; GAs need only the information about the objective function evaluation (the so-called fitness)for each member of the population; and GAs use probabilistic transition rules.

In any GA, the core operators are initialization (of the population), fitness evaluation, selection, mutation and crossover (also known as recombination). Other operators may also be included, such as local search, niching and decimation. Goldberg[7] described four reasons which make the application of GAs attractive: GAs can solve hard problems like many NP-hard problems; the interface building between GAs and existing models is generally simple; GAs are extensible.

To solve the mono-objective problem, a real coded GA is used. The Real-Biased Genetic Algorithm (RBGA) is a genetic algorithm with a different feature: the real biased crossover operator. The RBGA is defined as the successive application of the following operations: population evaluation and the fitness function computation; selection by roulette; real biased crossover; mutation; elitism. The real-biased crossover resembles a search following the direction of a tendency with an information that is similar to the one that is given by a kind of gradient vector evaluation, but with possibly long-range information validity (instead of the only local validity that is associated with a gradient). This operator only evaluates the objective function without any calculation of function derivatives, in the same way as the GA. If the parents are located near each other, then a step that implicitly uses a directional derivative information is executed. This procedure speeds up the local convergence to the optimal point. If the parents are far away from each other (maybe in different attraction basins), this procedure can be interpreted as a long-range trend information. The son individual is created taking into account such information. Further references of the algorithm can be found in Ref. 8.

It is important to highlight that due to the stochastic nature of the evolutionary techniques, it is necessary to perform any test over many executions of the algorithms to obtain a well-based judgment concerning their performance.

4. Tests and Results

Table 1 shows the parameter values used in this article. These parameters are the same used in Ref. 9. The parameter α is used as a decision variable of the optimization problem. Using the values presented in Table (1), and making $\alpha = 0$, the stable nontrivial equilibrium point is $A(0) = 8.3200$, $I(0) = 0.2773$, $F(0) = 5.5467$, $M(0) = 2.9120$ and $S(0) = 0$. This point is used as a initial conditions of the Initial Value Problem of the system (6).

Table 1.: Parameter values of the Dynamic System

r	ϕ	C	γ	β	β_s	μ_M	μ_A	μ_F	μ_S	μ_I
0.5	0.5	13	0.07	1	0.7	0.1	0.05	0.05	0.1	0.05

The RBGA algorithm was executed 30 times for each mono-objective problem, and it was started with the same basic parameters as listed below:

- Population size: 100 individuals
- Recombination rate: 0.9
- Mutation rate: 0.05
- Maximum Number of Generations: 100

Since our objective is only to solve the optimization problems with an efficient and previously tested Genetic Algorithm, we do not focus on the effect of the parameters over the RBGA performance.

The maximum number of generations was the only stop criterion. At the end of 30 executions of RBGA, we obtained the mean convergence line which corresponds to the mean value of the best individual throughout the 100 generations.

The evaluation of each individual in the proposed optimization problem involves the solution of a differential equation system and is time consuming. Therefore, to alleviate the computational time burden, we used a toolbox which allowed us to execute the algorithm in parallel - using 4 cores - , reducing the algorithm execution time.

Different scenarios are tested, which is exactly the same as used in Ref. 1:

- Scenario 1: $[c_1] = 1$, $[c_2] = 1$, $[c_3] = 1$ and $[c_4] = 1$;
- Scenario 2: $[c_1] = 1$, $[c_2] = 10$, $[c_3] = 1$ and $[c_4] = 1$;
- Scenario 3: $[c_1] = 1$, $[c_2] = 1$, $[c_3] = 100$ and $[c_4] = 1$;
- Scenario 4: $[c_1] = 1$, $[c_2] = 10$, $[c_3] = 100$ and $[c_4] = 1$;
- Scenario 5: $[c_1] = 10$, $[c_2] = 1$, $[c_3] = 1$ and $[c_4] = 1$;

To analyze the dynamic of each Dengue vector population, the plateau obtained using the optimal control theory[1] is chosen, called here as optimal control problem (OCP), and the mean value of all the executions of the Genetic Algorithm are obtained for three cases: constant control (CCP), simultaneous (SCP) and alternated (ACP) step-size control problems. Values less than 10^{-4} are set to zero.

Scenario 1

In this scenario, used as a reference value, the relative importance of each term on Dengue control cost is the same. Figure 1 represents the dynamic of the mating fertilized females population using the four control policies. The x-axis represents the time interval (in days) and the y-axis represents the number of individuals in each Dengue population. Table 2 shows the optimal values of the decision variables and of the objective function of the four policies considered in Scenario 1.

It is important to notice that all control policies, at the end of the time window, keep the mating fertilized females level as an acceptable plateau. It also can be seen that CCP, SSP and ASP present the same behavior during the beginning of the interval, decreasing the mating fertilized female level faster when compared with OCP. Observe that the rate of insecticide found

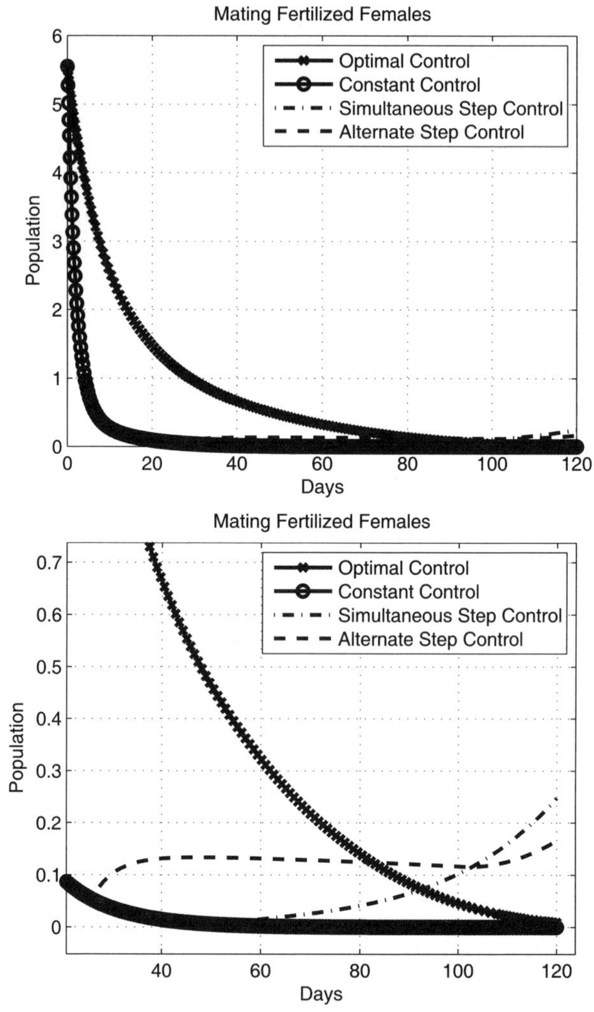

Figure 1.: SCENARIO 1: Dynamic of the mating fertilized females population on the: Optimal Control Problem, Constant Control Problem, Simultaneous Step Control Problem and Alternated Step Control Problem. The bottom figure is a zoom of the figure on top.

in OCP is smaller when compared with the obtained one in CCP, with a higher value of objective function. This is due to the reduction of the mating fertilized females in the OCP as can be seen in Figure 1. The SSP and ASP strategies show a different behavior from each other. The first

Table 2.: Optimal values for the decision variables and for the objective function of the four policies considered in Scenario 1 (Optimal Control Problem, Constant Control Problem, Simultaneous Control Problem and Alternated Step Control Problem).

SCENARIO	POLICY	u_1^*	u_2^*	u_3^*	u_4^*	J^*
1	OCP	0.0900	0.0455	120	120	459
1	CCP	0.4499	0.0001	120	120	136
1	SCP	0.4497	0.0001	53	63	129
1	ACP	0.4436	0.0488	26	102	117

one (SSP) proposes an application of insecticide for a longer period of time, decreasing the number of mating fertilized females to the same level as CCP. However, after the insecticide application, the number of mating fertilized females tends to increase while stabilizing to an acceptable level. The other strategy (ASP) proposes an application of insecticide for a smaller period of time and, after that period, the insertion of sterile males is able to keep a lower mating fertilized female level.

In this way, it is possible to state that the alternated step-size control is preferable over the constant control, since each control strategy is used during a smaller time interval while obtaining the smallest cost function value.

It might be noted that Genetic Algorithms generally find sub-optimal solutions, without a guarantee of optimality. For this, Figure 2 shows the convergence line for CCP, SSP and ASP optimization, in which the x-axes represent the generation and the y-axes represent the mean of the objective function value of the best individual, and It is possible to see that the algorithm is very robust and converges (possibly to a local optimum) after around 50 generations, being the ASP whose leads to a smallest objective function value.

Scenario 2

In fact, the production and release of sterile insects are more expensive than insecticide application, so to present a more realistic scenario, the cost of biological control is increased in ten times while maintaining the other values unchanged. Figure 3 represents the dynamic of the mating fertilized females population using the four control policies. The x-axis represents the

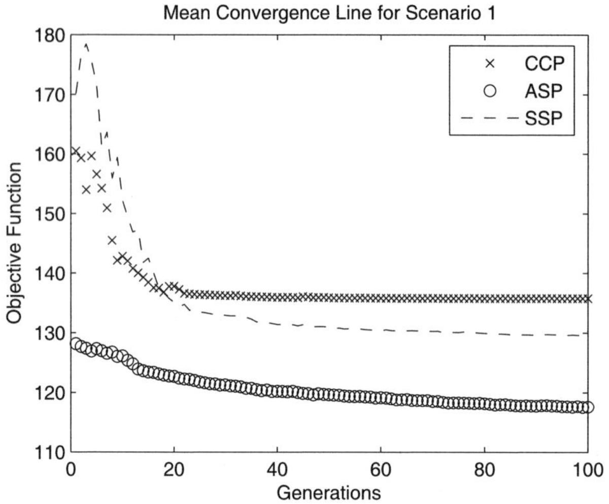

Figure 2.: Convergence line for the optimization of the Constant Control Problem, Simultaneous Control Problem and Alternated Step Control Problem in Scenario 1 using the RGBA.

time interval (in days) and the y-axis represents the number of individuals in each Dengue population. In the same way, all control policies keep the mating fertilized females level as an acceptable plateau. Table 3 shows the optimal values for the decision variables and for the objective function of the four policies considering Scenario 2.

Table 3.: Optimal values for the decision variables and for the objective function of the four policies considering Scenario 2 (Optimal Control Problem, Constant Control Problem, Simultaneous Control Problem and Alternated Step Control Problem).

SCENARIO	POLICY	u_1^*	u_2^*	u_3^*	u_4^*	J^*
2	OCP	0.1000	0.0051	120	120	361
2	CCP	0.4499	0.0003	120	120	136
2	SCP	0.4499	0.0003	52	62	129
2	ACP	0.4499	0.0000	53	0	129

It also can be seen that CCP, SSP and ASP present the same behavior during the beginning of the time interval, decreasing the mating

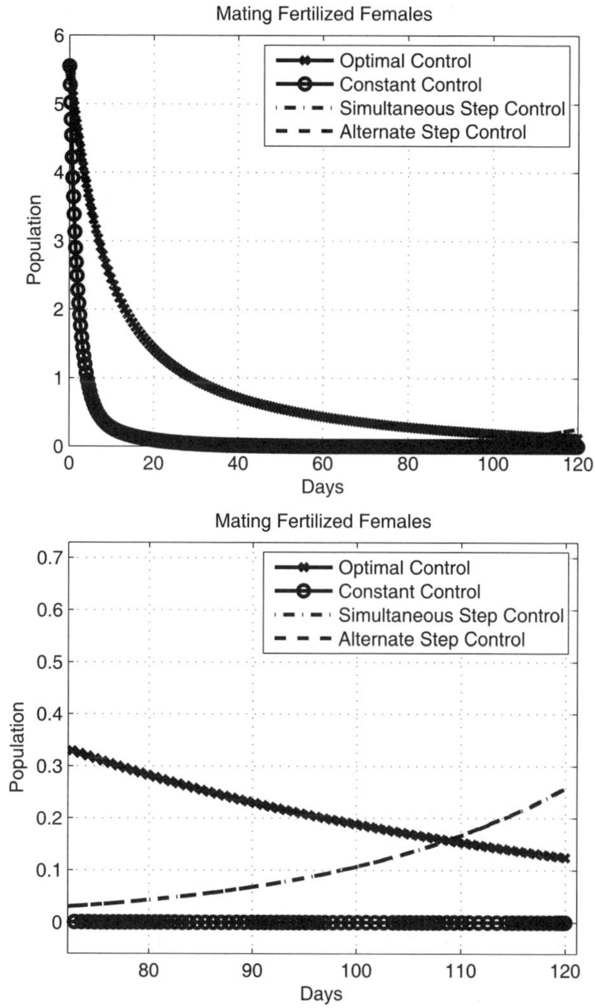

Figure 3.: SCENARIO 2: Dynamic of the mating fertilized females population on the: Optimal Control Problem, Constant Control Problem, Simultaneous Step Control Problem and Alternated Step Control Problem. The bottom figure is a zoom of the figure on top.

fertilized female level faster when compared with OCP. Moreover, the three proposed policies achieve a smaller number of females individuals, with a smaller objective function value. Since the cost of sterile males is higher in this scenario, the optimization procedure tends to propose optimal control policies that only use insecticides. In fact, the values of u_1 were almost the same in the three cases, and the value of the objective function in SPC is higher than in the SCP. This is due to the reduced number of days for which control is applied. According to Figure 3, the effect on the number of fertilized females is the same. Finally, ACP has chosen, smartly, not to use sterile males. Otherwise, the usage of sterile males would increase the cost function value without any reduction at the female population level. It can be highlighted that all the decision variables and the objective function values are the same in CCP, SCP and ACP in Scenarios 1 and 2. It means that the effect of sterile males is very small in these policies considering the two scenarios.

Scenario 3

The Dengue infection is, in general, a mild disease. However, in some cases, it can be evolved to the hemorrhagic form, which can be fatal. For this reason, the social cost must be very high. In this scenario, c_3 is set to 100 maintaining the other costs as in the reference case. Figure 4 represents the dynamic of the mating fertilized females population using the four control policies. The x-axis represents the time interval (in days) and the y-axis represents the number of individuals in each Dengue population. Table 4 shows the optimal values for the decision variables and for the objective function of the four policies considering Scenario 3. Note that the orders of magnitude in the J values are proportional to the increase in the value of c_3.

Table 4.: Optimal values for the decision variables and for the objective function of the four policies considering Scenario 3 (Optimal Control Problem, Constant Control Problem, Simultaneous Control Problem and Alternated Step Control Problem).

SCENARIO	POLICY	u_1^*	u_2^*	u_3^*	u_4^*	J^*
3	OCP	0.1000	0.0380	120	120	4.3857×10^4
3	CCP	0.4999	0.0005	120	120	1.2043×10^4
3	SCP	0.4999	0.0001	66	75	1.2061×10^4
3	ACP	0.4499	0.0778	33	116	1.2019×10^4

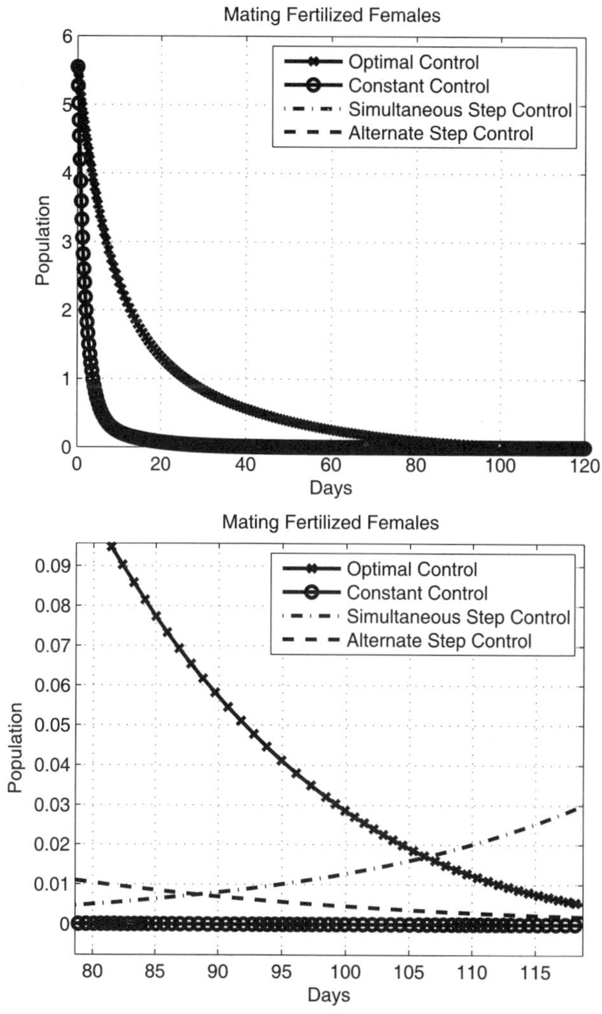

Figure 4.: SCENARIO 3: Dynamic of the mating fertilized females population on the: Optimal Control Problem, Constant Control Problem, Simultaneous Step Control Problem and Alternated Step Control Problem. The bottom figure is a zoom of the figure on top.

In the same way, all control policies keep the mating fertilized females level as an acceptable plateau. It also can be seen that CCP, SSP and ASP present the same behavior during the beginning of the time interval, decreasing the mating fertilized female level faster when compared with

OCP. It is noticeable that the number of fertilized females at the final of the time window, in SCP and ACP, is much smaller than the one in other scenarios (about one order of magnitude lower), as can be seen in Figure 4. This is due to the rate of male sterile released during the biological control phase. Also note that, at the end of the application of the insecticide in SCP, the number of fertilized females tends to increase. This indicates that the rate of sterile males being proposed by minimizing the objective function is not enough. What happens in the ACP is exactly the opposite, since it is able to find the optimal rate of sterile males to control the number of fertilized females.

Scenario 4

In this scenario, c_3 is set to 100, and c_2 is equal to 10, maintaining the other costs as in the reference case. This means that the sterile release is more expensive than the available insecticide and the high incidence of the disease requires elevated social cost.

Figure 5 represents the dynamic of the mating fertilized females population using the four control policies. The x-axis represents the time interval (in days) and the y-axis represents the number of individuals in each Dengue population. In the same way, all control policies keep the mating fertilized females level as an acceptable plateau. It also can be seen that CCP, SSP and ASP present the same behavior during the beginning of the time interval, decreasing the mating fertilized female level faster when compared with OCP. Table 5 shows the optimal values for the decision variables and for the objective function of the four policies considering Scenario 4.

Table 5.: Optimal values for the decision variables and for the objective function of the four policies considering Scenario 4 (Optimal Control Problem, Constant Control Problem, Simultaneous Control Problem and Alternated Step Control Problem).

SCENARIO	POLICY	u_1^*	u_2^*	u_3^*	u_4^*	J^*
4	OCP	0.1000	0.0180	120	120	4.1042×10^4
4	CCP	0.4999	0.0004	120	120	1.2042×10^4
4	SCP	0.4999	0.0003	114	94	1.2040×10^4
4	ACP	0.4499	0.0000	94	0	1.2034×10^4

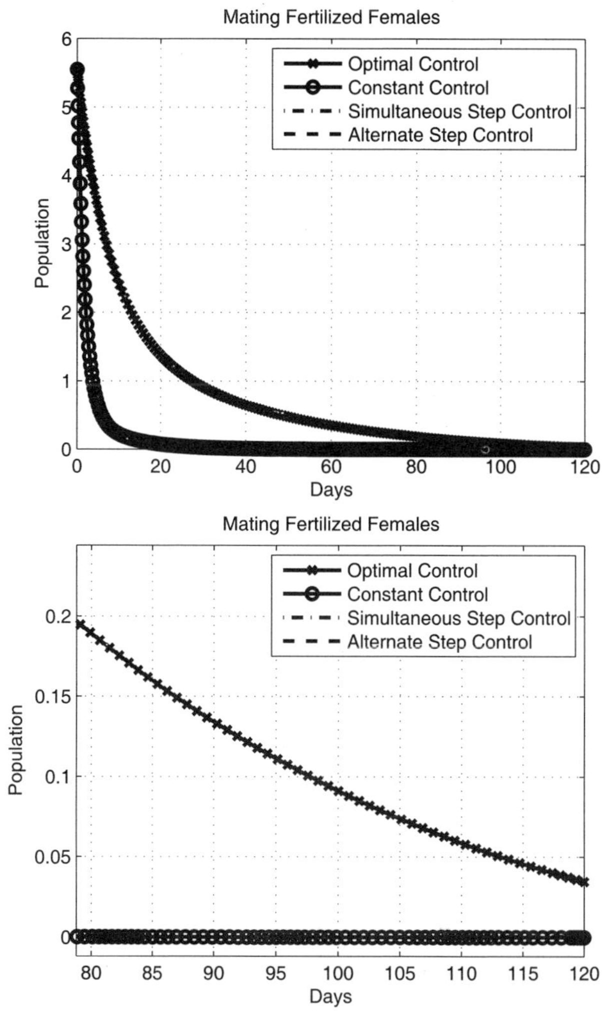

Figure 5.: SCENARIO 4: Dynamic of the mating fertilized females population on the: Optimal Control Problem, Constant Control Problem, Simultaneous Step Control Problem and Alternated Step Control Problem. The bottom figure is a zoom of the figure on top.

The values and results obtained with OCP and CCP were very similar for Scenarios 3 and 4, that is, the values were not affected by the high value of the cost of producing male sterile. Figure 5 shows that the application of insecticide last longer in SCP and ACP, leading to a higher reduction in the number of fertilized females, if compared to the results obtained in Scenario 3. In fact, the objective function values are similar, whereas the ACP did not need the use of sterile males. Despite the fact of the optimization algorithm has privileged the application of insecticides, it is believed that this scenario relates more the reality of the scenarios considered so far. The search for a more consistent scenario will lead, in future work, to consider a multiobjective approach.

Scenario 5

It can be observed an increasing trend of resistant mosquitoes to insecticide, which let to the development of new products. Since the cost of research, and development of new insecticides is very expensive, the chemical application cost is increased in ten times $c_1 = 10$, maintaining all other costs the same as the reference. Table 6 shows the optimal values for the decision variables and for the objective function of the four policies considering Scenario 5.

Figure 6 represents the dynamic of the mating fertilized females population using the four control policies. The x-axis represents the time interval (in days) and the y-axis represents the number of individuals in each Dengue population. In the same way, all control policies keep the mating fertilized females level as an acceptable plateau. It also can be seen that CCP, SSP and ASP present the same behavior during the beginning of the interval, decreasing the mating fertilized female level faster when compared with OCP.

Table 6.: Optimal values for the decision variables and for the objective function of the four policies considering Scenario 5 (Optimal Control Problem, Constant Control Problem, Simultaneous Control Problem and Alternated Step Control Problem).

SCENARIO	POLICY	u_1^*	u_2^*	u_3^*	u_4^*	J^*
5	OCP	0.0210	0.138	120	120	735
5	CCP	0.1294	0.0002	120	120	217
5	SCP	0.1381	0.1587	31	118	231
5	ACP	0.1392	0.1598	38	112	142

This scenario led, obviously, to policies with a much smaller rate of

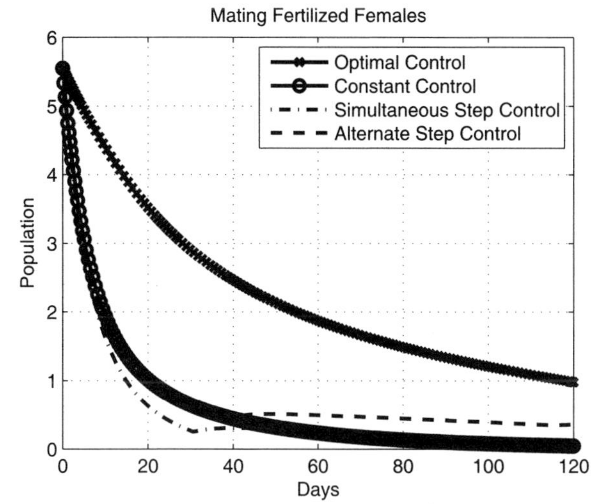

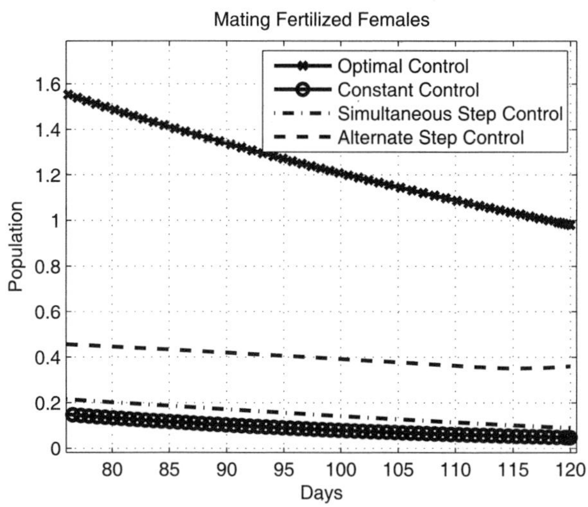

Figure 6.: SCENARIO 5: Dynamic of the mating fertilized females population on the: Optimal Control Problem, Constant Control Problem, Simultaneous Step Control Problem and Alternated Step Control Problem. The bottom figure is a zoom of the figure on top.

insecticides, compared to the previous scenarios, and to larger values of sterile males to be inserted. The problem is that the rate of sterile males is not able to reduce the number of fertilized females to acceptable levels after the time window, as shown in Figure 6. Compared to Scenario 1, the number of fertilized females at the end of the process is higher (an order of magnitude higher), as well as the values for the objective function. Nevertheless, the three proposed control strategies present objective function values smaller than the one in OCP, with a smaller total number of fertilized female maturation. The ACP presented the lowest objective function value, while presenting an overall number of females slightly higher. This can be seen as a tradeoff between cost control and the effect on the number of mosquitoes, which need to be considered according to what the government is willing to spend each year.

5. Conclusion

A mathematical model of the mosquitoes population is used to analyze the cost of a mixed control, using pesticides and sterile males, via mono-objective optimization schemes. Different approaches are tested to solve the control problem, using both constant and step-size continuous control. A parallel version of a genetic algorithm is used to solve each problem.

The studied case study shows that, considering both social and economic costs, it is possible to hint that: (i) constant control optimization policy is preferred that using a regression of the optimal control optimization into a plateau; (ii) step-size control policy is preferred that using constant control, since each control strategy can be used during smaller time intervals, even obtaining smaller cost function values; and (iii) the best of the four policies consists in applying control actions separately, using a small rate of chemical control just at the beginning of the season, in a few days, and to keep acceptable the mosquitoes level using biological control. Moreover, using the proposed optimization tool it is possible, more generally, in several scenarios, to find the optimal number of days in which each control strategy must be applied, and the respective rate of insecticide and sterile males that should be used.

As it is hard to adjust manually the constants of the objective-function without favoring the effect of one control variable over the another one, an alternative is to perform a multiobjective optimization, looking for compromise solutions related to the different parcels of the cost. the effect of temperature, rainfall and other factors in the dynamic system and simulations.

Acknowledgments

The authors would thank to the Brazilian Agencies CAPES, CNPq, FAPEMIG and CEFET-MG for financial support.

References

1. Thome, R.C.A. and Young, H.M. and Esteva, L. 'Optimal control of Aedes aegypti mosquitoes by the sterile insect technique and insecticide," Mathematical Biosciences, vol 223, p 12-23, 2010.
2. Barsante, L.S., Cardoso, R.T.N., Acebal, J.L. "Otimização Multiobjetivo no Controle de Gastos com Inseticidas e Machos Estéreis no Combate da Dengue," Proceeding of the XLIII Brazilian Symposium of Operational Research, Ubatuba, Brazil, 2011.
3. Esteva, L. and Young, H.M. "Control of Dengue Vector by the Sterile Insect Technique Considering Logistic Recruitment," TEMA Tendências em Matemática Aplicada e Computacional, vol 7, n 2, p 259-268, 2006.
4. Esteva, L. and Young, H.M. "Mathematical Model to Assess the Control of Aedes aegypti Mosquitoes by the Sterile Insect Technique," Mathematical Bioscience, vol 198, p 132-14, 2005.
5. Holland, J.H. "Concerning efficient adaptative systems," Self-Organizing Systems, In M.C. Yovits, G.T. Jacobi, and G. D. Goldstein editors, p 215-230, 1962.
6. Holland, J.H. "Outline for a logical theory of adaptive systems," Journal of the Association for Computing Machinery, v 9, p 297-314, 1962.
7. Goldberg, D.E. "Genetic Algorithms in Search, Optimization and Learning," Addison-Wesley, 1989.
8. Takahashi, R.H.C. and Vasconcelos, J.A. and Ramirez, J.A. and Krahenbuhl,L. "A multiobjective methodology for evaluation genetic operators," IEEE Transactions on Magnetics, vol 39, p 1321-1324, 2003.
9. Young, H.M. and Macoris, M.L.G. and Galvani, K.C. and Andrighetti, M.T.M. and Wanderley, D.M.V. "Assessing the effects of temperature on the population of Aedes aegypti, the vector of dengue," Epidem. Infect., vol 137, p 11881202, 2009.

DYNAMICS OF A VARROA-INFESTED HONEY BEE COLONIES MODEL

K. O. OKOSUN

Department of Mathematics, Vaal University of Technoloy, X021, Vanderbijlpark, 1900, South Africa
E-mail: kazeemo@vut.ac.za

In this paper a mathematical model for the dynamics of varroa-infested honey bees colonies is formulated in order to investigate their synergistic relationship in the presence of prevention hive and protection of forager. Firstly, the constant control case steady states is analyze and the basic reproduction number is calculated and investigate the existence and stability of equilibria. Two scenarios were considered, namely, (i) the dynamics of the endemic equilibrium of the honey bee population expressed in terms of the varroa mites and (ii) the dynamics of the endemic equilibrium of the varroa mites population expressed in terms of the infected honey bee. In the first case, it found that honey bee disease free equilibrium only exists in the absence of varroa mites and the infected bee population increases with increase in the varroa mites population. In the second scenario, the model exhibit multiple equilibria. From the numerical simulation, it is clear that for a infected free bee population, the varroa mites virus infected population can not exceed 2 units. Suggesting that intervention measures targetted towards preventing the bees from being infected even in the presence of varroa mites virus would not propogate the spread of the infestation of bee colonies. Secondly we incorporate time dependent controls, using Pontryagin's Maximum Principle to derive necessary conditions for the optimal control of the disease, the numerial results suggest that the comibination of prevention control (u_1) on hive and protection control on forager (u_2) is the most effective strategy to prevent Colony Collapse Disorder in honeybee colony.

1. Introduction

The global treat to insect pollinators of crops and wild plants are becoming alarming and their exstintion may have high impact on economic and environmental consequencies [11]. Not less than 15% of human food production relies on animal pollinators, most especially the bees. For instance, US fruit and vegetable growers requires the honey bees pollination services to generate $8-10 billion as farm income annually [5]. In addition to pollination, honey bees also play significant role as producer of honey and wax which in turn result in various nutritional and industrial uses.

Varroa mite is an external parasite which attacks the adult bees and the bee larvae. This parasite has been found to be the major reason for wintering loses of honey bee colonies [9]. Varroa mites have a huge economic impact on the beekeeping industry, they do not only feed on bees but also transmit deadly viruses to the bees. Studies has further shown that except these mites are effectively controlled honeybee colonies die out of existence [10].

Rafikov and Limeira [4] proposed a simple mathematical model of interaction between the sugarcane borer (*Diatraea saccharalis*) and its egg parasitoid *Trichogramma galloi.* In [3], a tritrophic model consisting of tea plant, pest and predator was considered to analyze different field observations. Wang and Song [2] proposed tow mathematical models concerning continous and impulsive pest control strategies respectively. Bhattacharyya and Ghosh [1] studied the role of predation dynamics in oscillation of pest population in insect ecology. Sun and Chen [12] formulated and investigated the pest control models in accordance with the mathematical theory of epidemiology.

Mathematical tools has been helpful in predicting infectious diseases spread. Much of the research efforts in this regard are mainly targeted towards diseases of human populations, not much of the disease dynamics of honeybee and varroa mite population has been studied in the literature. For example Olayiwola et al [5] presented a simple mathematical model to describe the infection of honeybee colony by the Acute Paralysis Virus (APV) carried by parasitic varroa mites. The authors in [7] studied a mathematical model of a colony collapse disorder stimulated by the dynamics of a healthy hive. While in [6] a compactmental model of bee population dynamics was presented to explore how food availability and bees death rates interact to determine colony growth and development. A simple SIR-Ross-MacDonald like model of the infestation of a honeybee (*Apis mellifera*) colony by the Acute Bee Paralysis Virus (ABPV) transmitted by parasitic varroa mite was presented in [8]. In this paper, the model in [7] is modified to include the varrao mites logistic growth model.

The paper is organized as follows: Section 2 is devoted to the model description and analyzes. In Section 3, we use Pontryagin's Maximum Principle to investigate analysis of control strategies and to determine the necessary conditions for the optimal control of the disease. In Section 4 we show and discuss the simulation results. Our conclusion is presented in Section 5.

2. Model formulation

The model is subdivided into Hive (H), Forager (virus free) (F), Infected forager (I), number of mites that infest the colony (M) and number of mites that carry the virus (m). A logistic growth model is adopted for the varroa mites population. Where β is the rate at which uninfected bees become infected, μ_1, μ_2, μ_3 are the mortality rates for uninfected, infected honeybees and mites respectively. Hive to forager transition rates are greatly affected by disorder [7], and Colony Collapse Disorder in honeybee colony (CCD), increases this rate, we adopt two recruitment rates, a healthy α and unhealthy γ rates. Also, ψ and ω are half saturation constant, λ is the rate at which mites that do not carry the virus acquire it and Q is the carrying capacity of the mites. The term r represent the maximum birth rate of the mite.

$$\begin{cases} \frac{d}{dt}H = L\left(\frac{H}{H+\omega}\right) - H\left(\frac{\alpha\psi}{F+I+\psi} + \gamma\right)\left(1-u_1\right) \\ \frac{d}{dt}F = H\left(\frac{\alpha\psi}{F+I+\psi} + \gamma\right)\left(1-u_1\right) - \mu_1 F - (1-u_2)\frac{\beta F m}{m+I} \\ \frac{d}{dt}I = (1-u_2)\frac{\beta F m}{m+I} - \mu_2 I \\ \frac{d}{dt}M = rM\left(1-\frac{M}{Q}\right)\left(1-m\right) - (1-u_2)\frac{\lambda I M}{F+I} - \mu_3 M \\ \frac{d}{dt}m = rmM\left(1-\frac{M}{Q}\right) + (1-u_2)\frac{\lambda I M}{F+I} - \mu_3 m \end{cases} \tag{1}$$

2.1. *Basic properties of the malaria model*

2.1.1. *Positivity and boundedness of solutions*

For the transmission model (1) to be epidemiologically meaningful, it is important to prove that all its state variables are non-negative for all time. In other words, solutions of the model system (1) with non-negative initial data will remain non-negative for all time $t > 0$.

Theorem 2.1. *Let the initial data* $H(0) > 0, F(0) \geq 0, I(0) \geq 0, M(0) \geq 0, m(0) \geq 0$. *Then the solutions* $H(t), F(t), I(t), M(t), m(t)$ *of the model (1) are non-negative for all* $t > 0$. *Furthermore*

$$\limsup_{t\to\infty} N_b(t) \leq L\left(\frac{H}{H+\omega}\right) - \mu_1 F - \mu_2 I$$

and

$$\limsup_{t\to\infty} N_m(t) \le rM\left(1-\frac{M}{Q}\right) - \frac{\beta F m}{F+I}$$

with,

$$N_b(t) = H(t)+F(t)+I(t)+M(t)+m(t),$$
$$N_m(t) = M(t)+m(t).$$

Proof. Let $t_1 = \sup\{t>0 : H(t)>0, F(t)>0, I(t)>0, M(t)>0, m(t)>0\}$.

Since

$H(0)>0, F(0)>0, I(0)>0, M(0)>0, m(0)>0,$

then, $t_1>0$. If $t_1<\infty$, then

$H(t), F(t), I(t), M(t), m(t)$

is equal to zero at t_1. It follows from the first equation of the system (1), that

$$\frac{H}{dt} = L\left(\frac{H}{H+\omega}\right) - H\left(\frac{\alpha\psi}{F+I+\psi}+\gamma\right)\left(1-u\right),$$

Thus,

$$\frac{d}{dt}\left\{H(t)\; exp\left[(H\left(\frac{\alpha\psi}{F+I+\psi}+\gamma\right)\left(1-u\right))t\right]\right\}$$
$$= \; L\left(\frac{H}{H+\omega}\right)\; exp\left[(H\left(\frac{\alpha\psi}{F+I+\psi}+\gamma\right)\left(1-u\right))t\right]$$

Hence,

$$H(t_1)\; exp\left[(H\left(\frac{\alpha\psi}{F+I+\psi}+\gamma\right)\left(1-u\right))t\right] - H(0)$$
$$= \int_0^{t_1} L\left(\frac{H}{H+\omega}\right)\; exp\left[(H\left(\frac{\alpha\psi}{F+I+\psi}+\gamma\right)\left(1-u\right))p\right] dp$$

so that,

$$H(t_1) = H(0)\; exp\left[-(H\left(\frac{\alpha\psi}{F+I+\psi}+\gamma\right)\left(1-u\right))t_1\right]$$
$$+\; exp\left[-(H\left(\frac{\alpha\psi}{F+I+\psi}+\gamma\right)\left(1-u\right))t_1\right]$$

$$\int_0^{t_1} L\left(\frac{H}{H+\omega}\right) exp\left[(H\left(\frac{\alpha\psi}{F+I+\psi}+\gamma\right)\left(1-u\right))p\right]dp$$
$$> 0.$$

It can similarly be shown that $H(0) > 0, F(0) > 0, I(0) > 0, M(0) > 0$ and $m(0) > 0$ for all $t > 0$.

For the second part of the proof, it should be noted that $0 < H(t) \leq N_b(t), 0 < F(t) \leq N_b(t), 0 < I(t) \leq N_b(t), M(t) \leq N_m(t), 0 < m(t) \leq N_m(t)$. Adding the equations for the Bee components and the mites component of the model (1) gives

$$\frac{dN_b(t)}{dt} = L\left(\frac{H}{H+\omega}\right) - \mu_1 F - \mu_2 I, \tag{2}$$

$$\frac{dN_m(t)}{dt} = rM\left(1-\frac{M}{Q}\right) - \frac{\beta F m}{F+I}.$$

Thus,

$$L\left(\frac{H}{H+\omega}\right) - \mu_1 F - \mu_2 I \leq \frac{dN_h(t)}{dt} \leq L\left(\frac{H}{H+\omega}\right) - \mu_2 I,$$

$$rM\left(1-\frac{M}{Q}\right) - \frac{\beta F m}{F+I} \leq \frac{dN_m(t)}{dt} \leq rM\left(1-\frac{M}{Q}\right).$$

Hence,

$$L\left(\frac{H}{H+\omega+\mu_1 F+\mu_2 I}\right) \leq \liminf_{t\to\infty} N_b(t) \leq \limsup_{t\to\infty} N_b(t)$$
$$\leq L\left(\frac{H}{H+\omega+\mu_1 F}\right),$$
$$\frac{rM(Q-M)(F+I)}{Q+\beta F m} \leq \liminf_{t\to\infty} N_m(t) \leq \limsup_{t\to\infty} N_m(t)$$
$$\leq rM\left(1-\frac{M}{Q}\right),$$

as required. □

2.1.2. *Invariant regions*

The model (1) will be analyzed in a biologically-feasible region as follows. Since the model (1) represents individuals, all parameters in the model

are non-negative and one can show that the solutions of the system are non-negative, given non-negative initial values.
Consider the feasible region

$$\Gamma = \Gamma_b \times \Gamma_v \subset \mathbb{R}^3_+ \times \mathbb{R}^2_+$$

with,

$$\Gamma = \left\{ (H, F, I) \in \mathbb{R}^3_+ : 0 \leq N_b \leq L\left(\frac{H}{H + \omega + \mu_1 F}\right) \right\}$$

and

$$\Gamma_m = \left\{ (M, m) \in \mathbb{R}^2_+ : 0 \leq N_m(t) \leq rM\left(1 - \frac{M}{Q}\right), \right\}.$$

The following steps are followed to establish the positive invariance of Γ (i.e., solutions in Γ remain in Γ for all $t > 0$). The rate of change of the total population is obtained in equations (2) it follows that

$$\begin{aligned} \frac{dN_b(t)}{dt} &\leq L\left(\frac{H}{H + \omega}\right) - \mu_1 F, \\ \frac{dN_m(t)}{dt} &\leq rM\left(1 - \frac{M}{Q}\right). \end{aligned} \tag{3}$$

A standard comparison theorem can then be used to show that $N_b(t) \leq \int_0^t [L\left(\frac{H}{H+\omega}\right) - \mu_1 F]dt$ and $N_m(t) \leq \int_0^t [rM\left(1 - \frac{M}{Q}\right)]dt$. In particular, $N_b(t) \leq \frac{\Pi_h}{\mu_h}$ and $N_m(t) \leq \frac{\Pi_v}{\mu_v}$ if $N_b(0) \leq \frac{\Pi_h}{\mu_h}$ and $N_m(0) \leq \frac{\Pi_v}{\mu_v}$ respectively. Thus, the region Γ is positively-invariant. Hence, it is sufficient to consider the dynamics of the flow generated by (1) in Γ. In this region, the model is epidemiologically and mathematically well-posed. Thus, every solution of the basic model (1) with initial conditions in Γ remains in Γ for all $t > 0$. Therefore, the ω-limit sets of the system (1) are contained in Γ. This result is summarized below.

Lemma 2.1. *The region $\Gamma = \Gamma_h \times \Gamma_v \subset \mathbb{R}^3_+ \times \mathbb{R}^2_+$ is positively-invariant for the basic model (1) with non-negative initial conditions in $\mathbb{R}^5_+$*

2.2. *Stability of the disease-free equilibrium (DFE)*

The model (1) has a DFE, obtained by setting the right-hand sides of the equations in the model to zero, given by

$$E_0 = (H^*, F^*, I^*, M^*, m^*) = \left(H^*, F^*, 0, M^*, 0\right) \tag{4}$$

The linear stability of E_0 can be established using the next generation operator method in Driessche and Watmough [14] on the system (1), the matrices F and Ψ, for the new infection terms and the remaining transfer terms, are, respectively, given by,

It follows that the reproduction number of the model (1), denoted by R_0, is given by

$$R_0 = \frac{2Q\lambda\mu_1\left(\sqrt{K} + (1-u_1)\gamma\omega + \psi\mu_1 - L\right)}{\beta\left(K - [L-(1-u_1)\gamma\omega - \psi\mu_1]^2\right)}, \tag{5}$$

$$K = (L-(1-u_1)\gamma\omega)^2 + \psi\mu_1[2(L-(1-u_1)(2\alpha+\gamma)\omega) + \psi\mu_1]$$

Further, using Theorem 2 in Driessche and Watmough [14], the following result is established. The DFE is locally asymptotically stable if $R_0 < 1$ and unstable if $R_0 > 1$.

2.3. *Existence of endemic equilibrium*

Proof. Calculating the endemic equilibrium point, we obtain,

$$\begin{cases} M^* = \dfrac{Q[(F^*+I^*)[(m^*-1)r-\mu_3] - I^*\lambda]}{r(m^*-1)(F^*+I^*)} \\ \\ H^* = \dfrac{L(F^*+\psi) - (1-u_1)[(F^*\gamma + (\alpha+\gamma)\psi]\omega}{(1-u_1)[F^*\gamma + (\alpha+\gamma)\psi]} \end{cases} \tag{6}$$

The endemic equilibrium satisfies the following two polynomial scenario. We consider the bee endemic population expressed in terms of the varrao mites virus population in the first case, that is,

$$P(I^*) = -{I^*}^2\mu_2 - I^*m^*\mu_2 + m^*F\beta(1-u_2) = 0 \tag{7}$$

(1) If $m^* = 0$ then there is disease free equilibrium (that is $I^* = 0$)
(2) If $m^* \neq 0$ then there exit one positive root, when $\mu_3 = 0, u_2 = 0$, see 1(a) .
(3) If $m^* \neq 0$ then there exit one positive root, when $\mu_3 \neq 0, u_2 \neq 0$. If $I^* = \frac{m^*\mu_2 \pm \sqrt{{m^*}^2\mu_2^2 + 4m^*F^*\mu_2\beta(1-u_2)}}{2\mu_2}$, see Figure 2(a). The Figure suggest that the infected bee population increases with increase in the varroa mites virus infected population.

The unique endemic equilibrium is given as

$$H^* = \frac{L(F^* + \psi) - (1 - u_1)[(F^*\gamma + (\alpha + \gamma)\psi]\omega}{(1 - u_1)[F^*\gamma + (\alpha + \gamma)\psi]}$$

$$I^* = \frac{m^*\mu_2 \pm \sqrt{m^{*2}\mu_2^2 + 4m^*F^*\mu_2\beta(1 - u_2)}}{2\mu_2} \tag{8}$$

$$M^* = \frac{Q[(F^* + I^*)[(m^* - 1)r - \mu_3] - I^*\lambda]}{r(m^* - 1)(F^* + I^*)}$$

The second case is the varroa mites endemic population expressed as a function of the number of infected bees, given by the polynomial

$$P(m^*) = z_1(m^*)^3 + z_2(m^*)^2 + z_3m^* + z_4 = 0 \tag{9}$$

where,

$$\begin{aligned}
z_1 &= r * (F^* + I^*)^2(2Qr - \mu_3),\\
z_2 &= 2F^{*3}r^2\beta + r(F^* + I^*)[2QI^*\lambda + (2+3Q)(F^* + I^*)\mu_3 - 4Qr(F^* + I^*)],\\
z_3 &= (F^* + I^*)[Qr[2r(F^* + I^*) - 3I^*\lambda] + \mu_3[QI^*\lambda\\
&\quad + Q(F^* + I^*)\mu_3 - (1 + 3Q)r(F^* + I^*]]),\\
z_4 &= QI^*\lambda[r(F^* + I^*) - \lambda I^* - \mu_3(F^* + I^*)].
\end{aligned} \tag{10}$$

(1) For $I^* = 0, \mu_3 = 0$ (e.i. infected free bee population), then the polynomial (9) becomes

$$P(m^*) = m^*\left(2r^2F^{*2}Qm^{*2} + (2F^{*3}r^2\beta - 4QrF^*)m^* + F_J\right) = 0. \tag{11}$$

Where $F_J = Qr^2F^{*2} - F^{*2}r(1 + 3Q)$ and $m^* = 0$ correspond to disease free equilibrium. There exist two positive roots (that is,

$$m^* = \frac{(4QrF^* - 2F^{*3}r^2\beta) \pm \Omega}{2F^2r^2Q},$$

where

$$\Omega = \sqrt{(2F^{*3}r^2\beta - 4QrF^*)^2 - 16Q^2r^4F^4},$$

see Figure 1(b)). From the numerical simulation, it is clear that for a infected free bee population ($I^* = 0$), the varroa mites virus infected population can not exceed 2 units. Suggesting that intervention

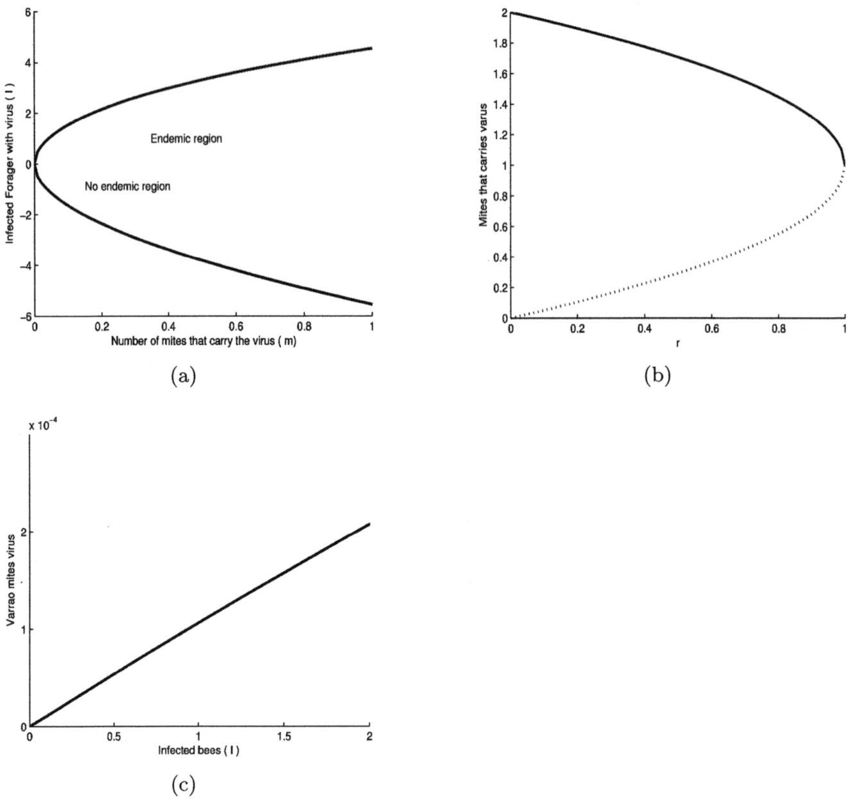

Figure 1. Simulations of the model showing the different scenarios: Using the following parameter values, $\mu_2 = 0.14, \beta = 0.8, F = 800, m = 700, 0 \leqslant u_2 \leqslant 1$

measures targetted towards preventing the bees from being infected even in the presence of varroa mites virus would not propogate the spread of the infestation of bee colonies. For $I^* \neq 0, \mu_3 \neq 0$, the Figure 2b, using λ, as bifurcation parameter, the model exhibits multiple endemic equilibria, transcritical bifurcation whenever $\lambda \leq 0.5$ and backward bifurcation whenever $2.9 \leq \lambda \leq 3.7$. The phase portrait is shown in 2c.

(2) For $I^* \neq 0$, see Figure 2b. □

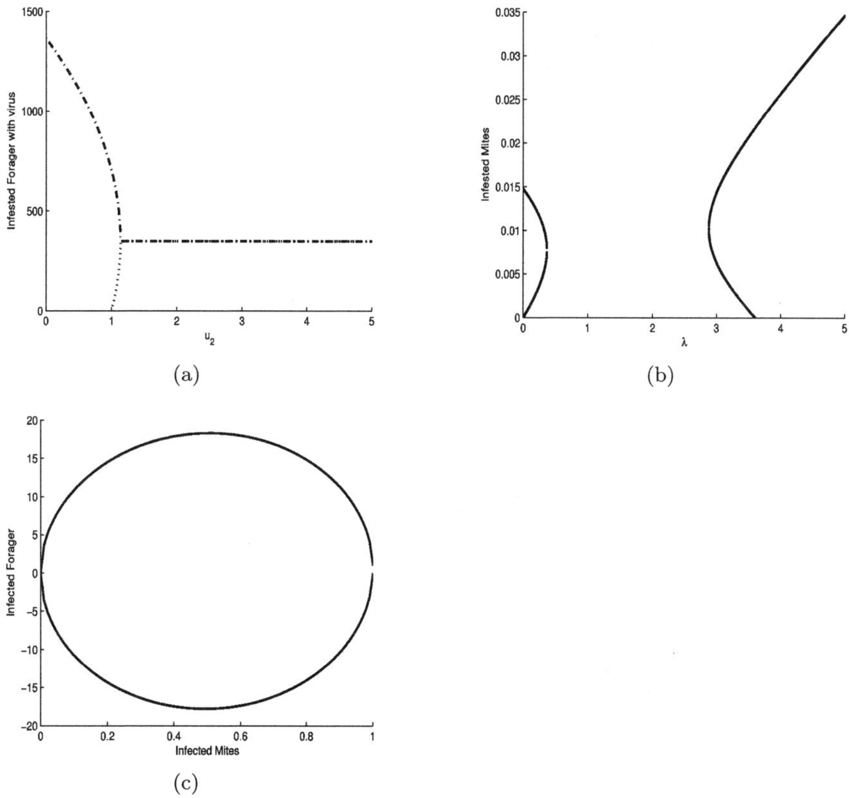

Figure 2. Simulations of the model showing the different scenarios: Using the following parameter values, $\mu_2 = 0.14, \beta = 0.8, F = 800, m = 700, 0 \leqslant u_2 \leqslant 1$

3. Analysis of optimal control

In the this section, we apply optimal control method using Pontryagin's Maximum Principle to determine the necessary conditions for the optimal control of the model. We consider time dependent controls (u_1 and u_2) in the model (1) to determine the optimal strategy for controlling the disease. Hence we consider the objective functional

$$J(u_1, u_2) = \int_0^{t_f} [AI + Bm + Cu_1^2 + Du_2^2]dt \tag{12}$$

The control function, u_1 and u_2 are bounded, Lebesgue integrable functions. The control $u_1(t)$ represents the efforts on preventing hive from being infested with mites. While the control $u_2t)$ represents the efforts on protect the forage from being infested with varroa-mites virus. The cost of applying preventions u_1 and u_2 are minimized subject to the system (1).

where t_f is the final time and the coefficients, A, B, C, D are balancing cost factors due to scales. We seek to find an optimal control, u_1^*, such that

$$J(u_1^*, u_2^*) = \min\{J(u_1, u_2)|u_1, u_2 \in \mathcal{U}\} \tag{13}$$

where $\mathcal{U} = \{(u_1, u_2)$ such that u_1 and u_2 are measurable with $0 \leq u_1 \leq 1$ and $0 \leq u_2 \leq 1$, for $t \in [0, t_f]\}$ is the control set.

The necessary conditions that an optimal solution must satisfy come from the Pontryagin et al [19] Maximum Principle. This principle converts (1)-(12) into a problem of minimizing pointwise a Hamiltonian H_a, with respect to u.

$$\begin{aligned}
H_a &= AI + Bm + Cu_1^2 + Dc_2^2 \\
&+G_H\left\{L\left(\frac{H}{H+\omega}\right) - H\left(\frac{\alpha\psi}{F+I+\psi} + \gamma\right)\left(1-u_1\right)\right\} \\
&+G_F\left\{H\left(\frac{\alpha\psi}{F+I+\psi} + \gamma\right)\left(1-u_1\right) - \mu_1 F - (1-u_2)\frac{\beta F m}{m+I}\right\} \\
&+G_I\left\{(1-u_2)\frac{\beta F m}{m+I} - \mu_2 I\right\} \\
&+G_M\left\{rM\left(1-\frac{M}{Q}\right)\left(1-m\right) - (1-u_2)\frac{\lambda I M}{F+I} - \mu_3 M\right\} \\
&+G_m\left\{rmM\left(1-\frac{M}{Q}\right) + (1-u_2)\frac{\lambda I M}{F+I} - \mu_3 m\right\}
\end{aligned} \tag{14}$$

where the G_H, G_F, G_I, G_M, G_m are the adjoint variables or co-state variables. The system of equations is found by taking the appropriate partial derivatives of the Hamiltonian (14) with respect to the associated state variable.

Theorem 3.1. *Given optimal controls u_1^*, u_2^* and solutions H, F, I, M, m of the corresponding state system (1)- (12) that minimize $J(u_1, u_2)$ over U. Then there exists adjoint variables G_H, G_F, G_I, G_M, G_m satisfying*

$$\frac{-dG_i}{dt} = \frac{\partial H_a}{\partial i} \tag{15}$$

where $i = H, F, I, M, m$ and with transversality conditions

$$G_H(t_f) = G_F(t_f) = G_I(t_f) = G_M(t_f) = G_m(t_f) = 0 \tag{16}$$

and

$$u_1^* = \min\left\{1, \max\left(0, \frac{H\left(\dfrac{\alpha\psi}{F+I+\psi}+\gamma\right)(G_F - G_H)}{2C}\right)\right\}, \tag{17}$$

$$u_2^* = \min\left\{1, \max\left(0, \left(\frac{\dfrac{\beta m F}{m+I}(G_F - G_H) + \dfrac{\lambda I M}{F+I}(G_m - G_M)}{2D}\right)\right)\right\}, \tag{18}$$

Proof: Corollary 4.1 of Fleming and Rishel [13] gives the existence of an optimal control due to the convexity of the integrand of J with respect to optimal pair u_1, u_2, a *priori* boundedness of the state solutions, and the *Lipschitz* property of the state system with respect to the state variables. The differential equations governing the adjoint variables are obtained by differentiation of the Hamiltonian function, evaluated at the optimal control. Then the adjoint equations can be written as

$$\begin{aligned}
-\frac{dG_H}{dt} &= \left[\frac{\alpha\psi}{F+I+\psi}+\gamma\right](1-u_1)(G_H - G_F) - G_H\frac{L\omega}{(H+\omega)^2}\\
-\frac{dG_F}{dt} &= (G_H - G_F)(1-u_1)\left[\frac{-\alpha\psi}{(F+I+\omega)^2}\right]\\
&\quad +(G_F - G_I)(1-u_2)\left(\frac{m\beta}{m+I}\right)\\
&\quad +G_F\mu_1 + (G_M - G_m)(1-u_2)\left(\frac{-\lambda I m}{(F+I)^2}\right)\\
-\frac{dG_I}{dt} &= (G_H - G_F)(1-u_1)\left[\frac{-\alpha\psi}{(F+I+\psi)^2}\right]\\
&\quad +(G_F - G_I)(1-u_2)\left[\frac{-mF\beta}{(m+I)^2}\right]\\
&\quad +(G_M - G_m)(1-u_2)\left[\frac{MF\lambda}{(F+I)^2}\right] + \mu_2 G_I\\
-\frac{dG_M}{dt} &= -G_M\left(r - \frac{2rM}{Q}\right)(1-m) + (G_M - G_m)(1-u_2)\frac{\lambda I}{F+I}
\end{aligned}$$

$$
\begin{aligned}
&-G_m\left(r-\frac{2rM}{Q}\right)m+\mu_3 G_M\\
&-\frac{dG_m}{dt}=(G_F-G_I)(1-u_2)\left(\frac{FI\beta}{(m+I)^2}\right) \qquad (19)\\
&+rM(G_M-G_m)\left(1-\frac{M}{Q}\right)+\mu_3 G_m
\end{aligned}
$$

Next, we discuss the numerical solutions of the optimality system and the corresponding results of varying the optimal control u, the parameter choices, and the interpretations from various cases.

4. Numerical results and discussions

In this section, we investigate numerically the effect of the optimal control strategy on the varrao-mites infestation on the Bee colony. The optimal control solution is obtained by solving the optimality system, which consists of the state system and the adjoint system. An iterative scheme is used for solving the optimality system. We start by solving the state equations with a guess for the controls over the simulated time using the fourth order Runge-Kutta scheme. Because of the transversality conditions (16), the adjoint equations are solved by the backward fourth order Runge-Kutta scheme using the current iterations solutions of the state equations. Then the controls are updated by using a convex combination of the previous controls and the value from the characterizations (18) - (19). This process is repeated and the iterations are stopped if the values of the unknowns at the previous iterations are very close to the ones at the present iteration [16], [17], [18], [19].

4.0.1. *Efforts on preventing Hive from mites u_1 only*

The Honey bee hive prevention control u_1 is used to optimize the objective function J while we set the other interventions (u_2) to zero. That is, only the hive prevention mechanisms is optimized without foragers being prevented from mites virus. We observed in Figure 3(a - c) that due to the control strategies, the number of infected foragers I and foragers (virus free) F shows no significant difference compared with cases without control. While the hive population H show significant improvement (difference). This strategy has no significant impact on controlling the infected forager population.

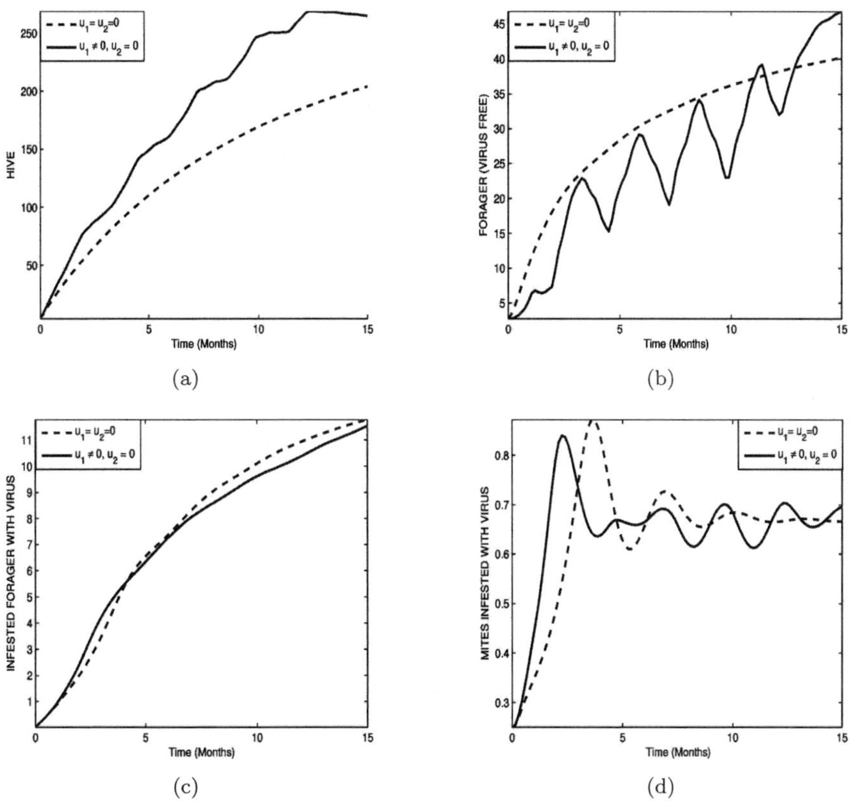

Figure 3. Simulations of the Honey Bee Colony

4.0.2. *Efforts on preventing Forager from mites infestation* u_2 *only*

The Forager prevention control u_2 is used to optimize the objective function J while we set the other interventions (u_1) to zero. That is, only the forager prevention mechanisms is optimized without the hive being prevented from mites. We observed in Figure 4(a - c) that due to the control strategies, the number of infected foragers I and foragers (virus free) F shows significant difference compared with cases without control. While the hive population H show little significant improvement (difference). This strategy has significant impact on controlling the infected forager population.

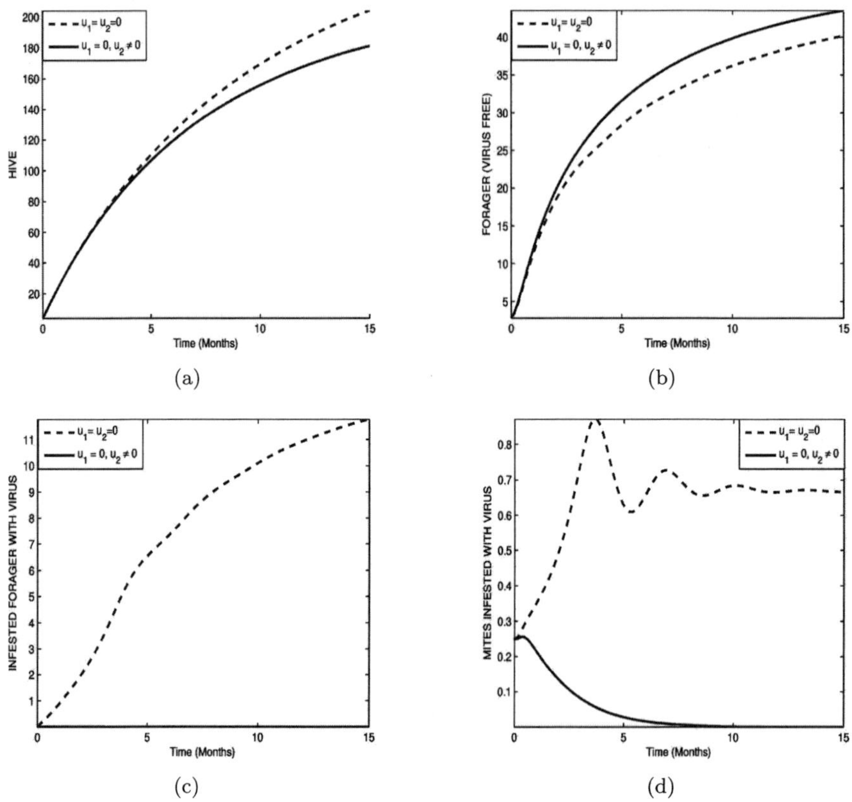

Figure 4. Simulations of the Honey Bee Colony

4.0.3. *Prevention of Hive u_1 and Protection of Forager u_2*

The Honey bee hive prevention control u_1 and control u_2 are used to optimize the objective function J.That is, both the hive prevention and forager protection mechanisms are optimized. We observed in Figure 5(a - c) that due to the control strategies, the number of infected foragers I and foragers (virus free) F shows significant difference compared with cases without control. While the hive population H also show significant improvement. It is clear that this strategy is more effective and effecient in preventing honey bee colony disorder.

In the Figures 6(a - b), we show the impact of parameter variation (β, r) and the phase portrait (Hive (H) against mites with virus (m)) in the dynamics of the colony.

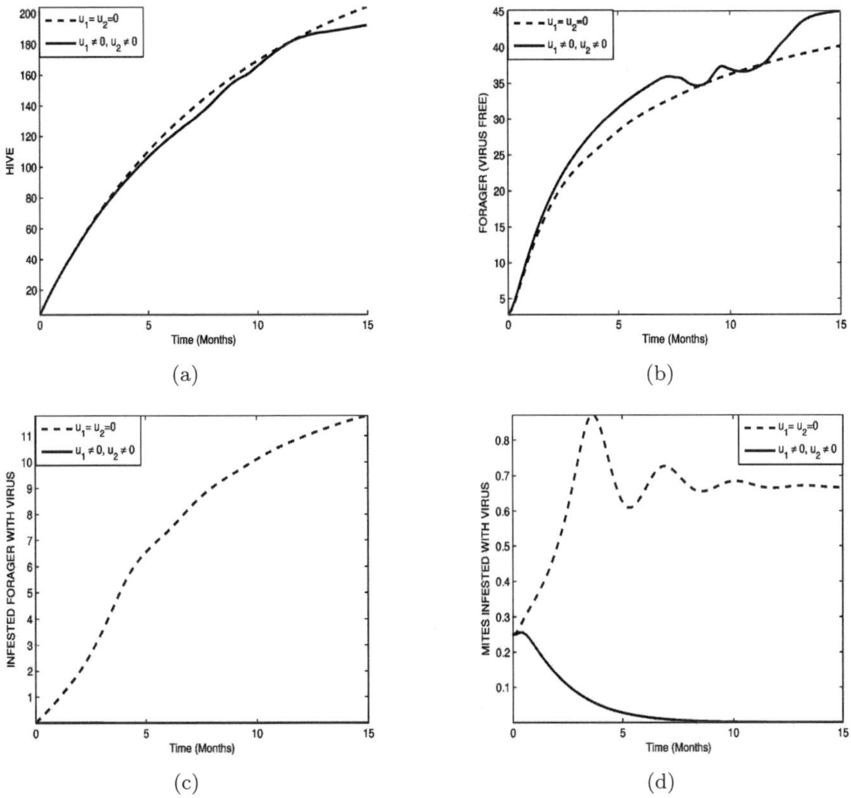

Figure 5. Simulations of the Honey Bee Colony

5. Concluding remarks

In this paper a mathematical model for the dynamics of varroa-infested honey bees colonies is formulated and analyzed. Two scenarios were considered, namely, (i) the dynamics of the endemic equilibrium of the honey bee population expressed in terms of the varroa mites, this is found that honey bee disease free equilibrium only exists in the absence of varroa mites and the infected bee population increases with increase in the varroa mites population and (ii) the dynamics of the endemic equilibrium of the varroa mites population expressed in terms of the infected honey bee, it is found that the model exhibit multiple equilibria. The numerical simulation show that for a infected free bee population, the varroa mites virus infected population can not exceed 2 units. We incorporate time dependent

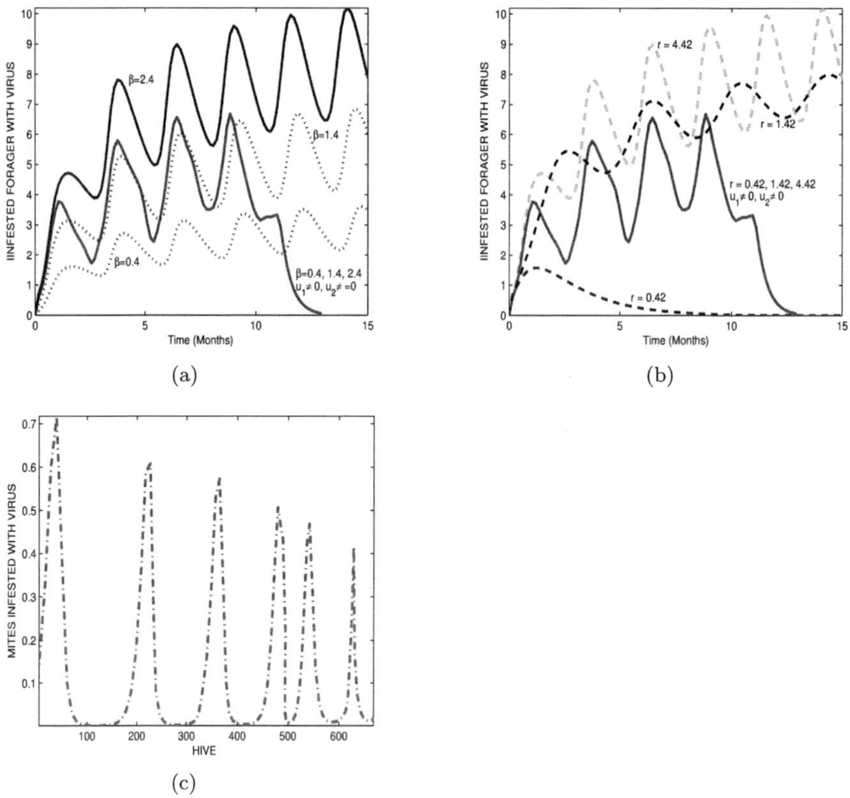

Figure 6. Simulations of the Honey Bee Colony

controls, using Pontryagin's Maximum Principle to derive necessary conditions for the optimal control of the disease, the numerial results suggest that the comibination of prevention control (u_1) on hive and protection control on forager (u_2) is the most effective strategy to prevent Colony Collapse Disorder in honey bee colony.

References

1. Bhattacharyya S and Ghosh S (2013). Oscillation in pest population and its management: A mathematical study. International Journal of Mathematics and Mathematical Sciences, volume 2013, Article ID 653080, 12 pages.
2. Wang X and Song X (2008). Mathematical models for the control of a pest population by infected pest. Computers and Mathematics with Applications, 56, 266 - 278.
3. Maiti A. Pal A.K. and Samata G.P. (2008). Usefulness of biocontrol of pest in tea: A mathematical model. Math. Model. Nat. Phenom. vol 3, no 4, 96-113.

4. Rafikov M and Limeira E. H. (2012). Mathematical modelling of the biological pest control of the sugarcane borer. International Journal of Computer Mathematics. vol 89, no 3, 390-401.
5. Olayiwola R.O. Asibor R.E. Mohammed A.A. Odebiyi O.A. and Kolo E.I. (2012). A mathematical model and simulation of virus epidemics in varroa-infested honey bee colonies. International Journal of Science and Technology. vol 2, no 12,807-816.
6. Khoury D.S. Barron A.B. and Myerscough M.R (2013). Modelling food and population dynamics in honey bee colonies. PLOS ONE, vol 8, issue 5, e59084, 1-7.
7. Dornberger L. Mitchell C. Hull B. Ventura W. Shopp H. Kribs-Zeleta C. Kojouharov H and Grover J. (2012). Death of the bees: A mathematical model of colony collapse disorder. Mathematics Preprint Series. Technical Report, The University of Texas, ARLINGTON. www.uta.edu/math/preprint.
8. Ratti V. Kevan P.G. and Eberl H.J. (2012). A mathematical model for population dynamics in honeybee colonies infested with varroa destructor and the Acute Bee Paralysis Virus.
9. Guzman-Novoa E. Eccles C.Y. McGowan J. Kelly P.G and Correa-Bentez A. (2010). Varroa destructor is the main culprit for the death and reduced populations of overwintered honey bee (*Apis mellifera*) colonies in Ontario Canada, *Apidologie*, vol 41, 443-450.
10. Seeley D. (1978). Iife history strategy of the honey bee *Apis mellifera*. Oecologia, vol 32, 109-118
11. Vanbergen A.J and the Insect Pollinators Initiative (2013). Threats to an ecosystem service: pressures on pollinators. Frontiers in Ecology and the Environment. The Ecological Society of America. doi:10.1890/120126, www.frontiersinecology.org.
12. Sun S and Chen L (2009). Mathematical modelling to control a pest population by infected pests. Applied Mathematics Modelling, vol 33, 2864 - 2873.
13. Fleming WH, Rishel RW. Deterministic and stochastic optimal control. Springer Verlag, New York (1975).
14. Driessche P.V.D., Watmough, J., 2002. Reproduction numbers and subthreshold endemic equilibria for compartmental models of disease transmission. *Math. Biosci.* 180: 29-48
15. Ozair M, Lashari A.A, Jung I.H, Okosun K.O (2012). Stability Analysis and Optimal Control of a vector-borne disease with nonlinear incidence. Discrete Dynamics inNature and Society.
16. Okosun, K.O., Ouifki, R., Marcus, N., 2013. Optimal control strategies and cost-effectiveness analysis of a malaria model. *BioSystems* 111, 83 - 101. doi:10.1016/j.biosystems.2012.09.008
17. Okosun, K.O., Makinde, O.D., 2013. Optimal control analysis of malaria in the presence of non-linear incidence rate. *Appl. Comput. Math.* 12(1), 20 - 32
18. Kirschner, D., Lenhart, S., Serbin, S., 1997. Optimal Control of the Chemotherapy of HIV. *J. Math. Biol.* 35: 775-792
19. Lenhart, S., Workman, J.T., 2007. Optimal Control Applied to Biological Models. *Chapman and Hall*

PROBABILITY DISTRIBUTIONS OF GC CONTENT REFLECT THE EVOLUTION OF PRIMATE SPECIES*

MARCO V. JOSÉ†

Instituto de Investigaciones Biomédicas,
Universidad Nacional Autónoma de México, CP 04510, México
E-mail: marcojose@biomedicas.unam.mx

QI LU

Centro Internacional de Ciencias A.C,
Campus UNAM UAEM, Col. Chamilpa, CP 62210,
Cuernavaca, Morelos, México

JUAN R. BOBADILLA

Instituto de Investigaciones Biomédicas,
Universidad Nacional Autónoma de México,
CP 04510, México

A novel statistics for measuring the (G+C)-content of a given chromosome, which represents the local average distance of GC dinucleotides was derived. We found that this statistics is log-linearly inversely related, for a broad range of scales, to the (G+C)-content in humans, chimpanzee, and rhesus macaque. We also determined the distribution of window sizes for which a given count of GC duplets is found along an entire chromosome. These distributions permit the visualization of how the GC duplet is distributed along a given chromosome and they capture simultaneously GC-poor, GC-intermediate, and GC-rich regions. Interestingly, these distributions followed a quasi-log normal distribution for all chromosomes in the three species and they were fitted to the Black-Scholes stochastic equation. Most deviations from the fittings occurred mainly at the tails that represent GC-poor and GC-rich regions. Some moments of these distributions were clearly different among some primate chromosomes, particularly the mean of several rhesus chromosomes. However, as a whole, the mean, variance, skewness, and kurtosis turned out not to be statistically different among the three primate species. The ratio of (G+C) over the size of each chromosome is practically constant for each chromosome in all the three primate species. This constant ratio imposes limits to the mean and variance of these distributions.

*This work is supported by PAPIIT; UNAM, Project IN107112
†Corresponding author.

1. Introduction

Almost three decades ago, it was demonstrated that comparison of Giemsa-banded karyotypes showed a very high degree of similarity between man, chimpanzee, gorilla, and orangutan[1]. The G-banded late-prophase chromosomes of these four species showed an extensive homology[1].

More recently, the whole genomes of *Homo sapiens*[2], *Pan troglodytes*[3], and *Rhesus macaque*[4] have been available. This information permits to carry out more accurate and systematic comparisons in order to understand the global structure, functional properties, and evolutionary aspects of primate genomes. Comparisons of DNA sequences between humans and the great apes showed that the African apes, especially the chimpanzees and the bonobos, but also the gorillas, are more closely related to humans than are the orangutans in Asia[5]. Thus, from a genetic standpoint, humans are essentially African apes. Our sense of uniqueness as a species was further shattered by the revelation that human DNA sequences differ by, on average, only 1.2% from those of the chimpanzees[6], as a consequence of humans and apes sharing a recent common ancestry. Explaining the evident biological traits that separate modern humans from our closest relatives, the chimpanzees, demands an explanation given the often cited 1.2-1.5% difference between orthologous nucleotide sequences[3,7]. This genetic distance is too small to account for their substantial differences. Regulatory changes[8], amino acid changes[3,9,10] and the detection of gene gain and loss since the split of humans from chimpanzees indicate that humans differ from chimpanzees by at least 6%. Another counterintuitive finding is that there seems to be more genes that have underwent positive selection in chimpanzee evolution than in human evolution[12]. Precise estimates of divergence times have also been obtained by comparisons of DNA sequences. It now seems clear that the human evolutionary lineage diverged from that of chimpanzees about 46 MYA, from that of gorillas about 68 MYA, from that of the orangutans about 1216 MYA[6,7], and that the rhesus macaque diverged from the ancestors of humans about 25 MYA[4]. Although the majority of regions in our genome are most closely related to chimpanzees and bonobos, a significant fraction is more closely related to gorillas[7]. A striking observation from the analysis of the human genome is the extent of DNA-sequence similarity among individuals from around the world: any two humans are thought to be about 99.9% identical in their DNA sequence[13]. The human species may be one of the most genetically uniform species on Earth. The common approach of most biologists is to look for differences among chromosomes of a given species and differences and similarities among species. The search of Single Nucleotide Polymorphism in the human genome is an example of this approach[14]. It is considered that the history of different chromosomes during evolution may not be the same[15,16] and the structure of a given chromosome can be very different from others[17,18,19]. The different evolutionary histories of each chromosome within a given species and among species raises the questions of how all primate genetic information is replicated in a coordinated manner before each cell division.

The sequence of each primate chromosome underpins an extremely broad range of evolutionary, genetic, medical problems and conservational issues. The analysis of statistical patterns of DNA sequences of whole genomes may provide more insights into the genetic differences and similarities between humans and non-human primates. Little attention has been put to comparisons of whole genomes that include non-coding RNA genes[20] and a bestiary of DNA sequences that are present in multiple copies in the genomes in which they reside, the so-called repetitive elements[21,22].

Statistical correlations in DNA sequences must not be a surprise given that chromosomes are complex systems which involve many different scales. DNA sequences are full of features at small, intermediate, and large scales: At short distances, there is strong 3-base periodicity signal in protein-coding regions which is absent in noncoding regions[23,24], and a weaker but ubiquitous 10-11 bases signal in many genomes[25]. The 3-base periodicity is a landmark of protein coding regions[26] and the periodicity of 10-11 bp reflects DNA bendability[25]. Long range correlations have been intensively studied[27,28]. In human DNA sequences the intermittent range, known as $1/f$ noise, is a universal feature, but interspersed repeats that represent about 50% of human genomic DNA sequences are not responsible for this pattern[28]. At intermediate length scales, there are *Alu* sequences of about 300 bases long[22] and nucleosome-forming sequences within 31-450 bases[29,30]. At large length scales, the most well-known features are the existence of alternating GC-rich and GC-poor isochores[31]. Heterogeneity of base composition within genomes from a variety of taxa ranging from virus to mammals was revealed several decades ago by thermal melting and sodium chloride gradient centrifugation[31,32,33]. The mammalian genome has been depicted as a discrete mosaic of long, fairly homogeneous genomic regions that are characterized by their guanine and cytosine (G+C)-content[33]. Isochores are large DNA segments, of average size well above 300 kb in mammals, which are characterized by an internal variation in GC well below that found in the entire mammalian genome. The human genome, for instance, is claimed to consist of five distinct isochore families: L1, L2, H1, H2, and H3, with (G+C) contents of $< 37\%$, $37\% - 42\%$, $42\% - 47\%$, $47\% - 52\%$, and $> 52\%$, respectively[33]. The origin of isochores and the source of variation of (G+C) content are still highly debated: are they the result of selection[34,35], or do they simply reflect variations in neutral substitutions patterns[36,37]. Several years ago, it was recognized that guanine and cytosine (G+C) content is an important parameter in considering evolution and phylogeny[38,39]. These Mb-scale variations in (G+C)-content result from variations of substitutions patterns that have affected both coding and non-coding regions. These genomic landscapes are correlated with many other important features (gene density, intron size, distribution of transposable elements, replication timing). Interestingly, this long-range variability of base composition is correlated with various genome features: GC-rich regions show a lower density of LINE and a higher density of *Alu* repeated elements, a higher level of methylation, a higher rate of recombination, and a much higher gene density[2,34,40,41,42,43]. These relationships demand that special attention should be paid to the evolution of (G+C)-content itself. Unraveling the origin of (G+C)-content (neutral evolution or selection) is essential to understand the functional significance (if any) of this peculiar genome organization. Thus, this long-range variability of (G+C)-content clearly reflects a fundamental feature of genome organization. It is puzzling that given the evolutionary distances considered, the temporal variations in (G+C)-content are negligible (human and chimpanzee orthologous sequences are 98% identical). A statistical description of the regional (G+C) variations along chromosomes is also a difficult problem. The isochore description assumes a distinction between locally (G+C)-rich and (G+C)-poor regions[45]. The sizes of these regions are larger than 100-300kb. The mere existence of isochores has been challenged[2,46]. The criticisms were that the transition from one region to another occurs within a relatively shorter length scale and that the random base model with sudden changes of bias does not describe appropriately the human genome. According to the global (G+C)-content, it is not expected that the human and non-human primates genomes do differ greatly among them. In one proposal for the historical evolution of (G+C)-content, GC-rich regions in ancestral amniote genomes decreased (G+C) to a level of those in the present-day mammals[47]. Another proposal is for the ancestral mammals[47] (therapsida) to have increased their (G+C) to those of present-day mammals. Some works indicate that isochores are vanishing[47] whereas other works suggest the opposite[48].

In this work we analyze the (G+C)-content of primate species via a novel indicator called *Local Average Distance of GC Dinucleotides* (*LADGC*). We show that this new indicator is log-linearly inversely related to the (G+C)-content along each whole primate chromosome. We also propose to model the probability distributions of (G+C)-content treating *LADGC* as a stochastic random variable. We show that its distribution for whole chromosomes follows a quasi-lognormal distribution consistent with the Black-Scholes model (related to the behavior of stock markets). Then with a single statistics we have a description of both the (G+C)-content and a single probability density function of the *LADGC* for each chromosome. We calculate the mean, variance, skweness, and kurtosis for all chromosomes of *Homo sapiens*, *Pan troglodytes*, and *Macaca mulatta*, (*Rhesus macaque*). We test the null hypothesis that all these statistical measures should be the same. Finally, we discuss our findings in the light of the present mathematical approach to further test the evolution of (G+C)-content in primate genomes.

2. Materials and methods

Human (*Homo sapiens*), chimpanzee (*Pan troglodytes*) and macaque (*Macaca mulatta*) complete genome sequences were downloaded from NCBI released, respectively, in March, October, and June of 2006. All the calculations on the complete genomes were performed with *ad hoc* programs in C programming language. Whenever we found an unidentified nucleotide (marked by "n") we randomly replace it by assigning a probability of 0.4 to G and C.

2.1. *Negative correlations between* LADGC *and GC content*

Instead of using the classic "sliding windows strategy"[49,51], which moves a fixed window size along the sequence and measures the (G+C)-content inside the window, we here propose to use the Local Average Distance of GC Dinucleotides First, we determine the distance series of the duplet GC along any chromosome. Then we fix an arbitrary number N of GC duplets at which we will perform the analysis. Once we found N GC dinucleotides, we record the distance L (bps) of every two neighbor GC dinucleotides. Thus,

$$LADGC = \langle GC \rangle_{LD} = \frac{\sum_{i=1}^{N-1} L_i}{N} \tag{1}$$

According to Eq. (1), $\langle GC \rangle_{LD}$ is a measure of the local average distance (bps) of N GC dinucleotides. On the other hand, if the local length $\sum_{i=1}^{N-1} L_i$ is given, it is also straightforward to calculate the local (G+C)-content. We have performed this calculation at small and large scales ($N = 60, 100, 1000, 2000$) for human, chimpanzee, and macaque. We found that there is a log-linear negative correlation between $LADGC$ and the (G+C)-content for all human and chimpanzee chromosomes and for most rhesus macaque chromosomes. In Fig. 1 (A, B, C) and Fig. 1 (D, E, F), we illustrate this relation for chromosomes 19 and X, respectively, for the 3 primates species. Then with a single statistical indicator we can directly read off the percentage of (G+C)-content along any single chromosome without the need to distinguish 5 types of different isochores. As all distances between GC duplets can be fit to a single straight line, all types of previously defined isochores seem to somehow obey the same rule as determined by our estimator. The 5 different types of isochores are included in a single log-linear plot. If we change to normal scale, the relationship is exponential and can be modeled as:

$$\langle GC \rangle_{LD} = ae^{(bX_{GC})} \tag{2}$$

where $a > 0$, $b < 0$ and a, b are two constants which depend on the chromosome and the local number of GC dinucleotides; X_{GC} represents the local (G+C)-content. The exponential form in Eq. (2), however, implies that correlations in GC-poor-regions are much stronger than those in GC-rich-regions. The explanation for this is as follows: If the local (G+C)-content is low (high), the probability to find GC dinucleotides in GC-poor (GC-rich) regions, also becomes small (high). Thus, the average distance of GC dinucleotides in GC-poor (GC-rich) regions tends to be large (short). In addition to the main negative trend at large-scales, we also found that in some chromosomes of macaque and few chromosomes of chimpanzee exhibit an extra structure at fine-scales (e.g. Fig. 1C). This structure indicates that at fine-scales there may be even stronger negative correlations between $LADGC$ and low (G+C)-content than the ones we observe at large-scales. Even when the scale is still large, for example for $N = 300$ (40-150kb), we can detect the decaying part in the macaque (not shown), but for chimpanzee even when the scale is "really fine", for which $N = 10$ (1-2kb), it just shows a slight decay (not shown). Further research shows that the extra negative correlation at fine scales for macaque is the strongest and for chimpanzee becomes very weak, whereas for human, it completely vanishes. Mathematically, we can model the fine-scale structure by the two-times exponential function,

$$\langle GC \rangle_{LD} = ae^{be^{(cX_{GC}+dX_{GC})}} \tag{3}$$

where $a > 0$, $b > 0$, $c < 0$, and $d < 0$ are constants, which depend on the particular chromosome and the scale used.

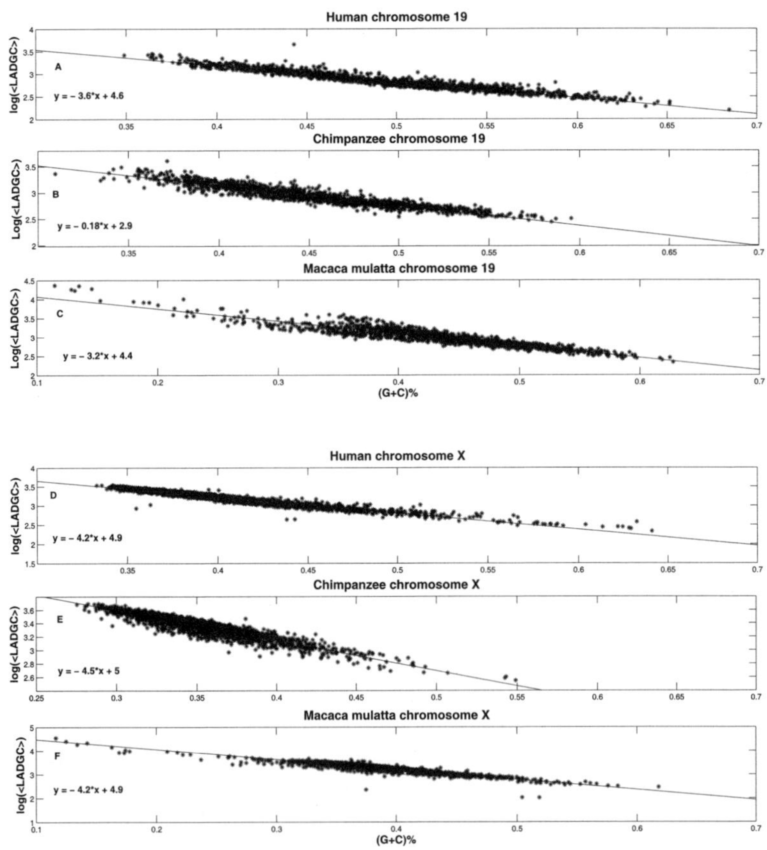

Figure 1. Linear plots of $\log\langle GC \rangle_{LD}$ vs (G+C)% for chromosome 19 in (A) human; (B) chimpanzee; (C) rhesus macaque; chromosome X in (D) human (E) chimpanzee (F) rhesus macaque. The local number of GC dinucleotides in each window is $N = 2000$. The fitted equation in each case is shown in each panel.

3. A mathematical evolutionary model for the evolution of (G+C)-content in primate species

In order to investigate the (G+C)-content evolution in different species following the foregoing concept, we employed the probability density function (PDF) of *LADGC* for all chromosomes of three primate species and found that the comparison of the PDFs among these species show very similar patterns and all PDFs can be approximately fitted to a log-normal distribution (Fig. 2). After comparing the distribution of the three species, we found that many of them exhibit different means but approximately the same variance. If we use a normal distribution to approximate the histogram of actual data, which means we neglect the effect of skewness and kurtosis, then, we can propose the following general stochastic equation to describe the evolutionary process:

$$dS = \mu(S,t)dt + \sigma(S,t)dW \tag{4}$$

where S represents the local average distance of GC ($LADGC$) and dW is a Wiener process. The idea behind Eq. (4) is that the evolutionary process is a random process which can be modeled by two factors, one is the mean which is the factor related to the trend and the other is the variance σ^2 which means noise. We can also make some simplifications. First, we suppose that noise in Eq. (4) is a small constant, in other words, the noise contribution is small when compared with the trend factor. The assumption is also consistent with the observation that the distributions of all the three species today have approximately the same variance. Otherwise, we could expect that the distributions would show large differences in its variance. Second, we suggest the function $\mu = V_S(t)t$ to be a candidate for the mean, where $V_S(t)$ is the speed of evolution of the different species at different times. We propose this function because the 3 distributions should be overlapped in their common ancestor. After that, they separated and evolved towards different directions and at the same time the distributions also began to separate from each order. In this circumstance, the simplest assumption is that the three species have different speeds of evolution. With these assumptions Eq. (4) is simplified to,

$$dS = V_S(t)tdt + \sigma dW \tag{5}$$

If we calculate the exponential transform in order to get the actual$LADGC(Y)$, and apply Itos law to $Y = e^S$, we get

$$\frac{dY}{Y} = \left(V_S(t) + \frac{\sigma^2}{2}\right)dt + \sigma dW \tag{6}$$

Eq. (6) is known as the Black-Scholes equation, originally formulated for modeling the behavior of stock markets[51]. This mathematical model is the most used in making financial decisions globally.

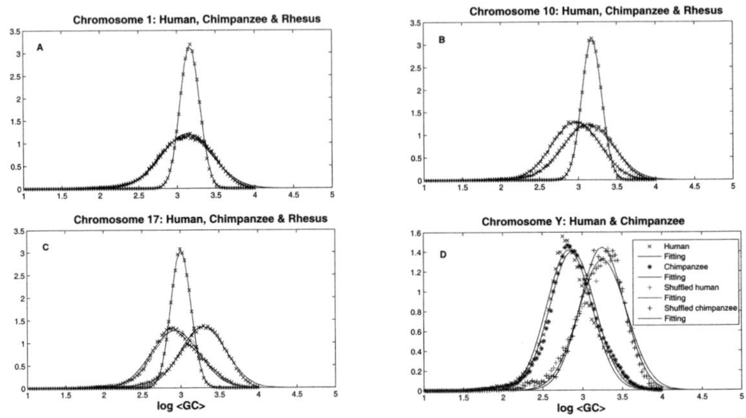

Figure 2. The PDFs of $\log\langle GC\rangle_{LD}$ ($N = 60$) for different chromosomes of human (crosses), chimpanzee (asterisks) and macaque (diamonds); human random control (pluses). All histograms were fitted to a log-Gaussian distribution (solid lines, solution of Eq. (6). In (A), (B), and (C) chromosomes 1, 10, and 17 of human, chimpanzee, and rhesus macaque, respectively; (D) chromosome Y of human and chimpanzee (the sequence of chromosome Y of rhesus macaque is not available); human (asterisks) and chimpanzee (pluses); the corresponding random controls are for human (asterisks) and for chimpanzee (crosses). In the Figure $\log\langle GC\rangle = \log\langle GC\rangle_{LD}$ is equivalent to $\log(X_{GC})$ and $\mu = \langle X_{GC}\rangle$.

3.1. *The probability distributions of (G+C)-content in primate species*

The negative correlations illustrated in Section 2.1 provide an important clue about how to measure the (G+C)-content evolution in a given species. The key idea is to use $\langle GC\rangle_{LD}$ as a bridge. For example, if we can estimate $\langle GC\rangle_{LD}$ in different species, by using its relationship with (G+C)-content (Eq. (2)), we can indirectly represent the (G+C)-content in different species by means of the distribution of $\langle GC\rangle_{LD}$. Thus, the probability distributions of $\langle GC\rangle_{LD}$ are equivalent to the probability distributions of (G+C)-content. This method enables us to compare the (G+C)-content evolution of different species. Additionally, the variable $\langle GC\rangle_{LD}$ can be easily and accurately quantified. In order to investigate the (G+C)-content evolution in different species following the foregoing concept, we calculated the probability density function (PDF) of $LADGC$ for all chromosomes of the three primate species (Eq. (6)).

In Fig. 2(A-D) the PDFs of $\langle GC \rangle_{LD}$ ($N = 60$) for different chromosomes of human, chimpanzee, and macaque are shown. In particular, we compare the PDFs for chromosomes 1, 10, 17 and Y for the three primate species together with their respective random controls. All the distributions for each chromosome follow approximately quasi-Gaussian distributions under the logarithm of $\langle GC \rangle_{LD}$ for each primate species. The fittings of the log-Gaussian distribution to the PDFs for all chromosomes for all species are statistically significant only if both tails are not included. This means that most of the $\langle GC \rangle_{LD}$ is the result of random variation whereas deviations of the log-Gaussian at the tails indicate long-range correlations at both GC-very-rich regions (left tails) and GC-very-poor regions (right tails). We found that the comparison of the PDFs among these species show similar patterns but there are both subtle and conspicuous differences. The deviations of the actual data from the log-Gaussian fittings are evolutionary relevant as they may be the outcome of positive and/or negative selection events. Note that the mean μ measures the average and σ^2 measures the variance of $\langle GC \rangle_{LD}$. Note from Eq. (2) Eq. (3) that $\langle GC \rangle_{LD}$ is a decreasing function of (G+C)-content and that generally, the bigger μ is, the lower the (G+C)-content will be in a chromosome. Thus, the average $\langle GC \rangle_{LD}$ and (G+C)-content of a chromosome are inversely related. Let us now assume that the evolution of the (G+C)-content is the result of positive or negative selection, then, we would like to further argue that for each chromosome the effect of μ comes from natural selection because it reflects a trend in the evolutionary process, while the variance should result mostly from neutral and/or nearly neutral selection because it is formed by random variability.

The mean of the PDF of some rhesus chromosomes may lie either at the right (Fig. 2C) or at the left (Fig. 2B) of the corresponding PDFs of both human and chimpanzee chromosomes. Herein rhesus macaque is regarded as an outgroup. When a PDF of rhesus chromosome lies at the left of the PDFs of human and chimpanzee, the distribution of GC dinucleotides is more clumped, whereas if they lie at the right, GC dinucleotides are more evenly distributed.

In regard to the Y chromosome both the mean and variance of the distribution are slightly larger in the chimpanzee than in humans. In contrast, the PDFs of $\log\langle GC \rangle_{LD}$ for chromosome X shows clearly that the mean in humans is greater than in chimpanzee and in turn the mean of the latter is greater than the one observed in macaque (not shown but see the slopes of Fig. 1D-E). This behavior clearly shows a trend of increasing the mean of this distribution. A similar behavior of having an ever greater mean can be observed when we compare the PDFs for autosome 19 but this is not the same for autosome 8. The PDF of chromosome 8 in chimpanzee is very similar to that found in humans, and both of them differ from the macaque. Still the mean for chromosome 8 is greater than the mean for chromosome 19 in macaque (not shown). Most distributions are undistinguishable from each other as can be seen in chromosome 1 (Fig. 1A). In order to test for the significance of these results, we randomized the position of each nucleotide for each chromosome and for each species and their corresponding distributions were also fitted to a log-Gaussian distribution. In these distributions the distances between GC dinucleotides are random. Interestingly, the three random distributions of each species overlapped for each chromosome (Fig. 2D). The random control follows a log-normal distribution in which the 3rd and 4th moments are zero. It is well known that the cumulants of a Gaussian process are identically zero for orders greater than two. In this case we have an identical and independent random variable. The variance of the random control distributions is always smaller than the variances of the actual distributions. The height is also higher in the control than the actual distributions. If the number N of GC dinucleotides is increased all distributions increase proportionally in size but the shape of the distributions remain unaltered. This is consistent with a hierarchical diffusion process but this observation is not further followed in this work.

We hypothesized that the overlapping of these random distributions would occur only if the ratio of (G+C) to the total size of each chromosome remains constant for each chromosome regardless of the primate species. We calculated for each chromosome and for each primate species the ratio of (G+C)/(A+T+G+C) and the expected value for all chromosomes of all species is0.41 ± 0.027. The mean and its standard deviation of (G+C)-content in human, chimpanzee, and rhesus macaque are 0.4158 ± 0.027, 0.4139 ± 0.0261, and 0.4133 ± 0.0263, respectively. These values are indeed practically identical.

In humans, chromosomes 4 (0.382) and 13 (0.385) and in chimpanzees also chromosomes 4 (0.381) and 13 (0.384), are the ones with the lowest (G+C)-content; in rhesus macaque the ones with the lowest (G+C)-content are chromosomes 5 (0.378) and 17 (0.384); in humans, chromosomes 19 (0.483) and 22 (0.479) and in chimpanzees also chromosomes 19 (0.479) and 22 (0.478), are the ones with the highest (G+C)-content; in rhesus macaque chromosomes 10 and 16 (0.455 each) and 19 (0.482) are the ones with the highest (G+C)-content.

Most actual PDFs already overlap the random distances of its respective control and both tails of the actual distributions lie below or above of the tails of the random control. In order to have a PDF at the right of the control the ratio of 0.4 must be overcome making the chromosomes mores AT-rich and/or the GC dinucletoides would be more homogeneously distributed. In other words, fixing the ratio of 0.4, the randomized control distributions are a reference barrier of homogenization of (G+C)-content. If the trend continues moving to the right, i.e. poorer GC-regions, we would have indeed a vanishing effect of the (G+C)-content. The average distances among GC dinucleotides would become more homogenized and this would produce a vanishing effect of the (G+C)-content. Then the distributions of distances of GC dinucleotides may be more clumped or dispersed in macaque than in chimpanzee and human. If the actual distributions lie at the right of the random controls then this implies that (G+C)-content has become further homogenized and this in turn is consistent with reports that isochores are vanishing[47]. However, if they lie at the left with the same GC-content will be distributed with more heterogeneity.

3.2. *Primate statistics of the PDFs*

In this section we present the results of the estimated mean, variance, kurtosis, and skewness for all chromosomes in each primate species (Fig. 3). An ANOVA test was unable to reject the hypothesis that all these statistical measures should be the same. Therefore, the fundamental structure of the distribution of (G+C)-content has remained essentially unchanged at the genomic level.

4. Discussion

In this work we have ushered in a novel statistics for analyzing the (G+C)-content of a given chromosome as reflected by the local average distance of GC dinucleotides that we denoted by $LADGC = \langle GC \rangle_{LD}$ We found that this statistics is log-linearly inversely related, for a broad range of scales, to the (G+C)-content in humans, chimpanzee and rhesus macaque. At low distances, several chromosomes of rhesus macaque and few of chimpanzee showed an additional exponential decay with very low levels of (G+C)-content (presumably a vestige of a primate paleogenome). We also

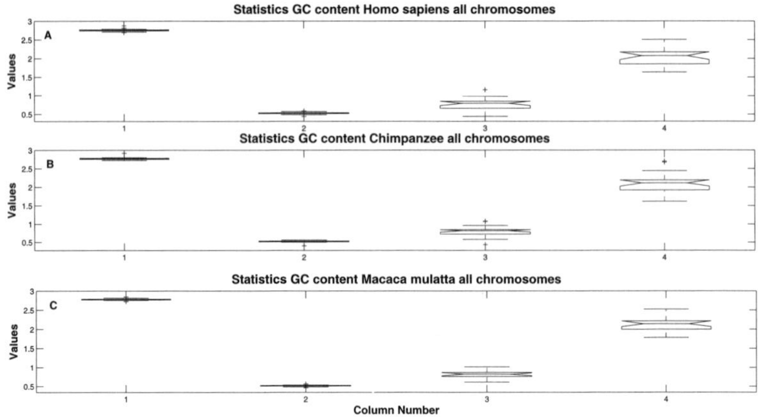

Figure 3. Box-whisker plots of the mean, variance, skewness and kurtosis of all PDFs of all chromosomes in (A) human; (B) chimpanzee, and (C) rhesus macaque. The center of error-bar is the mean value of local points. The maximum and minimum correspond to mean value standard deviation. The pluses are outliers.

determined the distribution of window sizes for which a given count of GC duplets is found along an entire chromosome. By doing so, we were able to visualize how the GC duplet is distributed along a given chromosome. This distribution is also directly related to the (G+C)-content. The probability distribution of (G+C)-content in all primate chromosomes follows a quasi-lognormal distribution that can be fitted to the Black and Scholes model (not including the tails). This model should be regarded as a null hypothesis of random variation of (G+C)-content. The potential use of this model is that several evolutionary hypothesis can be tested i.e., backward and/or forward predictions can be made. Then the evolution of these three species has been exponential in this regard (see Eq. (2)). Thus both the log-linear negative correlation between $LADGC$ and (G+C)% and the log-Gaussian distribution summarize GC-rich, GC-poor and GC-intermediate regions in a given chromosome. This type of distribution would be in agreement with neutral molecular evolution but the actual deviations from the log-normal as captured by long tails, skewness and kurtosis indicate strong positive natural selection effects. In contrast, the variance has remained approximately constant. There are not statistical differences between chromosomes of a given primate species (Fig. 3). Differences among chromosomes of different species lie mostly in the skweness, kurtosis and less often in the mean of the distributions. However, the most visible changes of the PDFs are shifts of the mean of some distributions, but this means significant changes in evolutionary terms. In turn this indicates that natural selection favors nonlinear relationships among the distance series of (G+C)-content. According to the shuffled lognormal control, the GC-intermediate regions of chromosomes are subjected mostly to neutral mutations. Our description of (G+C)-content is continuous which is in sharp contrast to the discrete description of mammalian genomes[33]. Analyses of the complete human genome[2,46] have already dismissed the underlying hypothetical picture of sharp boundaries between long homogeneous fragments: (G+C)-content turns out to vary continuously and somewhat erratically, along chromosomes; there are no clear boundaries between GC-poor and GC-rich regions, and isochores (notably, the GC-rich ones) are not as homogeneous as proposed initially[2,52]. However, a highly significant spatial autocorrelation of (G+C)-content was found, with most of the structure detectable at a large (300-kb) scale[2], indicating indeed a relative local homogeneity in base composition along mammalian chromosomes. In addition, sequence-based analysis of the human genome indicates that its compositional structure obeys scale invariant statistics[46].

It has been shown that the pattern of substitution tends to decrease the (G+C)-content in the human genome[47]. However, this process is extremely slow: since the divergence between human and chimpanzee (about 6 Myrs ago), regions with more than 50% of (G+C) lost about 0.2%[40]. A possible explanation is that these regions may be ever related to recombination hotspots. For human and chimpanzee, recombination hotspots have high recombination rates (10-1000 times higher) at fine scale[53] (1-2kb) and local pattern of recombination evolved rapidly[54]. Moreover, recombination drove the (G+C)-content evolution though substitution pattern[40]. Thus, if recombination ever transient drove the (G+C)-content evolution in human and chimpanzee, because the rate in hotspots is higher, then it would be straightforward to estimate the local length L, for which $L = Nae^{bX_{GC}}$. The value L measures whether the region is GC-rich or not. For example, if L is very big, it means that N GC dinucleotides are distributed in a large space, and then, it would be a GC-poor region.

Despite the fact that different chromosomes have experienced different evolutionary pressures and the structure of a given chromosome can be very different from others there are certainly constraints that each chromosome must satisfy. In order to test the direction of evolution, we also randomized the whole nucleotide sequence of all chromosomes. Most of the overlapping of these control distributions with the actual PDFs occurred at the middle. The reason behind this overlapping of the control distributions is due to the fact that the ratio of (G+C) over the size of each chromosome is practically constant for each species and throughout each of them. However, chromosome Y of both human and chimpanzees lie to the right of the control. The evolution of sexual chromosomes differs from that of autosomes because of differences in recombination rate, effective population sizes and mutation rates[17,55]. There are disputes about what leads to the variation of (G+C)-content in mammalian genomes, but the overall consensus is that the (G+C)-content in mammals is becoming homogenized[40,47]. Compared with investigations on (G+C)-content at large-scales ($10^3 \sim 10^4$ kb), the GC content at fine-scales ($10^0 \sim 10^1$ kb) has come into sight just recently[53,54]. In this context, the (G+C)-content evolution in a specific chromosome may be the result of several factors. In fact, the CpG $\rightarrow$ TpG transition (highly exponential) depends on local (G+C)-content[54] and CpG regional substitution is mainly affected by fine-scale neighbor (G+C)-content (1 2 kb)[53]. Genomic landscapes are not uniform across vertebrate chromosomes. Notably, the genomes of amniotes (reptiles, birds, and mammals) show a very strong heterogeneity of base composition along chromosomes. It has been shown that recombination is the major determinant of the evolution of (G+C)-content and therefore the selectionist models of isochore evolution have been dismissed. Different lines of evidence have suggested that (G+C)-content might be a consequence of the process of recombination[40]. Notably, analysis of the patterns of substitution in primate non-coding sequences have shown that recombination affects the relative rates of AT $\rightarrow$ GC and GC $\rightarrow$ AT substitutions[56,57,58]. This effect might result from the neutral process of biased gene conversion (BCG)[56,57,58]. According to this model, gene conversion (i.e. the copy/paste during meiotic recombination of one allele onto the other one at heterogozygous loci) is biased in favor of GC-alleles, which leads to an increase of probability of fixation of GC-alleles compared to AT-alleles). Thus, BGC should lead to an enrichment in (G+C)-content in genomic regions of high recombination compared to regions of low recombination. The effect of BGC is very similar to that of directional selection, and hence BGC can confound the tests that have been developed to detect selection in genomic sequences. The shift in (G+C)-content in the last 100 MY due to the current substitution pattern is less than 5%. Thus the regional (G+C)-content as measured in todays genomes reflects approximately those at about 100 MYA. However, given that recombination occurs essentially in hotspots that cover only 3% of the genome, that the BCG effect in hotspots is weak, and that hotspots have a short lifespan, is it possible that BGC drive the long-term evolution of the base composition of Mb-sequences. It has been shown that the impact of BGC in a given genomic fragment depends on the *local density* and intensity of recombination hotspots[40]. Both empirical data and theoretical calculations support the hypothesis that BGC has a strong impact on the evolution of (G+C)-content in amniotes. The BGC model[40] allowed to reject the alternative models for the evolution of isochores (mutation or selection). Thus, it seems that BGC may be the process at the origin and evolution of isochores. The impact of BGC on substitution patterns can be very strong, even in regions that are under selective pressure (coding sites or regulatory elements). In some cases, BGC overcomes purifying selection and leads to the fixation of deleterious AT $\rightarrow$ GC mutations[58]. Then, it is argued that along with mutation, selection and drift, BGC might be one of the major factors driving genome evolution and that it is essential to take this process into account if we want to be able to interpret sequences. Results on substitution patterns indicate that there is an erosion of the isochore structure of our genome[47]. However, the fact that BGC is presently weak in the human species does not exclude that BGC might have been more active in the past and might still be efficient in other species. We present results for 4 autosomes (chromosomes 1, 10, 17, and 19) and sexual chromosomes (X and Y). We could not analyze the Y chromosome in macaque because it has not been sequenced. Thus, the (G+C)-content at fine-scales can become relevant for different species despite the transition and substitution of CpG. We also argue that *LADGC* can shed some light on the problem of which sex, male or female, drove the (G+C)-content evolution. Our results do show that for (G+C)-content evolution in Y chromosome, negative selection is as important as positive selection. Several authors have proposed that GC-rich isochores might result from selection[35,36]. Other authors argue that the variation in mutation rate leads to this result[37,51] and the third suggests that biased gene conversion (BGC) is the main driven process[56,57,58]. It should be noted that the evolution of (G+C)-content affects all kind of sequences: exons, introns, intergenic regions, pseudogenes, transposable elements[2]. Thus, if selection is at work, this is not on the information content of genomic sequences, but simply on their (G+C)-content. However, the BGC process is presently too weak to maintain GC-rich isochores in the human genome. Further, the BGC model predicts the formation of very GC-rich isochores in species with higher effective population sizes or recombination rates. In other words, contrarily to the standard neutral mutational model, the BGC model predicts that the evolution of GC-rich isochores can be very rapid in species with large population size and high recombination rate. Thus, the BGC model provides a realistic explanation for the rapid origin of GC-rich isochores in the last common ancestor of amniotes, 310-350 Myrs ago[56,57,58]. It has been argued that recombination is the major determinant of the evolution of GC-content in primates[38]. Recombination affects the evolution of base composition in all kinds of non-coding sequences, transcribed or not. Thus, the human genome appears to be evolving toward a more homogeneous and less GC-rich base composition[47]. The (G+C)-content toward which the sequences are evolving is largely independent from the current (G+C)-content. Thus, the forces that have driven the evolution of isochores in mammalian genomes have changed both in intensity (these forces are not strong enough to maintain GC-rich isochores) and in localization along chromosomes. At the genomic scale of 1 Mb, recombination drives the evolution of (G+C)-content[40]. Thus the impact of recombination on sequence evolution is much stronger than the impact of (G+C)-content. Then, recombination has been driving the evolution of (G+C)-content in the human lineage during the last 6 MY. The substitutional pattern changed from an isochore-preserving to an isochore-degrading one. Isochores were established before the radiation of the eutherian mammals and have been subject to the process of homogeneization since then[40].

Final comment

In the XIX century, the hand was one of the symbols of the perfection of the human body, such as God has conceived it on the sixth day of Genesis. Although the Theory of Evolution finally prevailed, acknowledging mans simian lineage, some naturalists retained their view of the human being, merely shifting him from the status of the masterwork of Creation to that of the summit of evolution, its natural culmination. Evolution ceased to operate in the human species! Primates are all pentadactyls like their distant ancestors, but they present a derived feature: an opposable thumb on each limb. Evolution is not necessarily an improvement but simply a transformation. When human ancestors became bipedal, several MYA, the big toe evolved rapidly and lost its original mobility. The loss of its opposability is a derived characteristic that appeared late, linked together with the acquisition of bipedalism. Among other characteristics, it distinguishes man from his closest relative, the chimpanzee. Our most evolved finger is not our marvelous, archaic, opposable thumb, but our clumsy, recent big toe[58].

Acknowledgements

M.V.J. was financially supported by PAPIIT-IN107112, UNAM, México

References

1. J. J. Yunis, O. Prakash, The origin of man: A chromosomal pictorial legacy. *Science*, **215**:1525-1530 (1982).
2. E.S. Lander, et al. Initial sequencing and analysis of the human genome. *Nature*, **409**:860921 (2001).
3. T.S. Mikkelsen, L.W. Hillier, E.E. Eichler, M.C. Zody, D.B. Jaffe, et al. Initial sequence of the chimpanzee genome and comparison with the human genome. *Nature*, **437**:6987 (2005).
4. R.A. Gibbs, E.R. Mardis, K.A. Remington, R.L. Strausberg, J.C. Venter, J.M. Weinstock, R.K. Wilson, Evolutionary and biomedical insights from the *Rhesus Macaque* genome. *Rhesus Macaque* Genome Sequencing and Analysis Consortium. *Science*, **316**:22-234 (2007).
5. M.M. Miyamoto, J.L. Slightom, N. Goodman, Phylogenetic relations of humans and African apes from DNA sequences in the $\Psi\eta$globin region. *Science*, **238**:369-373 (1987).
6. F.C. Chen, W.H. Li, Genomic divergences between humans and other hominoids and the effective population size of the common ancestor of humans and chimpanzees. *Am. J. Hum. Genet.*, **68**:444456 (2001).
7. F.C. Chen, E.J. Vallender, H. Wang, C.S. Tzeng, W.H. Li, Genomic divergence between human and chimpanzee estimated from large-scale alignments of genomic sequences. *J. Hered.* **92**:481489 (2001).
8. M.W. Hahn, M.V. Rockman, N. Soranzo, D.B. Goldstein, G.A. Wray, Population genetic and phylogenetic evidence for positive selection on regulatory mutations at the factor VII locus in humans. *Genetics*, **167**:867877 (2004).
9. A. Boyko, S.H. Williamson, A.R. Indap, J.D. Degenhardt, R.D. Hernandez, et al. Assessing the evolutionary impact of amino acid mutations in the human genome. *PLoS Genet* **4**(5): e1000083 (2008).
10. C.D. Bustamante, A. Fledel-Alon, S. Williamson, R. Nielsen, M.T. Hubisz, et al. Natural selection on protein-coding genes in the human genome. *Nature*, **437**:11531157 (2005).
11. J.P. Demuth, T. De Bie, J.E. Stajich, N. Cristianini, M.W. Hahn, The Evolution of mammalian gene families. *PLoS ONE*, **1**(1): e85. doi:10.1371/ journal.pone.0000085 (2006).
12. M.A. Bakewell, P. Shi, J. Zhang, More genes underwent positive selection in chimpanzee evolution than in human evolution. *Proc. Nat. Acad. Sci. USA*, **104**:7489-7494 (2007).
13. S. Päävo, The mosaic that is our genome. *Nature*, **421**:409-412 (2003).

14. The International SNP Map Working Group, A map of human genome sequence variation containing 1.42 million single nucleotide polymorphisms. *Nature*, **409**:928-933 (2001).
15. N. Patterson, D.J. Richter, S. Gnerre, E.S. Lander, D. Reich, Genetic evidence for complex speciation of humans and chimpanzees. *Nature*, **441**:11031108 (2006).
16. A. Hobolth, O.F. Christensen, T. Mailund, M.H. Scierup, Genomic relationships and speciation times of human, chimpanzee, and gorilla inferred from a coalescent hidden Markov model. *PLoS Genetics*, **3**:0294-0304 (2007).
17. H. Skaletsky, et al., The male-specific region of the human Y chromosome is a mosaic of discrete sequence classes. *Nature*, **423**:825837 (2003).
18. S. Rozen, et al., Abundant gene conversion between arms of palindromes in human and ape Y chromosomes. *Nature*, **423**:873876 (2003).
19. S. Hughes, D. Zelus, D. Mouchiroud, Warm-blooded isochore structure in Nile crocodile and turtle. *Mol. Biol. Evol.*, **16**:15211527 (1999).
20. J.S. Mattick, I.V. Makunin, Non-coding RNA. *Human Mol. Gen.*, **15**:R17-R29 (2006).
21. S. Freeman, J. Herron, *Evolutionary Analysis.* Prentice Hall, Inc., Upper Saddle River, New Jersey (2001).
22. M.A. Batzer, P.L. Deininger, *Alu* repeats and human genomic diversity. *Nat., Rev. Gen.* **3**:370-379 (2002).
23. C.-K. Peng, S.V. Buldyrev, S. Havlin, M. Simons, H.E. Stanley, A. Goldberger, Mosaic organization of DNA nucleotides, *Phys. Rev. E*, **49**:16851689 (1994).
24. de Sousa M.V., Statistics of DNA sequences: a low frequency analysis, *Phys. Rev. E* **60**:59325937 (1999).
25. E.N. Trifonov, 3-, 10.5-, and 400-base periodicities in genome sequences. *Physica A*, **249**:511516 (1998).
26. I. López-Villaseñor, M.V. José, J. Sánchez, Three-base periodicity patterns and self-similarity in whole bacterial chromosomes. *Biophys. Biochem. Res. Comm.*, **325**:467-478 (2004).
27. C.-K Peng, S.V. Buldyrev, A.L. Goldberger, S. Havlin, F. Sciortino, M. Simons, H.E. Stanley, Long-range correlations in nucleotide sequences. *Nature*, **356**:168170 (1992).
28. W. Li, D. Holste, Universal $1/f$ noise, cross-overs of scaling exponents, and chromosome specific patterns of GC content in DNA sequences of the human genome. *Phys. Rev E*, **71**:041910:1-9 (2005).
29. I. Ioshikhes, S. Hosid, B.F. Pugh B.F., Variety of genomic DNA patterns for nucleosome positioning. *Genome Research*, **21**(11): 18631871 (2011).
30. D. Sosa, P. Miramontes, W. Li, V. Mireles, J.R. Bobadilla, M.V. José, Periodic distribution of a putative nucleosome positioning motif in human, nonhuman primates, and Archaea: Mutual information analysis, *Internat. J. Genomics*, http://dx.doi.org/10.1155/2013/963956.
31. G. Macaya, J.P. Thiery, G. Bernardi, An approach to the organization of eukaryotic genomes at a macromolecular level, *J. Mol. Biol.*, **108**:237254 (1976).

32. J. Filipski, J.P. Thiery, G. Bernardi, An analysis of the bovine genome by Cs2SO4-Ag density gradient centrifugation, *J. Mol. Biol.*, **80**:177197 (1973).
33. G. Bernardi, B. Olofsson, et al., The mosaic genome of warm-blooded vertebrates. *Science*, **228**:953958 (1985).
34. A. Eyre-Walker, Recombination and mammalian genome evolution. *Proc. Biol. Sci.*, **252**:237243 (1993).
35. A. Eyre-Walker, Evidence of selection on silent site base composition in mammals: Potential implications for the evolution of isochores and junk DNA. Genetics, 152:675683 (1999).
36. A. Eyre-Walker, L.D. Hurst, The evolution of isochores. *Nat. Rev. Genet.* **2**:549555 (2001).
37. K. Wolfe, P. Sharp, W. Li, Mutation rates differ among regions of the mammalian genome, *Nature*, **337**:283285 (1989).
38. N. Sueoka, Variation and heterogeneity of base composition of deoxyribonucleic acids: A compilation of old and new data, *J. Mol. Evol.*, **3**:31-40 (1961).
39. N. Sueoka, On the genetic basis of variation and heterogeneity of DNA base composition. *Proc. Nat. Acad. Sci. USA*, **48**:582-592 (1962).
40. L. Duret, P.F. Arndt, The impact of recombination on nucleotide substitutions in the human genome, *PLoS Genetics*, **4**(5):1-19 (2008).
41. D. Mouchiroud, G. DOnofrio, B. Aissani, G. Macaya, G. Gautier, et al., The distribution of genes in the human genome, *Gene*, **100**:181187 (1991).
42. L. Duret, D. Mouchiroud, Statistical analysis of vertebrate sequences reveals that long genes are scarce in GC-rich isochores, *J. Mol. Evol.*, **40**:308317 (1995).
43. K. Jabbari, G. Bernardi, CpG doublets, CpG islands and *Alu* repeats in long human DNA sequences from different isochore families, *Gene*, **224**:123128 (1998).
44. A.F. Smit, Interspersed repeats and other transposable elements in mammalian genomes, *Curr. Opin. Genet. Dev.*, **9**:657663 (1999).
45. M. Costantini, C. Oliver, F. Auletta, G. Bernardi, An isochore map of human chromosomes, *Genome Res.*, **16**:536-541 (2009).
46. N. Cohen, T. Dagan, L. Stone, D. Graur, GC Composition of the human genome: in search of isochores, *Mol. Biol. Evol.*, **22**(5):12601272 (2005).
47. L. Duret, M. Semon, G. Piganeau, D. Mouchiroud, N. Galtier, Vanishing GC-Rich isochores in mammalian genomes, *Genetics*, **162**:18371847 (2002).
48. W. Lee, L. Luo L., Periodicity of base correlation in nucleotide sequence, *Phys. Rev. E*, **56**(1):848-851 (1997).
49. G. Bernardi, Compositional constraints and genome evolution, *J. Mol. Evol.*, **24**:111 (1986).
50. N. Elango, S.H. Kim, NISC Comparative Sequencing Program, E. Vigota, S.V. Yi, Mutations of different molecular origins exhibit contrasting patterns of regional substitution rate variation, *PLoS Comput. Biol.*, **4**(2):1-10 (2008).
51. J.C. Hull, *Options Futures and Other Derivatives.* Prentice Hall. Chapter 11 (2000).
52. A. Nekrutenko, W.-H Li, Transposable elements are found in a large number of human protein-coding genes, *Trends Genet.*, **17**:619-621 (2001).

53. S.E. Ptak, D.A. Hinds, K. Koehler, B. Nickel, N. Pati, D.G. Ballinger, M. Przeworski, K.A. Frazer, S. Päävo, Fine-scale recombination patterns differ between chimpanzees and humans, *Nature Genetics*, **37**:429434 (2005).
54. W. Winckler, S.R. Myers, D.J. Richter, R.C. Onofrio, G.J. McDonald, et al., Comparison of fine-scale recombination rates in humans and chimpanzees, *Science*, **308**:107111 (2005)
55. G. Marais, Biased gene conversion: implications for genome and sex evolution, *Trends Genet.*, **19**:330338 (2003).
56. N. Galtier, G. Piganeau, D. Mouchiroud, L. Duret L., GC-content evolution in mammalian genomes: The biased gene conversion hypothesis, *Genetics*, **159**:907911 (2001).
57. N. Galtier, L. Duret L., Adaptation or biased gene conversion. Extending the null hypothesis of molecular evolution, *Trends Genet.* **23**:273-277 (2007).
58. N. Galtier, L. Duret, S. Glémin, V. Ranwez, GC-biased gene conversion promotes the fixation of deleterious amino acid changes in primates, *Trends Genet.* **25**:1-5 (2009).

MINING THE CONSTRAINTS OF PROTEIN EVOLUTION*

FERNANDO ENCINAS AND ANTONIO B. DE MIRANDA†

Laboratory of Computational and Systems Biology
Oswaldo Cruz Institute
Rio de Janeiro, RJ, 21040-360, Manguinhos, Brazil

The availability of different types of high-throughput data provides new opportunities for the identification of constraints that shape protein evolution; consequently, integrative computational approaches are essential to disclose the selective regimes that govern genomes. Combining text-mining analyses with other data mining techniques we have collected and analyzed data on various gene and protein characters to identify, classify and reveal existing associations between characters that may favor or hinder the rate at which proteins evolve. The use of latent constructs as an integrative procedure aimed to explain from a system perspective the relationships and the strength of these genome-wide characters allowed us to find that, at least for our data set, expression and structural constructs synthesize more the information in comparison to functional constructs. Overall, this work presents an integrative methodology intended to make the most of the available genomic data and describes an alternative framework to size the strength and links between determinants of protein evolution.

1. Introduction

The causes of variation in protein evolutionary rates have been a recurring topic of interest in the field of evolutionary biology [1,2,3]. Various comparative genomic analyses allowed the identification of individual factors, functional and structural, that favor or hinder the rate at which substitutions accumulate at nucleotide level [4,5,6]. Among these, although some examples against exist [7], gene expression has been indicated as major determinant of protein evolution [8,9].

The access to different types of biological information confirmed the complexity of organisms as living systems [10] and blurred the phenotypic boundaries at which selection operates [11]. Therefore, in the light of the

*This work was supported by the Institutional Cooperation of the Institute Oswaldo Cruz

†Corresponding author: antonio@fiocruz.br

ever-growing amount of high-throughput experimental data, there is a need to review the constraints that govern evolutionary change and to integrate related data to tackle protein evolution from an integrated perspective.

Integration of related data is particularly fruitful as it brings out the real value of individual data sets; however, to make this integration feasible and meaningful, it is necessary the application of advanced computational methods accompanied by mathematical and statistical approaches adequately braced with a theoretical framework [12].

Data mining as an applied science is a computer assisted process of analyzing large amounts of data to summarize it into valuable information [13]. Through a wide range of techniques, data mining approaches allow the recognition of patterns that are not instantly apparent and have the flexibility to offer both individual and system-level explanations [14].

In this work we present a combined methodology that, starting with a text-mining analysis, collected data on genome-wide variables that may constitute determinants of protein evolution. Hierarchical clustering and advanced factor analyses were used to explain the structure of the data set at a higher level.

2. Methods

2.1. *Text Mining*

Sixty one PDF research articles on protein evolution were manually downloaded from PUBMED and converted to plain text. An in-house code implemented in C language was used to process these plain texts by extracting sections of interest such as abstract, introduction and discussion. Resulting text files formed the document collection that was analyzed by the tm package [15] in R environment [16]. Text transformation, corpus construction and association between frequent terms were used to process the information from texts.

2.2. *Data collection*

We collected expression information including mRNA levels, translational efficiency and protein abundance for genes encoded in the genome of *Saccharomyces cerevisiae* for which comparative transcriptome/proteome analyses were conducted in [17]. Functional data consisting of dispensability and number of interactions were downloaded from (http://chemogenomics.stanford.edu/supplements/01yfh/files) and

Database of Interacting Proteins (http://dip.doe-mbi.ucla.edu/dip/) respectively. Structure-related information consisting of native structure classification, low complexity percentage and protein length were retrieved from Pedant Database (http://pedant.helmholtz-muenchen.de/genomes.jsp?cat egory=fungal). Finally, all genes were classified according to Gene Ontology classification using the Slim Mapper of Saccharomyces Genome Database (SGD) (http://www.yeastgenome.org/).

Pairs of orthologous genes between *Saccharomyces cerevisiae* and *Schizosaccharomyces pombe* were found using the stand-alone version of the InParanoid algorithm [18] and aligned using the ClustalW 2.0 [19] program with default parameters. Evolutionary rates, number of non-synonymous substitutions per synonymous site (dN) and synonymous substitutions per synonymous site (dS), between each ortholog pair, were estimated using the method of Nei and Gojobori implemented in MEGA 4 [20].

2.3. *Data Mining*

Summarization can be viewed as a compression of data into a smaller set of patterns retaining the maximum informative representation. We have used the following data mining techniques to summarize our data set:

2.3.1. *Hierarchical clustering of variables*

An ascendant hierarchical algorithm was used to arrange qualitative and quantitative variables in clusters of decreasing homogeneity. The homogeneity of a cluster is defined as the sum of correlation ratios (for qualitative variables) and the squared correlation (for quantitative ones) to a synthetic variable. The R package ClustOfVar [21] was used to implement the algorithm.

2.3.2. *Multiple Factor Analysis*

Multiple Factor Analysis (MFA) makes the synthesis of weighted Principal Component Analysis for quantitative variables and weighted Multiple Correspondence Analysis for qualitative variables making possible the analysis of variables structured into groups of related nature. Functions from the FactoMineR package [22] were used to perform MFA in six groups of variables arranged according to Table 1.

3. Results

3.1. *Genomic Variables derived from text identifiers*

An essential task, even for the simplest text mining analysis, is to find the terms that recur in a collection of documents. This enables the condensation of the whole content of information into a limited number of words. Frequent terms represent the identifiers of a collection therefore, finding significant associations between them (i.e., terms which co-occur) makes it possible to group and organize concepts to another level of valuable information. We have combined term frequency and term association analysis in a set of research articles to find new, potential constraints of protein evolution.

Thirty-one most frequent terms condensed the information of the texts and some of them visibly implied certain genomic information. In terms of co-occurrence counts, some terms presented significant correlations (Fig.1) that were very useful to support the intuitive attribution of one or more of them to a specific gene or protein character.

As a result, fourteen genomic variables, among gene and protein characters, were identified as prospective constraints of protein evolution and included as the focus of study in subsequent analyses. Table 1 presents the terms, the data type, nature and brief description of the genomic variables considered in the study.

3.2. *Pair/wise analyses reveal existing relationships between various genomic variables*

We collected or calculated the values of genomic variables listed in Table 1 for 442 protein-coding genes in the genome of the model organism *Saccharomyces cerevisiae* as detailed in Methods section.

We were especially interested in analyzing the behavior of genomic characters that might relate either to evolutionary variables or to expression variables. Thus, in our data set, a negative correlation (- 0.3307, p<9.55e-13) was found between the level of expression (mRNA level) and the number of non-synonymous substitutions (dN) and between translation efficiency and dn (- 0.2467, p<1.48e-07) (data not shown).

Later, turning into some of the "new" genomic characters we found in text-mining analysis, we also found that the instability index of a protein, a structure-related variable, presented high positive correlation with dN and a strong negative correlation with some expression variables such as

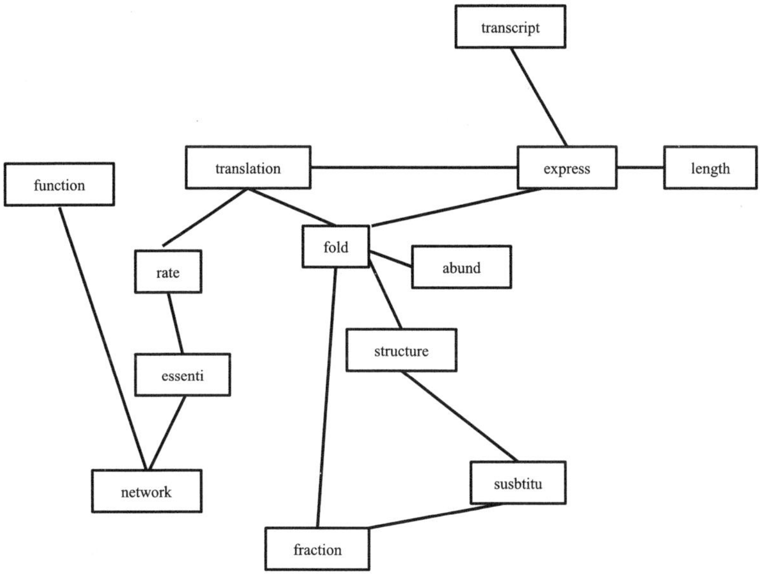

Figure 1. Term association network. Edges indicate significant co-occurrences between terms and provide the framework for concept classifications and collection identification.

translation efficiency. Although these preliminary results demonstrate the potential of text-mining approaches to generate novel information and reinforce the notion that more and strong genomic constraints do exist, they poorly contribute to our understanding on the evolution of proteins from an integrated perspective.

3.3. *Clustering of variables reveals the underlying structure of the data*

As clustering genomic variables in homogeneous groups would provide meaningful global information, we applied a hierarchical clustering algorithm based on agglomerative schemes to the mixture of quantitative and qualitative variables from our data set.

Aggregation levels demonstrated that four clusters would be enough to reveal the structure of the data. Thus, as depicted in the dendrogram of Fig. 2, most variables appeared to form clusters easily defined by the nature of the correlating variables.

Table 1. Detailed description of the origin, type and nature of genomic information.

Stem term	Gene/protein feature	Variable type	Nature
rate, substitu	number of synonymous substtitutions (dS)	continuous	Evolutive
rate, substitu	number of non-synonymous substitutions (dN)	continuous	Evolutive
express	mRNA level	continuous	Expression
abund	protein level	continuous	Expression
translation	translation efficiency	continuous	Expression
codon	codon adaptation index (CAI)	continuous	Expression
length	protein length	continuous	Structural
structure	native structure	categorical	Structural
structure	instability index	continuous	Structural
structure	stability	categorical	Structural
region, structure	low complexity percentage	continuous	Structural
network	number of interactions	continuous	Functional
essenti	essentiality	categorical	Functional
essenti	dispensability	continuous	Functional

In terms of homogeneity, low-complexity, instability index and stability, three structure-related variables, clearly grouped in the same cluster. Essentiality and number of interactions grouped together with native structure in a second cluster. Protein abundance, translation efficiency and protein length, all related to the translation machinery linked in a third cluster. Finally, evolutive variables dS and dN grouped together with a expression related variable, mRNA level.

3.4. *Latent constructs are useful to integrate genomic data and provide a descriptive system perspective*

Grouping genomic variables into clusters allowed us to grasp the underlying structure of our data set; nevertheless, no information is provided about the type or direction (positive or negative) of existing relationships between variables.

Aimed to analyze simultaneously multiple sets of variables, Multiple Factor Analyses (MFA) use an arrangement of variables in groups of related nature to evaluate the influence of each group and to reveal if there is any relationship between such groups. A descriptive concept or latent

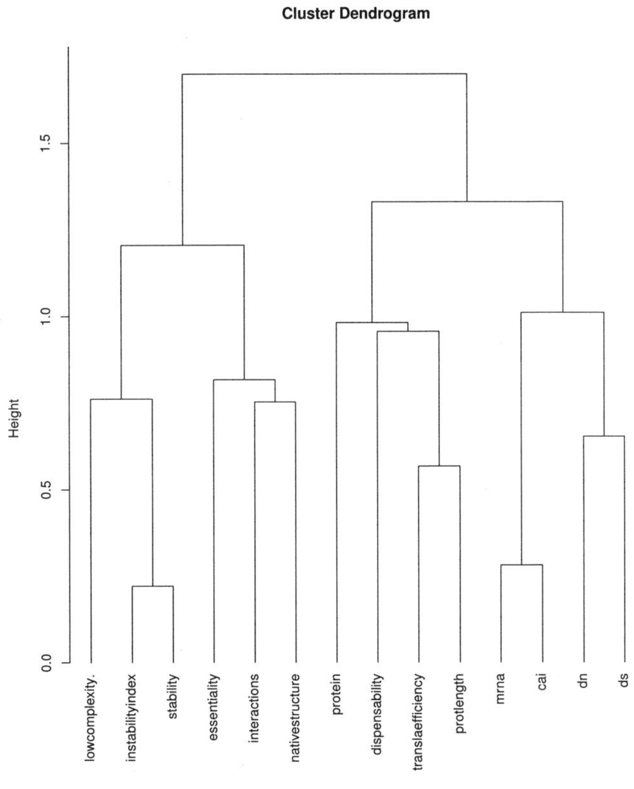

Figure 2. Hierarchical clustering of variables. Four groups of correlating variables reveal the structure of the data set.

construct can be associated to each group in order to attain a system-level interpretation.

Six groups of related genomic variables were created as detailed in Methods section and Table 1 to be analyzed by functions included in the package FactoMiner [22]. Figure 3 shows the quality representation of each group of variables clearly separated in the axes projection.

The distance between groups suggests, as we expected, that each of them represents distinctive but integrated information on three major determinants of protein evolution: structure, expression and function. Structural constructs (struct and structcat) appeared to have high coordinates on the first axis, whereas expression construct (express) had the highest coordi-

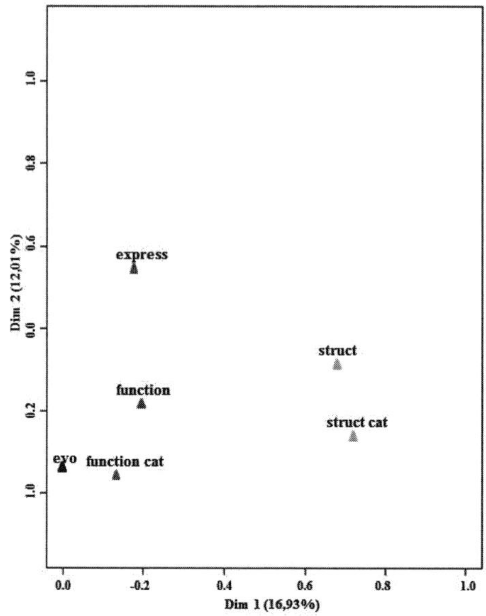

Figure 3. Quality representation of latent constructs. Related data on three major determinants of protein evolution can be integrated using latent constructs that synthesize distinct information reliably.

nates to the second axis. Both located distant from the evolution construct (evo), which has been set as supplementary group, and from the point of origin showing that these groups of variables helped the most in the synthesis of the information. Function constructs (function and functioncat) on the other hand, although separated equally, they presented low coordinates on the first two axes, consequently little power of discrimination.

Individual coordinates for members of each group provide the definitive descriptive system perspective proposed throughout the work. Figure 4 presents a plot of the factorial map of a correlation circle in which it is noticeable, on the one hand, the opposition between expression variables and the number of non-synonymous substitutions; on the other hand, the high correlation between structure-related variables (low complexity percentage and instability index) and finally, the positive association between translation efficiency and synonymous substitutions both opposing to the length of a protein and to its dispensability.

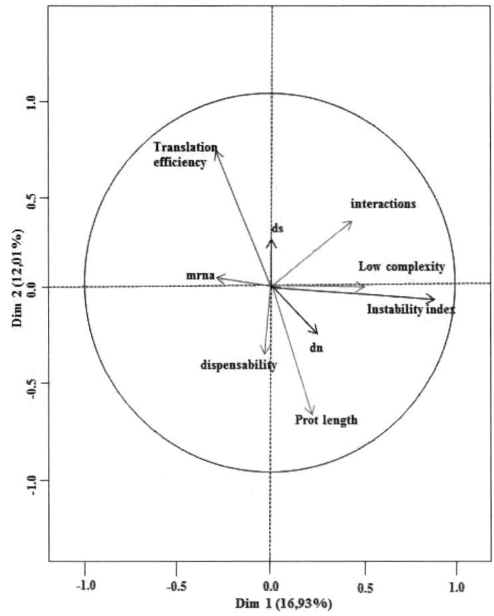

Figure 4. Circle of correlations. The individual coordinates show graphically the relationships between variables

4. Discussion

Text mining has emerged as a leading-edge technology that takes advantage of techniques of information retrieval, natural language processing and data mining, to cope with the non-trivial task of dealing with the ambiguity in language and the unstructured nature of written documents [23]. In biology, its applications vary from drug discovery [24] and disease-gene associations [25] to the systematic review of protocols and analysis of trends in molecular biology [26].

Assigning the identifiers of our text collection to gene or protein features, we have been able to distinguish variables that appear to be unacknowledged constraints of protein evolution. The instability index, the translation efficiency and percentage of low complexity regions in a protein strongly correlate with the number of non- synonymous substitutions (dN) accumulated.

In the same direction, our results showed that the level of activation of a gene, expressed by its mRNA level, also correlate negatively with dN, supporting the view that highly expressed genes tend to evolve at a slow rate. It has been suggested that evolution progresses through changes in protein expression rather than sequence [27]; therefore, gene expression constitutes the "key" element in our understanding of protein evolution.

While this "key" is generally interpreted as the unequivocal negative association between these variables (dN and mRNA level), it can be also argued that it holds a simplistic view of what gene expression really represents and especially that restricts the action of selection to a narrow margin. Gene expression can be explained by the level at which one exon is transcribed, by the number of translations per transcript or by the level of structurally functional proteins in the cell. Thus, transcription, translation and protein abundance might be important to different extents and selection may have a role at different stages accordingly [28].

Due to the requirement to form and maintain the definitive active (as in the case of enzymes) site that probably exerts a strong selective pressure on a protein to adopt just one stable and conserved fold, protein structures are generally regarded as "fossil records" of molecular evolution [29]. However, as more protein structures become available and more structural genomics projects are generating new and unprecedented information, a major biological question is how a system's physical properties influence its capacity to evolve.

On the one hand, it has been shown that contrary to the traditional view that protein function equates with a stable three-dimensional structure, many gene sequences in eukaryotic genomes encode large segments or even entire proteins that lack a well-structured three-dimensional fold and moreover, some of these regions can be highly conserved between species [30,31]. On the other hand, there is strong evidence that the capacity of one protein to evolve is enhanced by the mutational robustness conferred by extra stability [32].

As we see, the availability of different types of high-throughput biological data serves as evidence of the complexity that living organisms have reached in millions of years under the influence of selective forces that shaped their evolutionary history. However, the real informative value of individual data sets is truly appreciated only if these are combined or integrated in a single framework.

Data-mining techniques can provide such a framework and constitute an ideal option for the analysis of "different-but related" data sets. In this way, latent concepts play important roles in the theoretical work of many fields [33] and we took advantage of their virtue to act as components of both individual and system-level explanations to review the classic views of protein evolution in the light of the genomic data available.

The classic view would state that protein evolution is basically affected by selection acting on protein structure and function; moreover, mRNA level, as proxy for gene expression, has been pointed to have a major influence on the evolution of the corresponding gene. In contrast, our approach prioritized the quest for general over particular determinants of protein evolution.

5. Conclusions

Life sciences are facing the challenge of handling and analyzing biological information through the use of more innovative computational methods to respond the growing need of making sense of large amounts of experimental data. Integration of related data is useful to this purpose as it brings out the real value of individual data sets and, if linked to a theoretical framework, it provides the system-level perspective to review classical assumptions and test new hypothesis.

In this work, combining text-mining techniques with simple correlation analyses we have been able to identify genomic features that appear to be overlooked when studying the rates of protein evolution and the targets of selective forces. Translation efficiency, structural instability and low complexity regions showed strong correlation with the rate at which a protein evolves.

Latent constructs were used as an alternative to integrate related genomic information and to approach the evolution of biological organisms as systems formed by different components. We could recognize clearly distinct constructs that each in turn bring different information and found that, in general, expression and structural constructs explain more our data set in comparison to functional constructs. Overall, our results suggest that rather than taking mRNA levels as major determinants of protein evolution, other expression-related variables such as translation efficiency should be considered.

References

1. C. Pal, B. Papp and M.J.Lercher, *Nat. Rev. Genet.* **7**, 5 (2006)
2. J.I. Lucas-Lledo and M. Lynch, *Mol. Biol. Evol.* **26**, 5 (2009)
3. X. Du X, D.J. Lipman and J.L. Cherry, *Genome Biol. Evol.* **5**, 3 (2013)
4. S. Vieira-Silva, M. Touchon, S.S. Abby and E.P. Rocha, *Proc. Natl. Acad. Sci.* **108**,50 (2011)

5. J. Coulombe-Huntington and Y. Xia, *PLoS Comput. Biol.* **8**, 10 (2012)
6. S. Chakraborty, B. Kahali and T.C. Ghosh, *BMC Syst. Biol.* **12**,4 (2010)
7. I. Tirosh and N. Barkai, *Trends Genet.* **24**, 3 (2008)
8. D.A. Drummond, A. Raval and C.O. Wilke, *Mol. Biol. Evol.* **23**, 2 (2006)
9. J.F. Gout, D. Kahn and L. Duret, *PLoS Genet.* **6**, 5 (2010)
10. B. Berger, J. Peng and M. Singh, *Nat. Rev. Genet.* **14**, 5 (2013)
11. E. Koonin and Y. Wolf, *Nat. Rev. Genet.* **11**, 7 (2010)
12. P.V. Gopalacharyulu, E. Lindfors, C. Bounsaythip, T. Kivioja, L. Yetukuri, J. Hollmn and M. Oresic, *Bioinformatics* **21**, 1 (2005)
13. H. Bensmail and A. Haoudi, *J Biomed. Biotech.* **2**, (2005)
14. D. Rebholz-Schuhmann, A. Oellrich and R. Hoehndorf, *Nat. Rev. Genet.* **13**, 12 (2012)
15. I. Feinerer, K. Hornik and D. Meyer, *J. Stat. Soft.* **25**, 5 (2008)
16. R. Ihaka and R. Gentleman,*J. Comp. Graph. Stat.* **5**, 3 (1996)
17. V.L. MacKay, X. Li, M.R. Flory, E. Turcott, G.L. Law, K.A. Serikawa, X.L. Xu, H. Lee, D.R. Goodlett, R. Aebersold, L.P. Zhao and D.R. Morris, *Moll. Cell. Proteomics.* **3**, 5 (2004)
18. G. Ostlund, T. Schmitt, K. Forslund, T. Kostler, D.N. Messina, S. Roppa, O. Frings and E.L. Sonnhammer. *Nucleic Acids Res.* **38** (2010)
19. J.D. Thompson, D.G. Higgins and T.J. Gibson, *Nucleic Acids Res.* **22** (1994)
20. K. Tamura, J. Dudley, M. Nei and S. Kumar, *Mol. Biol. Evol.* **24**, 8 (2007)
21. M. Chavent, V. Kuentz-Simonet, B. Liquet and J. Saracco, *J. Stat. Soft.* **50**, 13 (2012)
22. S. Le, J. Josse and F. Husson, *J. Stat. Soft.* **25**, 1 (2008)
23. D. McDonald and U. Kelly, *JISC*(2012)
24. C. Plake, M. Schroeder, *Curr. Pharm. Biotechnol.* **12**, 3 (2011)
25. H. Al-Mubaid and R.K. Singh, *Int. J. Bioinform. Res. Appl.* **6**, 3 (2010)
26. M. Krallinger, R.A. Erhardt, A. Valencia, *DDT.* **10**, 6 (2005)
27. C. Bustamante, A. Fledel-Alon, S. Williamson, R. Nielsen, M.T. Hubisz, S. Glanowski, D.M. Tanenbaum, T.J. White, J.J. Sninsky, R.D. Hernandez, D. Civello, M.D. Adams, M. Cargill and A.G. Clark, *Nature* **437**, 7062 (2005)
28. E.P. Rocha, *Trends Genet.* **22**, 8 (2006)
29. A. Andreeva and A.G. Murzin, *Curr. Opin. Struct. Biol.* **16**, 3 (2006)
30. J. Nilsson, M. Grahn and A.P. Wright, *Genome Biol.* **12**, 7 (2011)
31. H.J. Dyson and P.E. Wright, *Nat. Rev. Mol. Cell. Biol.* **6**, 3 (2005)
32. J.D. Bloom, S.T. Labthavikul, C.R. Otey and F.H. Arnold, *Proc. Natl. Acad. Sci.* **103**, 15 (2006)
33. K. Bollen, *Annu. Rev. Psychol.* **53**, 605 (2002)

ENTROPY MEASURES BASED METHOD FOR THE CLASSIFICATION OF PROTEIN DOMAINS INTO FAMILIES AND CLANS

N. CARELS

Laboratory of Functional Genomics and Bioinformatics,
Oswaldo Cruz Institute, 21.045-900
Rio de Janeiro, Brazil

CECILIA F. MONDAINI

Institute of Mathematics,
UFRJ/CCMN, 21.941-909,
Rio de Janeiro, Brazil

R. P. MONDAINI*

Alberto Luiz Coimbra Institute,
UFRJ/CT/COPPE, 21.941-972
Rio de Janeiro, Brazil
E-mail: mondaini@cos.ufrj.br

Protein Function is conserved in Protein domains. These are able to perform independent folding and are approximately conserved among species. A major goal in the study of proteins is the identification and characterization of domains and domain families. Some algorithms have appeared in the scientific literature to obtain information on domains from three-dimensional structure of proteins, but many of these results do not agree with those defined by expert biologists. A time statistical analysis of aminoacid sequences should then be performed in order to improve this situation or to clarify the origin of disagreement.

1. Introduction

The universal concept of entropy is of fundamental significance for mathematical and biological sciences. The concept was introduced in the applications of Thermodynamics but it is generic enough to unveil the nature of

*Corresponding author.

physical processes occurring in systems from the galaxies to the biosphere of the Earth. Entropy is always transferred to the surrounding medium and an experimentalist gets information from the decrease of his/her entropy. This decrease of entropy is responsible for all creations of life, including sciences and art [1]. First of all, we have to note that the inexistence of a natural ordering or underlying nature should be circumvented in order to favour the conventional statistical calculations. In this note, we choose to introduce entropy measures in order to reach this aim [2]. We have also to note that the distribution of domains can be seen as one aspect of biological complexity in evolution. We usually think on protein domains as discrete structural units with 20 to 400 aminoacids. However, it is known that slightly less than one third of the domains are discontinuous. The domains are usually grouped into families and stored in databases [3]. This improves the accuracy of the functional inference as well as the clustering of new members. We work here with the domain sequences which have been grouped on a PFAM database (version 27.0 - March 2013) [4]. A probabilistic model which represents the aminoacid structure through hidden Markov models (HMM) [5,6] is responsible for the organization of this database. The PFAM database has currently $14 \cdot 10^3$ families and the number of domains on them goes from tens to thousands [4]. The size variation of these domains is between ~ 20 to > 200 aminoacids. It is also supposed that domain families, somehow related to their structure and probability of occurrence of aminoacids, can be gathered into "superfamilies" named clans. We work with 499 clans of this database. In the present work, we propose to set the scenario to question the biological relevance of the clan concept by using entropy measures based methods. Detailed statistical results based on these methods will be published elsewhere.

2. Entropy-based methods for the assessment of Protein databases

We look for families with a specified number of disjoint blocks. Each block has m lines and n columns, with a total of $m \cdot n$ aminoacids. We first consider m sequences with n_k aminoacids ($k = 1, \ldots, m$) each. In order to form a block as described above, we have to delete ($n_k - n$ aminoacids from each sequence, where n is given by

$$n = \min\{n_1, n_2, \ldots, n_m\}. \tag{1}$$

Let $n_j(a)$ be the number of times that the aminoacid a, $a = 1, \ldots, 20$, will appear in the jth column of the block. The probability of occurrence of this aminoacid is

$$p_j(a) = \frac{n_j(a)}{m}, \quad j = 1, \ldots, n. \tag{2}$$

It is easy to see from eq. (2) that only sequences with $m \geq 20$ will be admissible.

We then have

$$\sum_a p_j(a) = 1, \quad \forall j. \tag{3}$$

It is convenient to think on $p_j(a)$ as the components of a vector p_j of the probability space:

$$p_j^T = (p_j(1), \ldots, p_j(20)), \tag{4}$$

where p_j^T denotes the transpose of the vector p_j.

We choose a Sharma-Mittal type of entropy measure [7] to associate to this probability distribution:

$$h_j(r,s) = -\frac{1}{2s \ln 2} \sum_a [(p_j(a))^{r+s} - (p_j(a))^{r-s}]. \tag{5}$$

If a model probability distribution $q_j(a)$ is used to approximate $p_j(a)$, the information lost is given by

$$I_j(r,s) = \frac{1}{2s \ln 2} \sum_a q_j(a) \left[\left(\frac{p_j(a)}{q_j(a)} \right)^{r+s} - \left(\frac{p_j(a)}{q_j(a)} \right)^{r-s} \right]. \tag{6}$$

We also introduce the average p_j-entropy and average information lost as convenient entropy measures:

$$h(r,s) = \frac{1}{n} \sum_j h_j(r,s), \quad j = 1, \ldots, n, \tag{7}$$

$$I(r,s) = \frac{1}{n} \sum_j I_j(r,s), \quad j = 1, \ldots, n, \tag{8}$$

respectively.

The introduction of other probability distributions will lead to other entropy measures. We now consider the joint probability $P_{jk}(a,b)$ of finding the aminoacid a in column j and the aminoacid b in column k. It is convenient to think on $P_{jk}(a,b)$ as the elements of $n(n-1)/2$ square matrices P_{jk} of 20th order.

From eq. (3), we have

$$\sum_a \sum_b P_{jk}(a,b) = \sum_a P_{jk}(a) = \sum_a p_j(a) = 1, \tag{9}$$

for every j, k with $j < k$, $j = 1, \ldots, n-1$, $k = 2, \ldots, n$. The entropy measure corresponding to this joint probability is

$$H_{jk}(r,s) = -\frac{1}{2s\ln 2}\sum_a \sum_b [(P_{jk}(a,b))^{r+s} - (P_{jk}(a,b))^{r-s}]. \tag{10}$$

Another useful measure is the mutual information distance, which is a measure of dependence of the columns of the block $m \times n$:

$$M_{jk}(r,s) = \frac{1}{2s\ln 2}\sum_a \sum_b P_{jk}(a,b)\left[\left(\frac{P_{jk}(a,b)}{p_j(a)p_k(b)}\right)^{r+s-1} - \left(\frac{P_{jk}(a,b)}{p_j(a)p_k(b)}\right)^{r-s-1}\right]. \tag{11}$$

The corresponding average measures will be given by

$$H(r,s) = \frac{2}{n(n-1)}\sum_j \sum_k H_{jk}(r,s), \quad j < k,\ j = 1, \ldots, n-1,\ k = 2, \ldots, n. \tag{12}$$

$$M(r,s) = \frac{2}{n(n-1)}\sum_j \sum_k M_{jk}(r,s), \quad j < k,\ j = 1, \ldots, n-1,\ k = 2, \ldots, n. \tag{13}$$

From eqs. (10) and (11) we also introduce another entropy measure:

$$H_{jk}(r,s) - M_{jk}(r,s) \le H_{jk}(r,s) \tag{14}$$

and its normalized version, the Jaccard measure

$$0 \le J_{jk}(r,s) = 1 - \frac{M_{jk}(r,s)}{H_{jk}(r,s)} \le 1, \tag{15}$$

with the corresponding average Jaccard measure given by

$$J(r,s) = \frac{2}{n(n-1)}\sum_j \sum_k J_{jk}(r,s). \tag{16}$$

We now take the limits $r \to 1$, $s \to 0$ over the entropy measures given in eqs. (5) to (15). These limit forms will correspond to Shannon versions and we have, from eq. (5)

$$h_j(1,0) = -\sum_a p_j(a)\log_2 p_j(a). \tag{17}$$

With respect to these limits, eq. (6) turns into

$$(KL)_j = I_j(1,0) = \sum_a p_j(a) \log_2 \left(\frac{p_j(a)}{q_j(a)} \right), \tag{18}$$

which is known as Kullback-Leibler divergence in the scientific literature. This is not a true distance in the probability space, since it lacks symmetry.

Let us now introduce the last probability distribution to be considered in this work, the overall probability distribution, given by

$$q_j(a) = q(a), \quad \forall j, \tag{19}$$

where

$$q(a) = \frac{1}{n} \sum_j p_j(a), \quad j = 1, \ldots, n. \tag{20}$$

From eqs. (3) and (20), we get

$$\sum_a q(a) = 1, \quad a = 1, \ldots, 20. \tag{21}$$

We can now construct a true distance entropy measure. We use the overall probability distribution $q(a)$ to write

$$\frac{1}{n} \sum_j I_j(1,0) = h_q - h, \quad q_j(a) = q(a), \quad \forall j, \tag{22}$$

where

$$h_q = - \sum_a q(a) \log_2 q(a) \tag{23}$$

and

$$h = \frac{1}{n} \sum_j h_j(1,0). \tag{24}$$

Eq. (22) can now be written as

$$\frac{1}{n} \sum_j I_j(1,0) = h \left(\sum_j \frac{1}{n} p_j \right) - \sum_j \frac{1}{n} h_j \tag{25}$$

and this is the Jensen-Shannon distance

$$JS = h \left(\sum_j w_j p_j \right) - \sum_j w_j h_j \tag{26}$$

with constant weights $w_j = \frac{1}{n}$.

We then have

$$(JS)_{w_j=\frac{1}{n}} = \frac{1}{n}\sum_j I_j(1,0), \quad q_j(a) = q(a). \tag{27}$$

The normalized version of this entropy measure is given by

$$0 \leq (NJS)_{w_j=\frac{1}{n}} = 1 - \frac{h}{h_q} \leq 1. \tag{28}$$

The entropy corresponding to the probability $P_{jk}(a,b)$ and the mutual information distance, as defined in eqs. (10) and (11), turn into

$$H_{jk}(1,0) = -\sum_a \sum_b P_{jk}(a,b) \log_2 P_{jk}(a,b) \tag{29}$$

and

$$M_{jk}(1,0) = \sum_a \sum_b P_{jk}(a,b) \log_2 \left(\frac{P_{jk}(a,b)}{p_j(a)p_k(b)}\right), \tag{30}$$

respectively. We also obtain the following additional identity in this limit:

$$M_{jk}(1,0) \equiv h_j(1,0) + h_k(1,0) - H_{jk}(1,0). \tag{31}$$

The definitions given in eqs. (12), (13) and (16) remain true in this limit. In particular, we have for the Jaccard distance in the Shannon limit:

$$0 \leq J_{jk}(1,0) \equiv 1 - \frac{M_{jk}(1,0)}{H_{jk}(1,0)} \leq 1. \tag{32}$$

3. Methodology of *in silico* experiments

We choose to analyze 8 clans among the 499 clans of the database. On each clan we choose 8 families and on each family we take 10 blocks with $m = 100$ lines and $n = 100$ columns. This has been done by testing the robustness of some entropy measures by exhaustive calculations on blocks obtained from the same family (the intra family distances of columns). In Fig. 1 we can see that $n = 100$ columns are enough to show effects of stability on Shannon Entropy and Kullback-Leibler divergence for the shortened PF00003 protein family.

In other words, it is important for statistical consistency to know the size of the sequence at which we get the stabilization to zero of the entropy gain (KL). Otherwise, we could attribute the successful measures of entropy variation to trivial parameters without any correlation with protein structure.

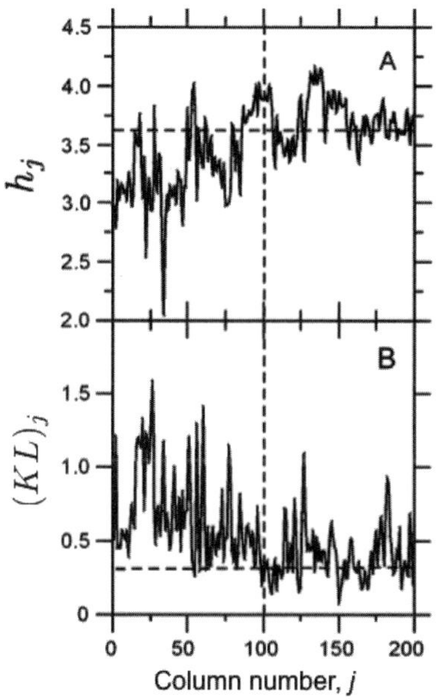

Figure 1. The robustness of entropy measures for blocks with $n \geq 100$ columns can be checked from this figure. The calculations have been done for the PF00003 protein family.

Since the size of the sequence for stabilization is now guaranteed and it is realized by choosing blocks of $10^2 \times 10^2$ aminoacids, we are free to test the behaviour of some entropy measures. In Fig. 2 we show the histograms for Jensen-Shannon distance, average mutual distance and average Jaccard distance, defined in eqs. (27), (12) and (16), respectively. We have used $5,827$ domain families of the PFAM database and we take one block per family. These domain families contain sequences of sufficient size and number.

We shall note that the average Jaccard distance looks very convenient due to the approximate symmetry of its histogram, to undertake any kind of exhaustive statistical analysis in the Shannon limit.

The statistical parameters of the distributions which we use are the average value and the standard deviation. We compare the intra-family distances (d_{intraf}), the inter-family distances of the same clan (d_{interfsc}) and the inter-family distances of different clans (d_{interfdc}). The usual idea of a clan would lead to

$$d_{\text{intraf}} < d_{\text{interfsc}} < d_{\text{interfdc}}. \tag{33}$$

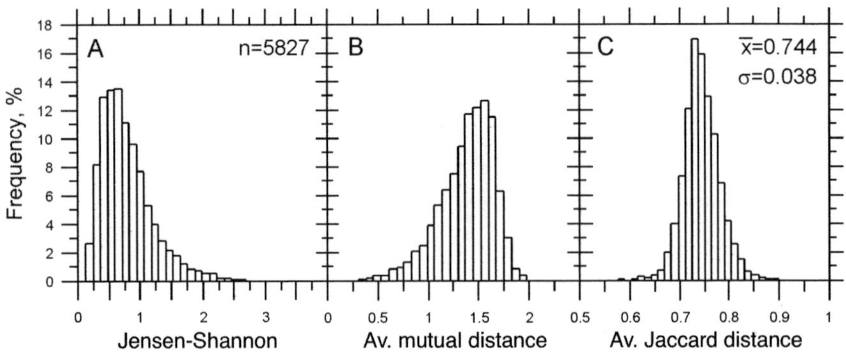

Figure 2. Histograms of some entropy measures corresponding to 5,827 blocks of size $10^2 \times 10^2$ chosen from 5,827 families; one block per family.

However, the results which have been used until now with the Jaccard distance do not confirm this, but

$$d_{\text{intraf}} < d_{\text{interfsc}} \approx d_{\text{interfdc}} \tag{34}$$

and this will correspond to challenge the relevance of the clan concept.

If this concept is really related to existence of the relationship with the three-dimensional structure of the families, we think that we should use another definition of entropy measures in order to circumvent this discrepancy. We hope that entropy measures with two parameters r, s could be expected to separate intra-family distances, inter-family distances of the same clan and inter-family distances of different clans by different pairs of values r, s or different parts in the rs-plane for each entropy measure. This will also be treated carefully in a forthcoming publication.

Appendix. The expression of Shannon Entropy measures by vectors and matrices of the probabilistic space

From eqs. (17) and (18), we can write:

$$h_j(1,0) = -p_j^T L_2 p_j$$

and

$$I_j(1,0) = p_j^T (L_2 p_j - L_2 q_j)$$

where p_j^T, $(L_2 p_j)^T$ and $(L_2 q_j)^T$ are the vectors

$$p_j^T = (p_j(1), \ldots, p_j(20))$$

$$(L_2p_j)^T = (\log_2 p_j(1), \ldots, \log_2 p_j(20))$$
$$(L_2q_j)^T = (\log_2 q_j(1), \ldots, \log_2 q_j(20))$$

Eqs. (29) and (30) can be written

$$H_{jk}(1,0) = -Tr(P_{jk}^T L_2 P_{jk})$$
$$M_{jk}(1,0) = Tr[P_{jk}^T (L_2 P_{jk} - L_2(p_j p_k^T))]$$

where P_{jk}, L_2P_{jk} and $L_2(p_jp_k^T)$ stand for $n(n-1)/2$ square matrices of 20th order with elements $P_{jk}(a,b)$, $\log_2 P_{jk}(a,b)$ and $\log_2(p_j(a)p_k(b))$, respectively.

An example with $n = 8$ and 3 aminoacids Q, D, N can be written as

$$P_{jk} = \begin{pmatrix} P_{jk}(Q,Q) & P_{jk}(Q,D) & P_{jk}(Q,N) \\ P_{jk}(D,Q) & P_{jk}(D,D) & P_{jk}(D,N) \\ P_{jk}(N,Q) & P_{jk}(N,D) & P_{jk}(N,N) \end{pmatrix}$$

$$p_jp_k^T = \begin{pmatrix} p_j(Q)p_k(Q) & p_j(Q)p_k(D) & p_j(Q)p_k(N) \\ p_j(D)p_k(Q) & p_j(D)p_k(D) & p_j(D)p_k(N) \\ p_j(N)p_k(Q) & p_j(N)p_k(D) & p_j(N)p_k(N) \end{pmatrix}$$

and analogous formulae for the corresponding matrices L_2P_{jk} and $L_2(p_jp_k^T)$.

There are $8(8-1)/2 = 28$ matrices of each form above and these are given by the arrays

$$\begin{matrix} P_{12} & P_{13} & P_{14} & P_{15} & P_{16} & P_{17} & P_{18} \\ & P_{23} & P_{24} & P_{25} & P_{26} & P_{27} & P_{28} \\ & & P_{34} & P_{35} & P_{36} & P_{37} & P_{38} \\ & & & P_{45} & P_{46} & P_{47} & P_{48} \\ & & & & P_{56} & P_{57} & P_{58} \\ & & & & & P_{67} & P_{68} \\ & & & & & & P_{78} \end{matrix}$$

$$\begin{matrix} p_1p_2^T & p_1p_3^T & p_1p_4^T & p_1p_5^T & p_1p_6^T & p_1p_7^T & p_1p_8^T \\ & p_2p_3^T & p_2p_4^T & p_2p_5^T & p_2p_6^T & p_2p_7^T & p_2p_8^T \\ & & p_3p_4^T & p_3p_5^T & p_3p_6^T & p_3p_7^T & p_3p_8^T \\ & & & p_4p_5^T & p_4p_6^T & p_4p_7^T & p_4p_8^T \\ & & & & p_5p_6^T & p_5p_7^T & p_5p_8^T \\ & & & & & p_6p_7^T & p_6p_8^T \\ & & & & & & p_7p_8^T \end{matrix}$$

and analogous arrays for the corresponding matrices L_2P_{jk} and $L_2(p_jp_K^T)$.

References

1. M. V. Volkenstein, *Entropy and Information*, Birkhäuser, Basel, Boston, Berlin (2009).
2. W. R. Atchley et al., *Correlations among aminoacid sites in bHLH Protein Domains: An Information Theoretic Analysis*, Mol. Biol. Evol. **17** (1), 164-178 (2000).
3. J. Lin and B. Rost, *Domains, motifs and clusters in the protein universe*, Curr. Opin. Chem. Biol. **7**, 5-11 (2003).
4. M. Punta et al., *The Pfam Protein Families Database*, Nucleic Acids Res. **40**, D290-D301 (2012).
5. R. Durbin, S. Eddy, A. Krogh and G. Mitchison, *Biological Sequence Analysis, Probabilistic Models of Proteins and Nucleic Acids*, Cambridge University Press (1999).
6. M. Borodovsky and S. Ekisheva, *Problems and Solutions in Biological Sequence Analysis*, 1st ed., Cambridge University Press (2006).
7. C. Beck, *Generalised information and entropy measures in physics*, Contemp. Phys. **50** (4), 495-510 (2009).

MODELLING OF POROUS ELASTIC AND VISCOELASTIC MEDIA AND ITS APPLICATION TO THE BRAIN

R. BEGG, J. MURLEY, AND M. KOHANDEL
Department of Applied Mathematics,
University of Waterloo,
200 University Ave. W.
Waterloo, ON N2L 2G1, Canada
E-mail: rbegg@staffmail.ed.ac.uk, jmurley@uwaterloo.ca, kohandel@uwaterloo.ca

S. SIVALOGANATHAN
Centre for Mathematical Medicine,
Fields Institute for Mathematical Sciences,
222 College Street
Toronto, ON M5T 3J1, Canada
E-mail: ssivaloganathan@uwaterloo.ca

In this paper we focus on continuum mechanics based models that can be used to study the mechanical response of brain tissue to high and low frequency loading. All of the models presented and reviewed are based on Biot's consolidation theory; however, we shall use homogenization methods and other approaches to develop generalizations of Biot's theory with a view to developing models that can be applied to brain biomechanics to study conditions such as hydrocephalus and trauma.

1. Introduction

Brain tissue is composed of primarily of neurons, glial cells and a significant extra-cellular space (estimated to occupy a volume fraction of 0.2[44]). The extracellular space is not empty, however, but is occupied by the extracellular matrix, composed mainly of various fibrous proteins andproteoglycans[47], which provide structural support for the brain cells, in addition to performing various other important functions.

Although the structure of the brain is inhomogeneous and non-isotropic, models of the brain have often assumed isotropy and homogeneity in order to make the problem of modelling the mechanics of the brain more tractable.

Biot's theory of consolidation[7] (originally intended to model soil consolidation under loading) has been widely used, either in it's basic or modified form, for the modeling of porous-elastic media. In recent years, poroelasticity theory has found significiant applications in brain biomechanics [25,35,50,55] in the context of modeling hydrocephalus. For example Kackzmarek *et al.*[25] take as their starting point, the application of Biot's theory to the case of a cylinder using the equations in a form developed by Kenyon[27].

The equations presented in Biot[7] describe the behaviour of a composite medium made up of two phases: a solid and a fluid. The elastic solid is filled with pores, like a sponge, so that a fluid such as water or air may occupy some space inside the solid skeleton and even move from one location to another, possible flowing from one side of the solid to the other. In the case of the brain the solid phase is made up of cells and intercellular matrix material and the fluid phase consists of cerebrospinal fluid (CSF). The original paper[7], deals with the modeling of the settlement of soils under load (consolidation).

Biot's theory has been further developed and elaborated upon in the following papers[8,9,10,11]. However, the extended theory is much more complicated and is therefore more difficult to deal with in applications. A simple model with fewer parameters is preferred in papers such as[35,50,55], since this reduces the number of unknowns that either need to be determined from data or arbitrarily assigned (for the purposes of investigating qualitative results) if the data is not available.

One extension to poroelastic theory, which has been proposed to model the behaviour of tissue, leads to the so-called poroviscoelastic models These introduce the possibility of a viscoelastic solid component of a porous material, and allow for a greater variety of deformation behaviours. In the case of the brain, Franceschini et al.[17] concluded from experimental data that there is a viscoelastic effect in the behaviour of brain tissue subjected to consolidation experiments. Mak[38] proposed a viscoelastic extension to poroelasticity by adding a convolution term, involving the rate of strain and a relaxation function, to the overall poroelastic stress/strain relationship. Biot[9] provides a quite general analysis of a porous viscoelastic solid.

Miller and Chinzei[41] performed an experiment involving several compression tests of cylindrical swine-brain samples, and interpreted their results using a single-phase viscoelastic model of the brain tissue. Miller[42] and Taylor and Miller[54] use the same model and data for the estimation of the elastic parameters of the brain in different situations. Taylor and Miller[54] for example focus on low strain-rate hydrocephalus, whereas Miller[42] focusses on higher strain-rates observed in surgical procedures and brain trauma.

In this paper, we focus on mechanical models of brain tissue, where the underlying solid has a porous structure. In particular, we assume that the brain is made up of an isotropic (porous) viscoelastic solid phase, with pores filled in with CSF. We assume also that the resulting macroscopic 'solid skeleton' is isotropic. The viscoelastic parameters of the solid skeleton are different from those of the constituent solid in this case–if an incompressible lump of rubber is filled with holes the resulting structure can be compressed with all volume change due to the holes expanding or contracting. In this paper we use a poroelastic/poroviscoelastic analogy and relate the parameters of the constituent solid to those of the overall solid using the formulae of Mackenzie[37]. Mackenzie's formulae are specialised instances of other approximation formulae presented in the literature (see the reviews [5,21,58] and other citations in Section 4). In this case we use Mackenzie's formulae because we assume that the brain is composed of a single, uniform solid phase (although this assumption is certainly too strong, it is necessary for mathematical tractability) with included pores or voids. The more general formulae apply when there are two or more solid phases present in the mixture. Our approach differs from that of Mak[38] in that we begin with a viscoelastic solid and then proceed to derive equations for the overall behaviour of the porous solid/fluid system, rather than adding a convolution term to the overall poroelastic equations. Here we add viscoelasticity *before* the effect of porosity (and solid/fluid interaction) is considered, in contrast to Mak who adds viscoelasticity *after* the effects of porosity have been taken into account.

2. Biot's Consolidation Theory

As a starting point we take the general equations governing consolidation from Biot's 1941 paper[7]. We also have an equation governing elastic equilibrium and the time-evolution of the pressure function within the material

$$\tau_{ij} = 2Ge_{ij} + \frac{2G\nu}{1-2\nu}e_{kk}\delta_{ij} - \alpha P, \tag{1}$$

$$G\nabla^2\mathbf{u} + \frac{G}{1-2\nu}\nabla\nabla\cdot\mathbf{u} - \alpha\nabla P = 0, \tag{2}$$

$$k\nabla^2 P = \alpha\frac{\partial}{\partial t}\nabla\cdot\mathbf{u} + \frac{1}{Q}\frac{\partial P}{\partial t}. \tag{3}$$

where G and ν are the shear modulus and Poisson's ratio of the solid (elastic) skeleton respectively, k is the permeability of the brain and α, $1/Q$ are constants representing the following quantities:

- α is the ratio of the volume of fluid squeezed out of the porous material to the volume change of the material if the material is compressed while allowing fluid to escape. According to[57] (Equation (3.8), page 49):

$$\alpha = 1 - \frac{K}{K_s'},$$

where K_s' is the so-called unjacketed bulk modulus and K is the overall bulk modulus of the solid skeleton. Under the assumption that the solid skeleton is composed of a single, uniform isotropic solid, K_s' is exactly the bulk modulus of the solid comprising the solid skeleton.

- $\frac{1}{Q}$ is a measure of the increase per unit pressure in volume fraction of fluid within the porous material if the volume of the material is kept constant. In Wang[57], this parameter is denoted as $\frac{1}{M}$ or S_ϵ. Equation (2.27) of[57] states that

$$S_\epsilon = \frac{\alpha}{K_u B},$$

where B is Skempton's coefficient (Page 64[57]), and K_u is the so-called undrained bulk modulus (Page 54). Skempton's coefficient is said to be close to 1.0 for fluid saturated soil. We shall assume that saturated brain tissue behaves similarly, and thus $B = 1$. Equation (3.74) of [57] gives the following expression for B:

$$B = \frac{1 - K/K_u}{1 - K/K_s'}.$$

The assumption that $B = 1$ thus gives $K_u = K_s'$ and therefore

$$\frac{1}{Q} = \frac{\alpha}{K_s'}.$$

The dimensions of the variables involved are given in Table 1.

Table 1. A table showing the variables in Biot's consolidation theory and their units.

Variable Name	Description	Dimensions
G	Shear Modulus	$\text{Pa} = \text{N m}^{-2} = \text{kg m}^{-1}\text{s}^{-2}$
Q	Biot Parameter	Pa
P	Fluid Pressure	Pa
k	Permeability	$\text{m}^2\ \text{Pa}^{-1}\ \text{s}^{-1} = \text{m}^4\ \text{N}^{-1}\ \text{s}^{-1} = \text{kg}^{-1}\ \text{m}^3\ \text{s}$
ν	Poisson's Ratio	dimensionless
α	Biot Parameter	dimensionless
$\mathbf{u}$	Displacement	m

3. A Poroviscoelastic Model and the Poroelastic/ viscoelastic analogy

In the case of an isotropic, porous, viscoelastic material, Biot[9] (Equation (6.18)) gives, in our notation:

$$\hat{\theta} = \alpha\hat{e} + \frac{\hat{P}}{Q}, \tag{4}$$

where θ is the increment in volume fraction of fluid, e is the trace of the strain tensor, and $\hat{\theta}$, $\hat{e}$, etc. represent the Laplace transform in time of the corresponding symbol. α and $1/Q$ in this case are functions of the transformed time variable s.

Again, with Laplace transforms, in the viscoelastic case Biot[9] (Equation (6.21)) obtains:

$$\hat{T}_{ij} = \lambda\hat{e}\delta_{ij} + 2G\hat{e}_{ij} - \alpha\hat{P},$$

where λ is Lamé coefficient (notation has been changed to be more in keeping with this paper). Writing this in terms of bulk modulus and shear modulus, we obtain:

$$\hat{T}_{ij} = K\hat{e}\delta_{ij} + 2G(\hat{e}_{ij} - \frac{1}{3}\hat{e}\delta_{ij}) - \alpha\hat{P}, \tag{5}$$

where, again, K and G are functions of s.

These equations are formally equivalent to the poroelastic model. We take the relationships between α, $1/Q$ and the elastic moduli K, K_s' from the elastic case and apply them here.

To the above, we add Darcy's law:

$$V = -k\nabla P,$$

where k is the permeability of the porous material and $V \cdot n$ is the volume of water per second flowing per unit area in the direction denoted by the unit vector n.

Now the increment in fluid volume θ changes according to the equation:

$$\frac{\partial \theta}{\partial t} = -\nabla \cdot V.$$

Thus we have

$$\frac{\partial \theta}{\partial t} = k\nabla^2 P.$$

Taking Laplace transforms of both sides of the above equation gives:

$$s\hat{\theta} - \theta_0 = k\nabla^2 \hat{P}, \tag{6}$$

where θ_0 is the initial increment in fluid content. However, we shall assume later that the solid constituent is a Kelvin-Voigt solid, which implies that there can be no instantaneous deformation of the skeleton, and hence $\theta_0 = 0$. Now, substituting $s\hat{\theta} = k\nabla^2 \hat{P}$ into (4), we obtain:

$$k\nabla^2 \hat{P} = s\alpha\hat{e} + s\frac{\hat{P}}{Q}. \tag{7}$$

The above equation will be used in the following sections.

Given the formal equivalence of the Laplace-transformed equations presented in this section to those of conventional poroelasticity, we propose that the parameters K and G, giving the overall viscoelastic bulk and shear moduli of the porous solid structure, may be related to the parameters of the constituent solid via existing formulae which relate the overall elastic (non-viscous) parameters of a composite material with those of its constituent parts. These kinds of formulae are dealt with in the following section.

4. Effective viscoelastic parameters

In order to estimate the overall parameters of the solid skeleton from those of the constituent solid, we take formulae from[37], which is a special case of existing effective medium theories for solids composed of a mixture of elastic materials. Note that we are using an elastic/viscoelastic analogy, so that both the elastic parameters of the solid constituent, and those of the solid skeleton, will be functions of s.

Mackenzie's formulae are given as follows (in our notation):

$$K = \frac{4K_sG_s(1-\phi)}{4G_s + 3K_s\phi} + O(\phi^3), \tag{8}$$

$$G = G_s - \phi_f G_s \frac{15K_s + 20G_s}{9K_s + 8G_s} + O(\phi^2), \tag{9}$$

where K and G are the effective bulk and shear moduli of the solid skeleton, while K_s and G_s are those of the constituent solid. We drop the error terms to obtain our estimates for K and G.

Given that brain tissue is not composed of a single type of solid, but rather a mixture of solids, we shall discuss some sources describing more elaborate formulae which deal with these more complicated situations, although below we will use the assumption that the solid part is a single, uniform material for the sake of simplicity.

Reviews of such theory in the literature include those of Berryman[5] and Zaoui[58]. An older review by Hashin[21] has a large list of references.

One of the early examples of such a theory is due to Einstein[14] (English translation in the 1956 Dover publication[15]), who estimated the effective viscosity of a solution consisting of dilute rigid spheres suspended in a viscous fluid.

Mackenzie[37] estimated the elastic constants of an elastic solid containing spherical voids by what Hashin[21] refers to as the generalised self-consistent scheme, where a spherical void is contained in a spherical shell of the solid comprising the matrix; this solid shell is in turn surrounded by a solid possessing the desired effective elastic constants. The effective elastic constants must be chosen such that the inclusion of isolated (small) composite spheres as shown in Figure 1 does not effect the overall response of the material to bulk and shear stress.

Eshelby[16] studied the effect of ellipsoidal inclusions in an elastic matrix, deriving formulae for effective elastic constants of an elastic medium containing spherical inhomogeneities with differing elastic moduli to the surrounding matrix. These are generalizations of Mackenzie's results and reduce to Mackenzie's estimates when the elastic moduli of the inhomogeneities are reduced to zero.

Kröner[31,32] and Kröner and Koch[30] discuss how effective elastic constants can be derived from random media.

For the case of an isotropic effective medium composed of N isotropic phases, if we consider each phase as distributed throughout the reference medium as spherical inclusions, the estimates for the effective bulk modulus

Figure 1. The scheme presented by Mackenzie[37]: A central spherical void (white) is surrounded by a spherical shell (black) of the constituent solid, which is in turn surrounded by a spherical shell (grey) of the effective material, composed of both solid and pores. The radius of the outer shell is taken as much larger than that of the inner shell. Applying bulk or shear stress and equating the displacements at the outer boundary of the composite sphere shown and a homogeneous sphere of effective material gives expressions for the effective bulk and shear moduli in terms of the bulk and shear moduli of the constituent solid and the pore volume fraction.

K and shear modulus G are given by[4]:

$$K = \sum_{p=1}^{N} \phi_p K_p \left(1+\alpha_0\left(\frac{K_p}{K_0}-1\right)\right)^{-1} \times \left[\sum_{p=1}^{N} \phi_p \left(1+\alpha_0\left(\frac{K_p}{K_0}-1\right)\right)^{-1}\right]^{-1} \tag{10}$$

$$G = \sum_{p=1}^{N} \phi_p G_p \left(1+\beta_0\left(\frac{G_p}{G_0}-1\right)\right)^{-1} \times \left[\sum_{p=1}^{N} \phi_p \left(1+\beta_0\left(\frac{G_p}{G_0}-1\right)\right)^{-1}\right]^{-1}, \tag{11}$$

where

$$\alpha_0 = \frac{3K_0}{3K_0+4G_0}; \qquad \beta_0 = \frac{6(K_0+2G_0)}{5(3K_0+4G_0)}.$$

Mackenzie's estimates derive from the above when $N = 2$, $K_2 = G_2 = 0$ and $K_0 = K_1$. Eshelby's estimates for spherical inclusions derive from the above when $N = 2$ and K_0, G_0 are equal to either of K_1, G_1 or K_2, G_2. The case where K_0, G_0 are equal to K_p, G_p for some $p = 1, \ldots, N$ has been referred to as the generalised self-consistent scheme, as mentioned above, but is also known as the Mori-Tanaka scheme[43] and signifies that one component acts as the matrix, while the other is embedded in the first as isolated inclusions.

A notable alternative choice for K_0 and G_0 is $K_0 = K$ and $G_0 = G$. This gives the so-called self-consistent[22,30,32] estimate for the elastic moduli K and G. In contrast to the generalised self-consistent scheme, here one can imagine many spherical inclusions of the constituents of the effective material embedded directly in the effective medium, rather than surrounded by a spherical shell of some other constituent. This choice is appropriate for perfectly disordered mixture of constituents, so that none can be said to play the role of matrix or inclusion, for example.

5. Specialisation of the model for a one-dimensional consolidation experiment

Franceschini et al.[17] studied the consolidation of cylindrical brain tissue specimens under constant load. The experimental conditions allowed drainage of fluid from the top and bottom of the specimen. We shall now specialise the above model to one-dimension and try to reproduce the qualitative behaviour observed in this work[17].

Assumptions: Assume that the brain is composed of a porous, isotropic Kelvin-Voigt solid. So that the solid bulk and shear moduli ,K_s and G_s, are given by

$$K_s = K_1 + sK_2; \qquad G_s = G_1 + sG_2,$$

and α and $1/Q$ are therefore given by

$$\alpha = 1 - \frac{K}{K_s}; \qquad \frac{1}{Q} = \frac{\alpha}{K_s},$$

where K and G are given by (8) and (9) respectively.

The two equations which must be specialised to one-dimension are (5) and (7).

The experimental setup of Franceschini et al.[17] allows only for displacement in the z direction, while radial expansion is prevented by a surrounding, impermeable sheath. Furthermore the stress throughout the cylinder is constant, so that the Laplace transform (in time) of the stress tensor is $\hat{T}_{ij} = \frac{T_{ij}}{s}$. Thus, (5) reduces to the following equation involving stress and strain in the z direction only:

$$\frac{\sigma_z}{s} = \frac{1}{a}\frac{\partial \hat{w}}{\partial z} - \alpha \hat{P}, \tag{12}$$

where $\sigma_z < 0$ is the stress-compenent in the z direction and is equal to the negative of the applied load L, w is the displacement in the z direction and

$$a = \frac{1}{K + \frac{4}{3}G}.$$

Equation (7) reduces to:

$$k\frac{\partial^2 \hat{P}}{\partial z^2} = s\alpha\frac{\partial \hat{w}}{\partial z} + s\frac{\hat{P}}{Q}.$$

Solving (12) for $\frac{\partial \hat{w}}{\partial z}$ and substituting this into the above equation then gives:

$$\frac{\partial^2 \hat{P}}{\partial z^2} - \frac{s}{k}\left[\alpha^2 a + \frac{1}{Q}\right]\hat{P} = \frac{\alpha a \sigma_z}{k}. \tag{13}$$

In the experiments of Franceschini et al., fluid is allowed to flow out from the top and bottom of the specimen as it contracts. Thus, the boundary conditions are given as

$$\hat{P}(z = 0) = \hat{P}(z = h) = 0, \tag{14}$$

where h is the height of the specimen (in metres). Note that if we multiply (13) by s and let $s \to \infty$, then using the final value theorem, in conjunction with the boundary conditions above, gives

$$\lim_{t\to\infty} P = 0.$$

The solution to the above equation, along with the given boundary conditions, is given by:

$$\hat{P} = \frac{A}{B}\left[(e^{\sqrt{B}z} - 1) - (e^{\sqrt{B}h} - 1)\frac{\sinh(\sqrt{B}z)}{\sinh(\sqrt{B}h)}\right], \tag{15}$$

where

$$A = \frac{s}{k}\left[\alpha^2 a + \frac{1}{Q}\right]; \qquad B = \frac{\alpha a \sigma_z}{k}.$$

The strain throughout the cylinder can be calculated from (12):

$$\frac{\partial \hat{w}}{\partial z} = a\alpha\hat{P} + \frac{a\sigma_z}{s}. \tag{16}$$

If we denote the total deflection of the top of the cylinder by w_0, then:

$$\hat{w_0} = \int_{z=0}^{z=h} \frac{\partial \hat{w}}{\partial z}\, dz = \int_{z=0}^{z=h} a\alpha\hat{P} + \frac{a\sigma_z}{s}\, dz. \tag{17}$$

We must apply an inverse laplace transform to $\hat{w_0}$ in order to find how the total deflection of the top of the cyclinder changes with time.

5.1. *Numerical inverse Laplace transform*

The function $\hat{w}_0$ is a very complicated function of s and is difficult to invert analytically. In order to invert $\hat{w}_0$ we use the Gaver-Stehfest algorithm (Abate and Whitt[1], page 52), implemented in MAPLE, to evaluate w_0 at any given value $t > 0$.

6. Curve fitting

Franceschini et al.[17] fitted two different models to consolidation data obtained from cylindrical brain specimens. We cannot compare our model directly with the data from[17], so instead we fitted our model to the one-dimensional Gibson-Lo[18] poro-viscoelastic model, which approximates the data in Franceschini et al.[17] to a reasonable degree.

The Gibson-Lo model consists of a solid skeleton which behaves as a hookean spring in series with a Kelvin-Voigt element The model is formulated from the start at the macro-scale in one-dimension, so the elastic parameters all represent effective parameters of the solid skeleton and not those of the constituent solid.

Franceschini et al. give a presentation of the Gibson-Lo model which differs slightly from the notation that Gibson and Lo used. To fit the data, Franceschini et al. determine two parameters, α and c_v, to obtain a least-squares curve-fit, using the Mathamatica function `FindFit`. Apart from an error in the formula (the sum that Franceschini et al. present should range over all odd n, not all n), from the presentation that Franceschini et al. give you cannot determine a curve from the values of α and c_v only. However, if one assumes that $E_1 = E_2$, so that $E_* = 2$, the curves one obtains look very similar to those from[17]. It is our opinion that this assumption was made, but not stated in the paper[17]. Regardless, it is an assumption that we make here in order to obtain a theoretical curve to which we may fit the model presented above.

Where Franceschini et al. use minutes and millimeters as their units of time and length respectively, we use seconds and meters.

We used the `NonlinearFit` command from the Statistics package of Maple to find a best fit for some curves presented in Franceschini et al.[17], produced by the Gibson-Lo[18] model. As a restriction we took the final strain to be -5%, which is in line with that specified in Franceschini et al.[17]. This means that G_1 may be determined from K_1 and ϕ, since the final strain is given by La_∞ where L is the load applied to the tissue-sample and a_∞ is the parameter a evaluated at $s = 0$:

$$a_\infty = \frac{1}{K_1^h + \frac{4}{3}G_1^h}; \quad K_1^h = \frac{4K_1G_1(1-\phi)}{4G_1 + 3K_1\phi}; \quad G_1^h = G_1 - \phi G_1 \frac{15K_1 + 20G_1}{9K_1 + 8G_1}. \tag{18}$$

We solve $La_\infty = 0.05$ in order to obtain G_1. There is a unique positive solution for G_1 as long as $\phi \leq 0.4$ (see Appendix Appendix A), which we shall assume, given that the extra-cellular space in the brain is estimated to occupy a volume fraction of around 0.2.

The experiments undertaken in Franceschini et al.[17] include some pre-compression, so it is likely that the pore volume fraction is less than 0.2.

7. Concluding remarks

The results from Section 6 indicate a significant viscous component in the solid part of the brain tissue in the experiments of Franceschini et al.. The experiments, however, may be significantly affected by the preloading: It is stated in Franceschini et al.[17] that each tissue sample was subjected to the weight of the measuring device itself, and allowed to settle, before any further weight was applied. The pre-loading weight was 1.3N, which translates to a pre-loading of around 1.8 kPa, given the specified diameter of the samples. If the brain tissue is initially soft and very porous, this could have a large effect both on the porosity and on the elastic/viscoelastic parameters of the tissue itself.

Appendix A. Determining G_1 from K_1 and ϕ

The requirement that the final strain of the sample be -5% reduces the number of free parameters by one. In this section we show how.

The final strain can be found by applying the final value theorem to Equation (16):

$$\lim_{s\to 0} s\frac{\partial \hat{w}}{\partial z} = \lim_{s\to 0} a\alpha s\hat{P} + \lim_{s\to 0} \frac{a\sigma_z}{\quad}.$$

Now, it can be seen that α and a tend to definite limits α_∞ and a_∞ as $s \to 0$, where a_∞ in particular is given by (18). Moreover, it can also be shown that $\lim_{s\to 0} s\hat{P} = 0$, so that we obtain:

$$0.05 = La_\infty$$

when the desired final strain is -5%. It shall now be shown that this equation has exactly one positive solution for G_1 as long as ϕ, $K_1 > 0$ and ϕ is small enough.

Rearranging the above equation, we obtain:

$$20L = \frac{1}{a_\infty}.$$

We shall show the following:

- $\frac{1}{a_\infty} = 0$ when $G_1 = 0$.
- $\frac{1}{a_\infty}$ is strictly increasing for $G_1 \geq 0$.

These two points show that there is a unique solution $G_1 > 0$.

The first point follows easily from the expanded expression for $\frac{1}{a_\infty}$:

$$\frac{1}{a_\infty} = \frac{4K_1G_1(1-\phi)}{4G_1 + 3K_1\phi} + \frac{4}{3}G_1 - \frac{4}{3}\phi G_1 \frac{15K_1 + 20G_1}{9K_1 + 8G_1}.$$

Substituting $G_1 = 0$ into the above gives $\frac{1}{a_\infty} = 0$.

For the second point we calculate the derivative of $\frac{1}{a_\infty}$ with respect to G_1 and obtain, after some calculation:

$$\frac{\partial}{\partial G_1}\frac{1}{a_\infty} = \frac{12K_1^2(1-\phi)\phi}{(4G_1 + 3K_1\phi)^2} + \frac{4}{3(9K_1 + 8G_1)^2}$$
$$\times \left\{(81 - 135\phi)K_1^2 + (144 - 360\phi)K_1G_1 + (64 - 160\phi)G_1^2\right\}.$$

It can be seen that

$$\frac{12K_1^2(1-\phi)\phi}{(4G_1 + 3K_1\phi)^2} > 0,$$

for all $K_1, G_1 > 0$ and $0 < \phi < 1$, while

$$\frac{4}{3(9K_1+8G_1)^2}\left\{(81-135\phi)K_1^2+(144 - 360\phi)K_1G_1+(64-160\phi)G_1^2\right\} \geq 0,$$

as long as $K_1, G_1 > 0$ and $0 < \phi \leq 0.4$.

References

1. J. Abate and W. Whitt, *Queueing Systems* **10**, 5 (1992).
2. C. G. Armstrong, W. M. Lai, and V. C. Mow, *J. Biomech. Eng.* **106**, 165 (1984).
3. A. Bensoussan, J. L. Lions, and G. Papanicolau, *Asymptotic Analysis for Periodic Structures*. North-Holland: Amsterdam (1978).
4. O. Bernard, F.-J. Ulm, and E. Lemarchand, *Cem. Concr. Res.* **33**, 1293 (2003).
5. J. G. Berryman, *J. Eng. Mech.* **131**, 928 (2005).
6. S. Beurthey, and A. Zaoui, *Eur. J. Mech. A/Solids* **19**, 1 (2000).
7. M. A. Biot, *J. Appl. Phys.* **12**, 155 (1941).

8. M. A. Biot, *J. Appl. Phys.* **26**, 182 (1955).
9. M. A. Biot, *J. Appl. Phys.* **27(5)**, 459 (1956).
10. M. A. Biot and D. G. Willis *J. Appl. Mech., Trans. A. S. M. E.* **24(79)**, 594 (1957)
11. R. Burridge and J. B. Keller, *J. Acoust. Soc. Am.* **70(4)** 1140 (1981)
12. T. Clopeau, J. L. Ferrín, R. P. Gilbert, and A. Mikelić, *Math. Comput. Model.* **33** 821 (2001).
13. J. M. Drake, O. Mostachfi, G. Tenti, and S. Sivaloganathan, *Can. J. Neurol. Sci.* **23** S5 (1996).
14. A. Einstein, *Ann. der Phys.* **19** 289 (1906).
15. A. Einstein, *Investigations on the theory of brownian movement.* Dover (1956).
16. J. D. Eshelby, *Proc. Roy. Soc. Lond. A* **204** 376 (1957).
17. G. Franceschini, D. Bigoni, P. Regitnig, and G. A. Holzapfel, *J. Mech. Phys. Solid.* **54** 2592 (2006).
18. R. E. Gibson and K. Y. Lo, *Acta Poly. Scand.* **10** 1 (1961).
19. R. P. Gilbert and A. Mikelić, *Nonlinear Anal.* **40** 185 (2000).
20. D. H. Griffel, *Applied functional analysis.* Wiley: Toronto (1981).
21. Z. Hashin, *J. Appl. Mech.* **50** 481 (1983).
22. A. V. Hershey, *ASME J. Appl. Mech.* **21** 236 (1954).
23. R. Hill, *J. Mech. Phys. Solids* **11** 357 (1963).
24. M. Hrapko, J. A. W. van Dommelen, G. W. M. Peters, and J. S. H. M. Wismans, *J. Biomech. Eng.* **130** 031003 (2008).
25. M. Kaczmarek, R. P. Subramaniam, and S. R. Neff, *Bull. Math. Biol.* **59** 295 (1997).
26. J. B. Keller, in *Nonlinear partial differential equations in engineering and applied science.* **54**. New York: Dekker. 429 (1980).
27. D. E. Kenyon, *J. Appl. Mech.* **43** 594 (1976).
28. Y. K. Kim, H. B. Kingsbury, *Experimental Mechanics* **19(7)** 252 (1979).
29. S. G. Kou and H. Wang, *Adv. Appl. Prob.* **35** 504 (2003).
30. E. Kröner and H. Koch, *Solid Mech. Arch.* **1** 183 (1976).
31. E. Kröner, *J. Mech. Phys. Solids* **25** 137 (1977).
32. E. Kröner, *J. Phys. F: Metal Phys.* **8** 2261 (1978).
33. Q. V. Le, F. Meftah, Q.-C. He, and Y. Le Pape, *Mech. Time-Depend. Mater.* **11** 309 (2007).
34. C. K. Lee and C. C. Mei *Int. J. Engng. Sci.* **35** 329 (1997).
35. D. N. Levine, *Bulletin of Mathematical Biology* **61** 875 (1999).
36. Y.-B. Lu, K. Franze, G. Seifart, C. Steinhäuser, F. Kirchoff, H. Wolburg, J. Guck, P. Janmey, E.-Q. Wei, J. Käs, and A. Reichenbach, *Proc. Natl. Acad. Sci.* **103** 17759 (2006).
37. J. K. Mackenzie, *Proc. Phys. Soc. B* **63** 2 (1950).
38. A. F. Mak, *Biorheology* **23** 371 (1986).
39. A. Meirmanov, *Euro. J. Appl. Math.* **19** 259 (2008).
40. H. Metz, J. McElhaney, and A. K. Ommaya, *J. Biomechanics* **3** 453 (1970).
41. K. Miller and K. Chinzei, *J. Biomechanics* **30** 1115 (1997).
42. K. Miller, *J. Biomechanics* **32** 531 (1999).

43. T. Mori and K. Tanaka, *Acta Metall.* **21** 571 (1973).
44. C. Nicholson and E. Syková, *Trends Neurosci.* **21** 207 (1998).
45. H. J. Reulen, R. Graham, M. Spatz, and I. Klatzo, *J. Neurosurg.* **46** 24 (1977).
46. T. Roose, P. A. Netti, L. L. Munn, Y. Boucher, and R. K. Jain, *Microvascular Research* **66** 204 (2003).
47. E. Ruoslahti, *Glycobiology* **6** 489 (1996).
48. A. Sahar, G. M. Hochwald, and J. Ransohoff, *J. Neurol. Sci.* **11** 81 (1970).
49. J. E. Santos, C. L. Ravazzoli, and J. Geiser, *Int. J. Solid. Struct.* **43** 1224 (2006).
50. A. Smillie, I. Sobey, and Z. Molnar, *J. Fluid. Mech.* **539** 417 (2005).
51. I. Stakgold, *Green's Functions and Boundary Value Problems.* Wiley: Toronto (1998).
52. R. L. Sternberg, A. J. Kalinowski, and J. S. Papadakis, *Nonlinear Partial Differential Equations in Engineering and Applied Science.* Marcell Dekker, Inc.: New York (1979).
53. T. Roose, P. A. Netti, L. L. Munn, Y. Boucher, and R. K. Jain, *Microvascular Research* **66** 204 (2003).
54. Z. Taylor and K. Miller, *J. Biomechanics* **37** 1263 (2004).
55. G. Tenti, S. Sivaloganathan, and J. M. Drake, *Canad. Appl. Math. Quart.* **7(1)** 93 (1999).
56. K. Terada, T. Ito, and N. Kikuchi, *Comput. Methods Appl. Mech. Engrg.* **153** 223 (1998).
57. H. F. Wang, *Theory of Linear Poroelasticity with Application to Geomechanics and Hydrogeology.* Princeton University Press: Princeton, New Jersey (2000).
58. A. Zaoui, *J. Eng. Mech.* **128** 808 (2002).
59. M.-Y. Zhou and P. Sheng, *Phys. Rev. B* **39(16)** 27 (1989).

THE MATHEMATICS OF LIVER TRANSPLANTATION

FRANCISCO ANTONIO BEZERRA COUTINHO*, ELEAZAR CHAIB, MARCOS AMAKU, MARCELO NASCIMENTO BURATTINI AND EDUARDO MASSAD

University of São Paulo, School of Medicine
LIM 01 and LIM 37, HCFMUSP
Avenida Doutor Arnaldo - 455
Sao Paulo - SP, CEP: 01246-903, Brazil
**Presenting author: coutinho@dim.fm.usp.br*

São Paulo is the pioneer Brazilian state in transplantation surgery. The Brazilian model for organ sharing has developed a system that prioritizes liver transplantion (LT) for patients who have the highest risk of wait-list mortality. In this paper we present a brief review of the application of mathematical models to the problem of how to improve the speed with which patients with liver failure can receive a liver graft and how alternative criteria for prioritizing patients in the waiting list can be proposed.

Key words: Liver transplantation, hepatocellular carcinoma, expanded criteria, model.

1. Introduction

Liver failure occurs when large parts of the liver become damaged beyond repair and the liver is no longer able to function. The most common causes of chronic liver failure (where the liver fails over months to years) include Hepatitis B, Hepatitis C, long term alcohol consumption, Cirrhosis, hemochromatosis (an inherited disorder that causes the body to absorb and store too much iron) and malnutrition. The causes of acute liver failure, when the liver fails rapidly, however, are often different. These include acetaminophen (Tylenol) overdose, viruses including hepatitis A, B, and C (especially in children), reactions to certain prescription and herbal medications, aflatoxin (*Aspergillus flavus*) and ingestion of poisonous wild mushrooms.

Liver transplantation or **hepatic transplantation** is the replacement of a diseased liver with a healthy liver from another person (allograft). The most commonly used technique is **orthotopic transplantation**, in which the native liver is removed and replaced by the donor organ in the same anatomic location as the original liver. Liver transplantation is a viable treatment option for end-stage liver disease and acute liver failure. Figure 1 shows a schema of orthotopic liver transplantation.

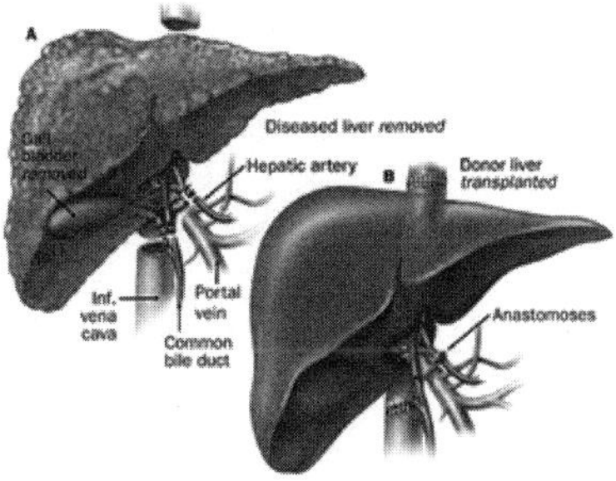

Figure 1. Schema of orthotopic liver transplantation. Available at http://218.17.222.227/xhnk/UploadFiles/upload/201103141158168927.png

In this paper we present a brief review of the application of mathematical models to the problem of how to improve the speed with which patients with liver failure can receive a liver graft and how alternative criteria for prioritizing patients in the waiting list can be proposed.

2. Waiting list dynamics in the state of São Paulo, Brazil

The aim of this study is to demonstrate the performance of our state liver transplantation program and analyze when the number of liver transplantation will meet the demand of our waiting list.

The data related to the actual number of liver transplantation, *Tr*, the incidence of new patients in the list, *I*, and the number of patients who died in the waiting list, *D*, in the State of So Paulo since 1997 as in Table 1.

We then fitter a continuous function to the number of transplantation carried out in each year. The equation has the form:

$$Tr = 107.07 \ln(year) + 72.943 \tag{1}$$

The result of the fitting is seen in Figure 2.

We next projected the size of the waiting list, L, by taking into account the incidence of new patients per year, I, the number of transplantations carried out in that year, Tr, and the number of patients that died in the

Table 1. Actual number of liver transplantation, Tr, the incidence of new patients in the list, I, and the number of patients who died in the waiting list, D, in the State of So Paulo since 1997.

$Year$	Tr	I	D
1997	63		
1998	160	553	321
1999	188	923	414
2000	238	1074	548
2001	244	1248	604
2002	242	1486	725
2003	289	1564	723
2004	295	1500	671

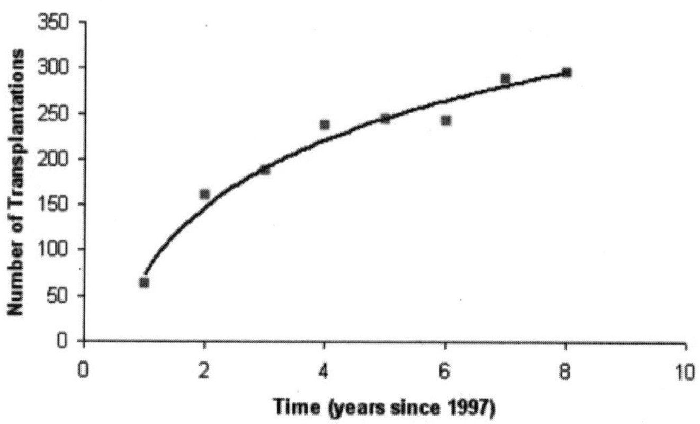

Figure 2. Fitting Eq. 1 to the number of liver transplantation performed in the period between 1997 and 2004.

waiting list, D. The dynamics of the waiting list is given by the difference equation:

$$L_{t+1} = L_t + I_t - D_t - Tr_t \tag{2}$$

that is, the list size at time $t + 1$ is equal to the size of the list at the time t, plus the new patients getting into the list at time t, minus those patients who died in the waiting list at time t, and minus those patients who received a graft at time t. The variables I, and D, from 2004 onward were projected by fitting an equation by maximum likelihood, in the same way that we did for Tr. Figure 3 shows the result of the projection.

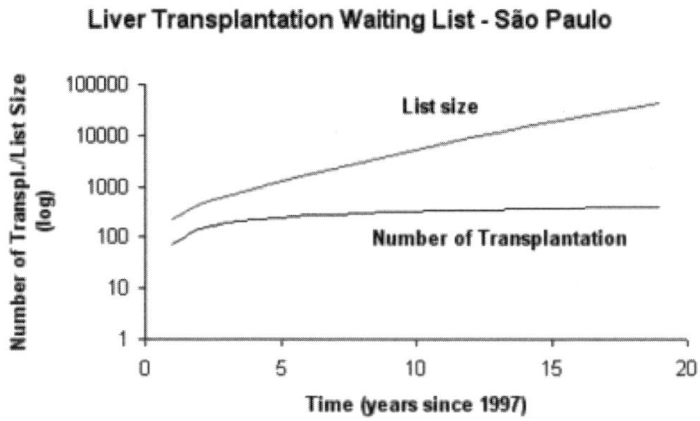

Figure 3. Projections of the list size and the number of liver transplantations in the State of São Paulo, Brazil.

Note that, unless the number of liver transplantation increases dramatically in the next decades the list size will steadily grow.

3. The potential impact of using ABO-compatible living donor liver transplantation (LDLT) on the liver transplantation program and waiting list in the state of São Paulo, Brazil

The aim of this work was to analyze through a mathematical equation, the potential impact of using ABO-compatible LDLT on both our liver transplantation program and waiting list in the state of Sao Paulo. We calculated the probability that a giving receptor has a compatible parent. For this we need first the frequency distribution of each blood type in the population of the State of Sao Paulo. We calculated the frequency distribution of each blood group alleles by applying the Hardy-Weinberg equation for 3 alleles, that is:

$$(p + q + r)^2 = p^2 + q^2 + r^2 + 2pq + 2pr + 2qr \tag{3}$$

We then calculated the frequency of each allele from the frequency of each blood in the Brazilian population. From this we estimated the probability of having a parent or a sibling compatible who could be potential donors for liver transplantation. The results are shown in Table 2.

Table 2. Frequency of ABO groups and probability of having a compatible parent or sibling in the State of São Paulo, Brazil.

Blood Group	Prevalence in the Population	Probability of having a compatible parent	Probability of having at least one compatible sibling
O	0.5067	0.154	0.350
A	0.3217	0.169	0.122
B	0.1345	0.059	0.052
AB	0.0371	0.029	0.029

Now, assuming that on average 25% of list is comprised by children, who should receive a liver from a parent and 75% is comprised by adults, who should receive a liver from a brother or sister, we can calculate the impact of LDLT in the waiting list by writing:

$$\begin{aligned} L_{t+1} &= L_t + I_t - D_t - Tr_t - \\ &= [(0.5076 \times 0.154) + (0.3217 \times 0.169) + \\ &\quad (0.1345 \times 0.059) + (0.0371 \times 0.029)] \times 0.25L_t - \\ &\quad [(0.5076 \times 0.350) + (0.3217 \times 0.122) + \\ &\quad (0.1345 \times 0.052) + (0.0371 \times 0.029)] \times 0.75L_t \end{aligned} \tag{4}$$

with $L_0 = 218$, the size of the list in 1997. The result of the impact of including the above potential donors in the transplantation program can be seen in Figure 4.

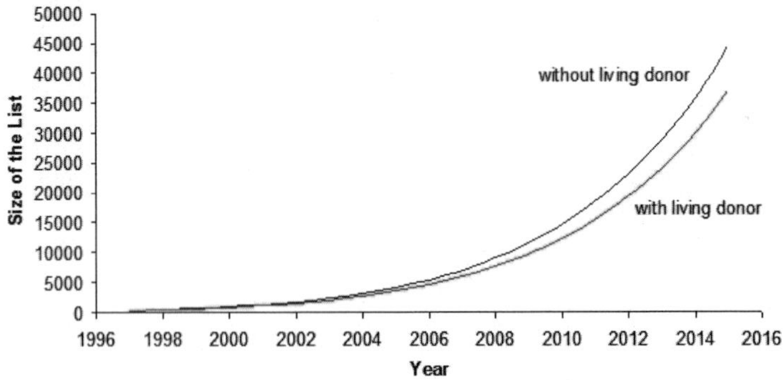

Figure 4. Impact of including living donors on the size of the waiting list.

Note that the impact of including relative donors on the size of the waiting list is not remarkable but it increases with time.

4. The potential impact of non-heart beating donors on the liver-transplantation waiting list of São Paulo, Brazil

We applied a mathematical model to analyze the potential impact of using Non-Heart Beating Donors (NHBD)policy on the liver transplantation waiting list in the State of So Paulo, Brazil. The model shows that, assuming only 10%, 480 additional liver transplantations. This additional transplantations figure when applied in the model results in a relative reduction of 37% in the size of the waiting list and a projected number of averted deaths of about $41,500$ in 20 years. The impact of including NHBD on the number of patients who die in the waiting list is shown in Figure 5.

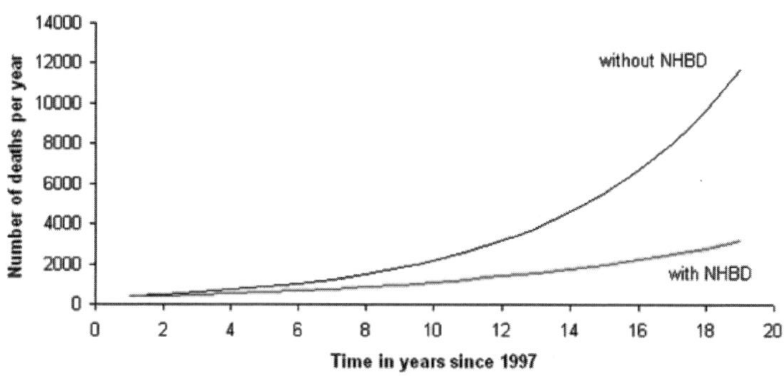

Figure 5. Impact of including NHBD on the number of patients who die every year in the waiting list.

Note that the number of patients who die every year in the waiting list is significantly reduced after 10 year of adopting such a policy [1,2,3,4,5,6,7,8].

5. Theoretical impact of an anti-HCV vaccine on the annual number of liver transplantation

The development of an effective vaccine for hepatitis C is of paramount importance, given the global disease burden and its public health impact. We simulated a theoretical vaccine with 98% efficacy and coverage of 95% of the susceptible population, an achievable program. The simulated period of vaccination varied from 0 to 70 years and we calculated, through a mathematical model, the reduction in the number of liver transplantations carried out each year. The model's structure is shown in Figure 6.

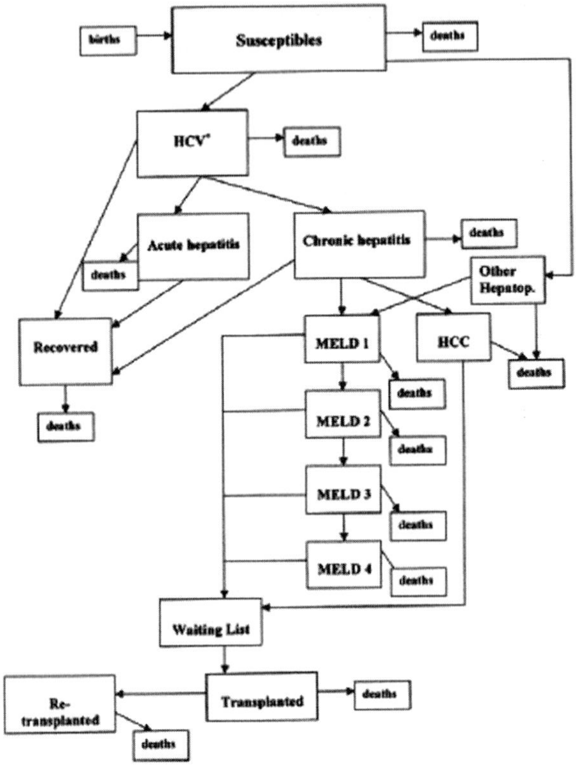

Figure 6. Model's structure to the study of the impact of a theoretical vaccine against hepatitis C on liver transplantation.

The corresponding set of equation is:

$$\frac{dS_1(t)}{dt} = \Lambda - \mu\beta_1 HCV(t)\frac{S_1(t)}{N} - \mu\beta_2 Ac(t)\frac{S_1(t)}{N} - \mu\beta_3 T_A(t)\frac{S_1(t)}{N}$$
$$-\mu\beta_4 Chr(t)\frac{S_1(t)}{N} - (\mu + \omega + \xi)S_1(t) \tag{5}$$
$$\frac{dS_2(t)}{dt} = \xi S_1(t) - \beta_1 HCV(t)\frac{S_2(t)}{N} - \beta_2 Ac(t)\frac{S_2(t)}{N} - \beta_3 T_A(t)\frac{S_2(t)}{N}$$
$$-\beta_4 Chr(t)\frac{S_2(t)}{N} - (\mu + \omega)S_2(t) \tag{6}$$

$$\frac{dHCV(t)}{dt} = \beta_1 HCV(t)\left(\frac{S_1(t)+S_2(t)}{N}\right) - \beta_2 Ac(t)\left(\frac{S_1(t)+S_2(t)}{N}\right) - \beta_3 T_A(t)\left(\frac{S_1(t)+S_2(t)}{N}\right) - \beta_4 Chr(t)\left(\frac{S_1(t)+S_2(t)}{N}\right) - (\gamma_1+\sigma_1+\sigma_2+\mu)HCV(t) \quad (7)$$

$$\frac{dAc(t)}{dt} = \sigma_1 HCV(t) - (\delta_1+\gamma_2+\alpha_1+\mu)Ac(t) \quad (8)$$

$$\frac{dR(t)}{dt} = \gamma_1 HCV(t) + \gamma_2 Ac(t) + \gamma_4 Ac(t) + \gamma_4 Chr(t) - \mu R(t) \quad (9)$$

$$\frac{dChr(t)}{dt} = \delta_1 Ac(t) + \sigma_2 HCV(t) - (\delta_2+\theta+\alpha_2+\gamma_3)Chr(t) \quad (10)$$

$$\frac{dMELD_1(t)}{dt} = \delta_2(Chr(t)+Oth(t)) - (\phi_1+\epsilon_1+\alpha_3+\mu)NT_cM_1(t) \quad (11)$$

$$\frac{dMELD_2(t)}{dt} = \epsilon_1 NT_cM_1(t) - (\phi_2+\epsilon_2+\alpha_4+\mu)NT_cM_2(t) \quad (12)$$

$$\frac{dMELD_3(t)}{dt} = \epsilon_2 NT_cM_3(t) - (\phi_3\epsilon_3+\alpha_5+\mu)NT_cM_3(t) \quad (13)$$

$$\frac{dMELD_4(t)}{dt} = \epsilon_3 NT_cM_3(t) - (\phi_4+\alpha_6+\mu)NT_cM_4(t) \quad (14)$$

$$\frac{dHCC(t)}{dt} = \theta Chr(t) - (\phi_5+\alpha_7+\mu)HCC(t) \quad (15)$$

$$\frac{dWL(t)}{dt} = \sum_{i,j=1}^{4} \phi_i NT_CM_j + \phi_5 Chr(t) - (\tau_1+\alpha_8+\mu)WL(t) \quad (16)$$

$$\frac{dTx_1(t)}{dt} = \tau_1 WL(t) - (\phi_6+\alpha_9+\mu)Tx_1(t) \quad (17)$$

$$\frac{WLTx(t)}{dt} = \phi_6 Tx_1(t) - (\tau_2 + \alpha_{10} + \mu)WLTx(t) \tag{18}$$

$$\frac{Tx_2(t)}{dt} = \tau_2 WLTx(t) - (\alpha_{11} + \mu)Tx_2 \tag{19}$$

$$\frac{Oth(t)}{dt} = \omega(S_1(t) + S_2(t)) - (\delta_2 + \alpha_{12} + \mu)Oth(t) \tag{20}$$

The impact of such a vaccine on reducing the number of liver transplantation is shown in Figure 7.

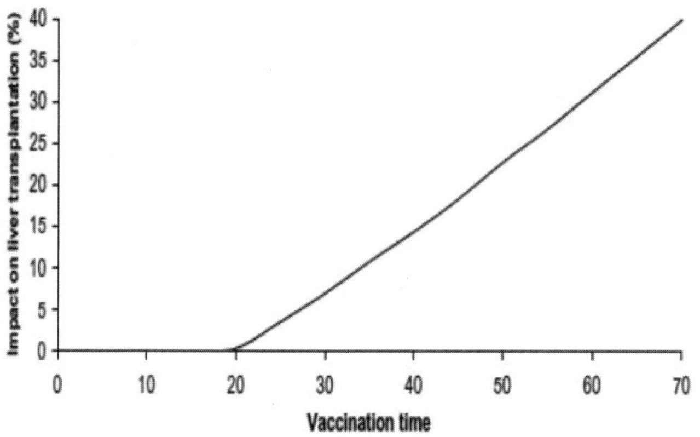

Figure 7. Impact of a theoretical vaccine against hepatitis C on liver transplantation.

Note that the the program is entirely inefficient until 20 years of vaccination and its impact rises linearly with time, reaching a maximum of 40% reduction. The model assumes that approximately 50% of all the liver transplantation carried out in our population are due to HCV infection. Therefore, the maximum reduction in the number of transplantation attained after 70 years is 10% less of the theoretical optimum. This is due to the 2% of primary vaccination failure plus the 5% in the coverage failure, which leaves a small proportion of susceptible individuals who will catch the infection and evolve to liver failure.

6. Does anti-HBV vaccine make any difference in long-term number of liver transplantation?

In the absence of any previous study comparing population treated and non-treated with respect to the number of liver failure due to HBV we have decided to apply a model previously proposed to study the projected impact of vaccination against hepatitis C on liver transplantation to the case of hepatitis B as a cause of liver transplantation. The strucuture of the studied model is very similar to the one related to the vaccine against hepatitis C and is shown in Figure 8.

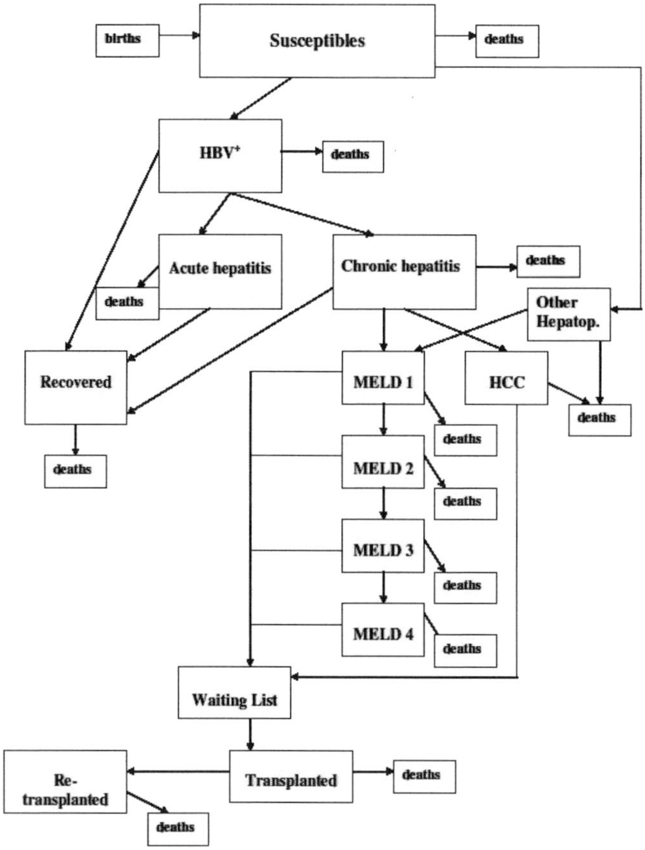

Figure 8. Model's structure to the study of the impact of the vaccine against hepatitis B on liver transplantation.

The corresponding set of equations is also very similar to the one studied in the case of hepatitis C:

$$\frac{dS_1(t)}{dt} = \Lambda - \mu\beta_1 HBV(t)\frac{S_1(t)}{N} - \mu\beta_2 Ac(t)\frac{S_1(t)}{N} - \mu\beta_3 T_A(t)\frac{S_1(t)}{N}$$
$$-\mu\beta_4 Chr(t)\frac{S_1(t)}{N} - (\mu + \omega + \xi)S_1(t) \qquad (21)$$

$$\frac{dS_2(t)}{dt} = \xi S_1(t) - \beta_1 HBV(t)\frac{S_2(t)}{N} - \beta_2 Ac(t)\frac{S_2(t)}{N} - \beta_3 T_A(t)\frac{S_2(t)}{N}$$
$$-\beta_4 Chr(t)\frac{S_2(t)}{N} - (\mu + \omega)S_2(t) \qquad (22)$$

$$\frac{dHBV(t)}{dt} = \beta_1 HBV(t)\left(\frac{S_1(t)+S_2(t)}{N}\right) - \beta_2 Ac(t)\left(\frac{S_1(t)+S_2(t)}{N}\right)$$
$$-\beta_3 T_A(t)\left(\frac{S_1(t)+S_2(t)}{N}\right) - \beta_4 Chr(t)\left(\frac{S_1(t)+S_2(t)}{N}\right)$$
$$-(\gamma_1 + \sigma_1 + \sigma_2 + \mu)HBV(t) \qquad (23)$$

$$\frac{dAc(t)}{dt} = \sigma_1 HBV(t) - (\delta_1 + \gamma_2 + \alpha_1 + \mu)Ac(t) \qquad (24)$$

$$\frac{dR(t)}{dt} = \gamma_1 HBV(t) + \gamma_2 Ac(t) + \gamma_4 Ac(t) + \gamma_4 Chr(t) - \mu R(t) \qquad (25)$$

$$\frac{dChr(t)}{dt} = \delta_1 Ac(t) + \sigma_2 HBV(t) - (\delta_2 + \theta + \alpha_2 + \gamma_3)Chr(t) \qquad (26)$$

$$\frac{dMELD_1(t)}{dt} = \delta_2(Chr(t) + Oth(t)) -$$
$$(\phi_1 + \epsilon_1 + \alpha_3 + \mu)NT_cM_1(t) \qquad (27)$$

$$\frac{dMELD_2(t)}{dt} = \epsilon_1 NT_cM_1(t) - (\phi_2 + \epsilon_2 + \alpha_4 + \mu)NT_cM_2(t) \qquad (28)$$

$$\frac{dMELD_3(t)}{dt} = \epsilon_2 NT_cM_2(t) - (\phi_3 + \epsilon_3\alpha_5 + \mu)NT_cM_3(t) \qquad (29)$$

$$\frac{dMELD_4(t)}{dt} = \epsilon_2 NT_cM_2(t) - (\phi_4 + \alpha_6 + \mu)NT_cM_4(t) \qquad (30)$$

$$\frac{dHCC(t)}{dt} = \theta Chr(t) - (\phi_5 + \alpha_7 + \mu)HCC(t) \quad (31)$$

$$\frac{dWL(t)}{dt} = \sum_{i,j=1}^{4} \phi_i NT_C M_j + \phi_5 Chr(t) - (\tau_1 + \alpha_8 + \mu)WL(t) \quad (32)$$

$$\frac{dTx_1(t)}{dt} = \tau_1 WL(t) - (\phi_6 + \alpha_9 + \mu)Tx_1(t) \quad (33)$$

$$\frac{WLTx(t)}{dt} = \phi_6 Tx_1(t) - (\tau_2 + \alpha_{10} + \mu)WLTx(t) \quad (34)$$

$$\frac{Tx_2(t)}{dt} = \tau_2 WLTx(t) - (\alpha_{11} + \mu)Tx_2 \quad (35)$$

$$\frac{Oth(t)}{dt} = \omega(S_1(t) + S_2(t)) - (\delta_2 + \alpha_{12} + \mu)Oth(t) \quad (36)$$

The impact of the vaccine against hepatitis B is shown in Figure 9.

Our analysis suggests that increasing the vaccination coverage against HBV in the State of So Paulo would have a relatively low impact on the number of liver transplantation. In addition, this impact would take several decades to materialize due to the long incubation period of liver failure due to HBV.

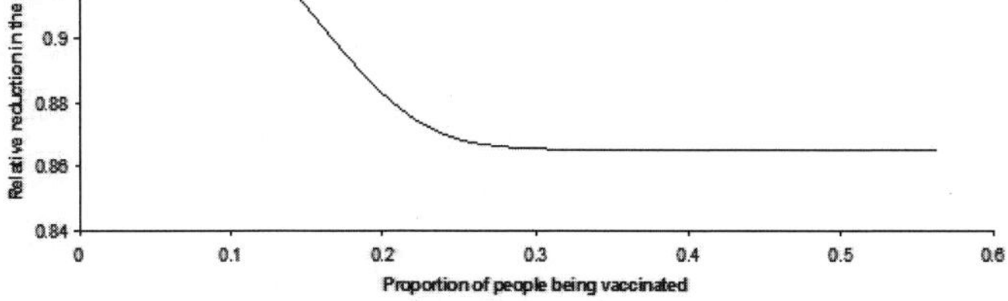

Figure 9. Impact of a vaccine against hepatitis B on the relative reduction in the number liver transplantations.

7. A model for optimizing the indications of liver transplantation in patients with hepatocellular carcinoma in the state of São Paulo, Brazil

The Milan Criteria, MC, is defined by the presence of a single nodule up to 5 cm, up to three nodules none larger than 3 cm, with no evidence of extra-hepatic spread or macrovascular invasion. The Brazilian law allows patients only within MC to be evaluated and considered for LT. This police implies that some patients with HCC slightly more advanced than those allowed by the current strict selection criteria will be excluded, even though LT for these patients might be associated with acceptable long-term outcomes.

In this study we propose a mathematical approach to study the consequences of relaxing the MC for patients with HCC that do not comply with the current rules for inclusion in the transplantation candidate list. We consider overall 5-years survival rates compatible with the ones reported in the literature. We simulate our model in order to reproduce what is known about the survival of the two groups of patients (those who comply with the strict MC and those who do not) and calculate the best strategy that would minimize the total mortality of the affected population, that is, the total number of people in both group that dies after 5 years of the implementation of the strategy, either by post-transplantation death or by death due to the basic HCC.

The model is based in the following assumptions:

- the mortality of both transplanted and non-transplanted HCC patients is a monotonically increasing function of tumor size at presentation (tumor size is, therefore, taken as an indication of gravity).
- the number of available livers to be grafted is limited and always less than the total number of HCC who have transplantation indication; and finally,
- the tumor size is distributed in the HCC population according to an exponential distribution, such that the probability that a given HCC patient has tumor size, s, is described by the probability density function (p.d.f.):

$$f(s, \lambda) = \lambda e^{\lambda s} \tag{37}$$

Assuming an average tumor size of 5 cm implies in l equal to 0.2, with variance 25 cm^2. We will simulate a population of $1,500$ HCC patients, which will result in tumor sizes averaging 5 ± 1.27 cm, which is in accord to what is observed in a real situation.

Total Mortality after T years is given by:

$$M(s_0) = N - F\int_0^{s_0} \lambda e^{-\lambda s} e^{-\alpha_t(s)T} ds -$$
$$N\int_0^{s_0} \left(1 - \frac{F}{N}\right) \lambda e^{-\lambda s} e^{-\alpha_{nt}(s)T} ds -$$
$$\int_{s_0}^{\infty} \lambda e^{-\lambda s} e^{-\alpha_{nt}(s)T} ds \quad (38)$$

where $t(s)$ is the mortality rate of transplanted patients and $nt(s)$ is the mortality rate of non-transplanted patients. We illustrate the above analysis for a simulation of a theoretical population of $1,500$ HCC patients with tumor size parameter distribution of l equal to 0.2, with variance 25 cm^2. As mentioned above, this implies in tumor sizes averaging 5 ± 1.27 cm, which is in accord to what is observed in a real situation. The total number of available livers to be grafted was assumed to be 500. With this, we simulated the total number of deaths in both transplanted and non-transplanted HCC patients after 5 years as a function of the tumor size of transplanted patients. The result is shown in Figure 10.

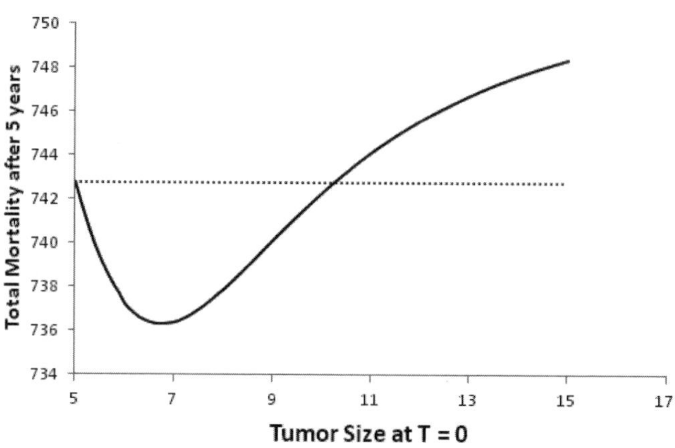

Figure 10. Total mortality of patients in the waiting list as a function of tumor size five years after transplantation, according to Eq. 38.

Figure 10 shows the total mortality in the HCC patients cohort, including those transplanted and those non-transplanted as well. Note that it is possible to include patients with tumor size up to 10 *cm* without increasing the total mortality of this cohort.

8. Final comments

The aim of this paper was to show how mathematical models can be useful in helping and/guiding the designing of optimization of a serious public health problem, namely, the growing size of the waiting list of liver transplantation.

In the last 10 years approximately 1300 patients are included every year in the liver transplantation waiting list in the State of So Paulo, of which only about 500 receive a graft. In the same period, an average 400 patients die every year in the waiting list. As seen in Figure 1 the prospects are dire indeed. Even the adoption of the Model for End-Stage Liver Disease (MELD) score (which comprises some laboratory exams and that is used as a surrogate of gravity) for an entrance criteria in the list, although resulting in a significant improvement in the first few years, will solve the problem.

Of course, the optimal solution of this serious public health problem involves preventative measure of the causes of liver failure and ways to increase the number of donors.

We showed that the use of mathematical models can help decision makers to improve their policies of list entrance criteria, as well as to quantify the impact of preventive measures like vaccination against the three viruses of hepatitis.

Much remains to be done but we hope that this brief review can stimulate young and better prepared mathematicians to contribute to the amelioration of this and other epidemiological problems related to human and animal health.

References

1. H. B. de Carvalho, F. Mesquita, E. Massad, R. C. Bueno, G. T. Lopes, M. A. Ruiz, M. N. Burattini, *J Acquir. Immune Defic. Syndr. Hum. Retrovirol.* **12(1)**, 84 (1996).
2. E. Chaib and E. Massad, *Transplant. Proc.* **37(10)**, 4329 (2005).
3. E. Chaib and E. Massad, *Transplantation* **84(9)** 1209 (2007).
4. E. Chaib and E. Massad, *Transpl. Int.* **21(3)** 290 (2008).
5. E. Chaib and E. Massad, *Liver Transpl.* **14** 1732 (2008).
6. E. Chaib, C. Fridman and E. Massad E, *Transplantation Proceedings* **41** 3775 (2009).

7. E. Massad, F. A. B. Coutinho, E. Chaib and M. N. Burattini, *Epidemiol. Infect.* **137(2)** 241 (2009).
8. E. Chaib, M. C. de Oliveira, F. H. Galvão, F. D. Silva, L. A. D'Albuquerque and E. Massad. *Med. Hypotheses* **75(3)** 324 (2010).

COMPLEXITY OF MOLECULAR SIGNALING NETWORKS FOR VARIOUS TYPES OF CANCER AND NEUROLOGICAL DISEASES CORRELATES WITH PATIENT SURVIVABILITY

D. BREITKREUTZ

Department of Oncology, University of Alberta, Edmonton, Alberta, T6G 1Z2, Canada

E. A. RIETMAN

Center of Cancer Systems Biology, St. Elizabeth's Medical Center, Tufts University School of Medicine, Boston, MA, 02135, USA

P. HINOW

Department of Mathematical Sciences, University of Wisconsin - Milwaukee, Milwaukee, WI, 53201-0413, USA

M. HEALEY

Department of Physics, University of Alberta, Edmonton, Alberta, T6G 2E1, Canada

J. A. TUSZYNSKI

Department of Physics, University of Alberta, Edmonton, T6G 2E1, Canada
Department of Oncology, University of Alberta, Edmonton, T6G 1Z2, Canada

As the population ages, the number of those affected by both cancer and neurodegenerative diseases will continue to increase. In order to offset the social, economic, and personal destruction that these diseases cause new methods of analyzing the disease states and identifying potential drug targets are becoming imperative for the treatment of these diseases. We had previously studied the correlation between cancer survival statistics and cancer pathway network properties, specifically the degree-entropy, and found that there was a correlation between degree-entropy and 5-year survival statistics. Additionally, we used the betweenness centrality measure to identify potential protein targets for new drugs. Recently, we examined other network properties for the cancer pathways, specifically relative automorphism group size and cyclomatic number. Furthermore, we extended the same strategy of network analysis to neurodegenerative conditions. As with the cancer pathways, we observed a correlation between the complexity of the networks with the mortality of patients after diagnosis, and identified potential target proteins.

1. Introduction

In recent years the systems biology approach to understanding cell biology has advanced. Information about the interactions between proteins has been determined by experimental techniques including yeast two-hybrid, affinity pull-down mass spectrometry, biochemical techniques, and next-generation sequencing[1,2]. This interaction information has been made available in online databases[3,4]. It is our goal to apply this systems biology knowledge so as to better understand human diseases.

The initial input to our approach are protein signaling pathways associated with specific human disease states. To obtain the pathway information, we have relied on the Kyoto Encyclopedia of Genes and Genomes (KEGG) PATHWAY database[5,6,7,8,9]. KEGG PATHWAY currently contains about 500 pathways, including more than 70 pathways for human diseases[10]. The KEGG pathway maps are manually drawn pathway maps that summarize the current knowledge concerning the protein interaction networks that are disregulated in a disease state. The nodes of these pathways include both specific proteins (for example p53 and alpha-synuclein) and also higher level cell and tissue processes (for example apoptosis and neuronal injury). Both extracellular and intracellular interactions are included. These signaling pathways are primarily protein-protein interaction (PPI) networks[11], therefore we discard the non-protein interactions. Once we have obtained the protein signaling pathways, our next step is to convert the pathways into mathematical graphs (also known as networks), and then to calculate network properties both for the overall graph (such as degree-entropy) and for specific proteins. In our graph representation specific proteins are nodes (nodes are also known as vertices), and the interactions between the proteins are represented as the connections between nodes (these connections are also known as edges). Furthermore, the information classifying a particular interaction as an activation or an inhibition is ignored; only the connections are considered as important.

We have investigated the correlation between several network properties and patient mortality statistics[12]. To obtain patient mortality statistics for different types of cancer we used the Surveillance Epidemiology and End Results (SEER) Program database[13]. We have also calculated measures for individual nodes (such as betweenness centrality), which we suggest can be used to identify new target proteins for drug discovery research[12].

In previous work we focused on the pathway maps for cancer[12,14]. In this paper we will review that previous work, and also present new work applying similar methods to neurodegenerative diseases. For both cancer[12] and for neurodegenerative diseases we have observed a correlation between degree-entropy and patient survival statistics.

2. Cancer pathways

2.1. *Cancer patient survival and degree-entropy*

Our initial conjecture was that diseases with more "robust" networks would be more resistant to drug treatment and would have worse prognosis. By robustness we meant networks that were more able to continue normal functioning after the removal of random nodes. To mathematically describe the concept robustness we tried several different measures, and we found that degree entropy had the best correlation with patient survival[12]. Degree-entropy, H, is defined as:

$$H = -\sum_{k=1}^{N-1} p(k) \log p(k) \qquad (1)$$

where N is the total number of nodes in the network and $p(k)$ is the degree (number of edges) of node k[15]. Degree-entropy is a measure of the complexity of a network, with more complex networks having a higher degree entropy. To further understand degree-entropy measure and its relationship to the concept of robustness, we calculated the entropy as random edges are removed. Figure 1 shows the change in entropy as random edges are removed from the acute myeloid leukemia (AML) network. The figure shows that the entropy tends to decrease as more edges are removed.

Figure 2 shows a plot of 5-year survival vs. degree-entropy for 13 cancer types (sites of the primary tumor)[12]. Note that prostate cancer was an outlier and was removed from the analysis. As Figure 2 shows, a lower frequency of 5-year survival is correlated with a higher degree-entropy. In other words, cancer types that have more complex KEGG pathways tend to have a worse prognosis.

2.2. *Cancer protein targets and betweenness centrality*

A second strand of the network analysis sought to identify specific proteins that would be the best targets for inhibition in order to disrupt the signaling pathway associated with the disease. The idea is that such proteins would be good targets for future drug discovery research projects.

For each of the cancer pathways we calculated the betweenness centrality measure for each node[12]. Betweenness centrality is a measure of the extent

Figure 1. Change in entropy of the acute myeloid leukemia network due to random edge removal.

that a node is situated on the paths between other nodes[16]. A node with higher betweenness centrality will tend to have a greater effect on a network if it is removed. Betweenness centrality is defined by:

$$c_B(V) = \sum_{s,t \in V} \frac{\sigma(s,t|v)}{\sigma(s,t)} \tag{2}$$

where $\sigma(s,t)$ is the number of shortest paths between two nodes (s,t) and $\sigma(s,t|v)$ is the number of those paths passing through nodes other than the (s,t) pair[16].

Table 1 gives the top 3 proteins in betweenness centrality for the types of cancer that have KEGG pathways. The most frequent protein in the combined top 3 lists is KRAS. This gene is a member of the Ras family and is frequently mutated in cancer cells[17]. Since it is an essential protein in normal cells, inhibiting its function with a systemic drug may not be a good strategy; however a drug that targeted only mutant KRAS might be useful. There are no currently approved anticancer drugs that directly target KRAS[18].

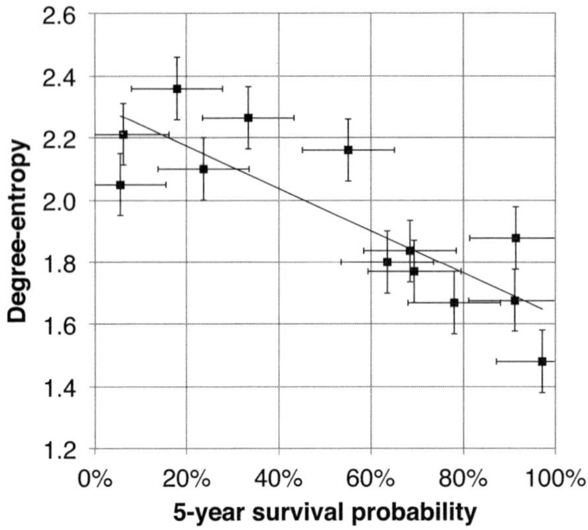

Figure 2. Scatter plot showing the correlation between cancer 5-year survival probability and degree-entropy. Data points are shown for 13 cancer sites, including all of the cancer types with KEGG pathways, except prostate cancer. The x-axis is the 5-year survival probability for the cancer site, from the SEER database[13]. The y-axis is the degree-entropy (H) for the cancer site, calculated from the KEGG pathway[10]. The line is a linear regression fit, with $R^2 = 0.7$. The error bars are set at fixed estimates of the error, with the x-axis error bars fixed at $\pm 10\%$ and the y-axis error bars fixed at ± 0.1. Figure reproduced from Breitkreutz *et al*[12].

3. Networks in Neurodegenerative Diseases

Many neurodegenerative diseases (NDD) result in the gradual decline of physical or psychological abilities, or a combination of physical and psychological abilities. The current treatment regimens for these diseases focus on symptomatic reduction, or on combatting the secondary disorders that are common in people suffering from NDD's. These diseases typically result in decreased lifespan relative to the normal population, while affecting older age groups. For example, the prevalence of Alzheimer's in those younger then 60 years is practically non-existent. Table 2 summarizes some of the important current prevalence statistics. These numbers are expected to increase as the mean age of our population increases, for example it is estimated that Alzheimer's will affect 1 in 85 people worldwide by the year 2050[20].

It is well known that today's developed nations hold a larger proportion of their general population in older age brackets. This refers to, of course,

Table 1. Cancer survival probabilities and network statistics for 14 cancer types. The columns B1, B2, and B3 give the HGNC gene symbols[19] for the top three betweenness centrality nodes. Table reproduced from Breitkreutz *et al*[12].

Cancer Type	B1	B2	B3
Acute myeloid leukemia	FLT3	SPI1	JUP
Basal cell carcinoma	GSK3B	CTNNB1	GLI1
Bladder cancer	MAP2K2	MAP2K1	MAPK3
Chronic myeloid leukemia	GRB2	MDM2	GAB2
Colorectal cancer	KRAS	RALGDS	CTNNB1
Endometrial cancer	GRB2	ADARB2	PDPK1
Glioma	EGFR	IGF1R	PDGFRB
Melanoma	NRAS	MAP2K1	MAPK3
Non small-cell lung cancer	KRAS	RASSF1	CCND1
Pancreatic cancer	KRAS	RALBP1	JAK1
Prostate cancer	GRB2	GSK3B	AKT1
Renal cell carcinoma	GAB1	RFC1	MAPK1
Small cell lung cancer	NFKBIA	PTK2	CCND1
Thyroid cancer	HRAS	NRAS	KRAS

Table 2. Summary of the epidemiological data for the key neurodegenerative diseases.

Parkinson's Disease	0.3% of people in industrialized nation[21]
Amyotrophic Lateral Sclerosis	Affects 1 (or 2) per 100 thousand people[22]
Huntington's Disease	5–10 cases per 100 thousand persons[23]
Alzheimer's Disease	4.5 million US citizens[24]

the current "baby boomer" generation. This age distibution places a burden on current health care systems due to age related diseases such as NDD's. In order to combat the effect that NDD's will have on the well-being of persons suffering from the disease, as well as on those responsible for their care, new methods to understand the disease states and to identify potential drug targets are needed.

3.1. *Cancer protein targets and betweenness centrality*

We analyzed the molecular signaling pathways in a similar manner as was previously performed for the analysis of cancer related signaling pathways[12]. In this portion of the research, we however narrowed our focus to those diseases that were both considered lethal and relatively common. This lead us to focus on the following NDD's:

(1) Alzheimer's Disease – A neurodegenerative condition characterized by the gradual loss of ability to form new memories, as well as the loss of previously obtained memories. This is also accompanied by psychological changes such as paranoia, anxiety, etc[25].
(2) Parkinson's Disease – A disease associated with the degeneration of motor neurons. This subsequently leads to the loss of voluntary and involuntary muscle coordination[26].
(3) Amyotrophic Lateral Sclerosis (ALS) – A degenerative motor neuron disease that causes degeneration in the nerves controlling muscles and coordination[27].
(4) Huntington's Disease – An inherited neurodegenerative disease associated with cognitive, emotional, and physical decline leading to incapacitation and eventually death[28].

In order to develop the network models of the pathways associated with each disease, the KEGG pathways were downloaded from the KEGG database in the KGML file format[5,9]. These were then converted to adjacency matrices using the KEGGgraph library[29]. For this study, we analyzed both the directed and undirected forms of these graphs. In the case of the undirected graphs, the determination of the network entropy was performed as previously mentioned. To determine the entropy of the directed graphs we applied the constraint that the entropy of the vertices entering a node was zero, whereas we calculated the entropy determined by assessing the probability of a vertex exiting the node. Where unconnected nodes were found in some networks, the analysis also lead to determining the variation in degree-entropy (Eq. 1) amongst networks both if the unconnected nodes were removed and if the nodes were not removed. Furthermore the betweenness centrality measure (Eq. 2) was reported for the top three scores in each network to determine the relative importance of the node to the network.

Due to the age dependent reduction in life span associated with NDD's, our comparative measure was assessed using the crude mean standardized mortality ratio (SMR) for each disorder. The SMR is defined as the ratio of the observed number of deaths to the expected number of deaths for an age or age group. This is typically represented as,

$$SMT = \frac{O}{E} = \frac{LE_{\text{Ref}}}{LE_{\text{Study}}} \tag{3}$$

where E and O are the expected and observed number of deaths respectively, and LE_{Ref} and LE_{Study} are the remaining life expectancy the reference population and the study population, respectively. The SMR can also be thought of as the relative probability of deaths between diseased individuals and the general population.

The values for SMR were found in the literature. The typical manner of reporting moratility rate for a range of ages is to use a stratified distribution of ages, i.e. if the age of the study group is between 20 years and 60 years, one might divide them into groups separated in 10 year increments. The SMR of each age bracket was then determined by using the expected remaining life of the middle term in the bracket (assessed from a worldwide mortality database) versus the life expectancy of the bracket itself. The crude mean SMR was then taken as the mean of the SMR's for all age brackets in the study. In cases where the distribution of ages was very sharp, the mean age and median survival times were used to estimate SMR's. Or in some cases when Hazard Ratios (HR) were reported, the unadjusted HR was used as an estimate of the SMR. In certain instances, where it was more difficult to find an adequate single study, multiple studies were used to determine the mean SMR values.

3.2. *Results*

Upon comparing the network entropy for both directed and undirected networks, as well as for networks both with and without the removal of unconnected nodes, we found that in all of the cases the SMR of patient groups increased linearly with increasing network entropy. This correlation was most significant in cases where the directed networks were used. The removal of unconnected nodes had little effect on the overall correlation between entropy and SMR values. The results for the directed network with no nodes removed is shown in Figure 3. The results for all cases are given in Table 3. We also ranked the top three betweenness centrality values, shown in Table 4.

3.3. *Discussion*

Using the KEGG pathways for neurodegenerative diseases, we have shown a strong relationship between the complexity of the disease (as measured by its network entropy) and a poor disease prognosis as measured by SMR.

Furthermore, the measure of betweenness centrality was used to identify the important nodes in each disease state. For example, betweenness

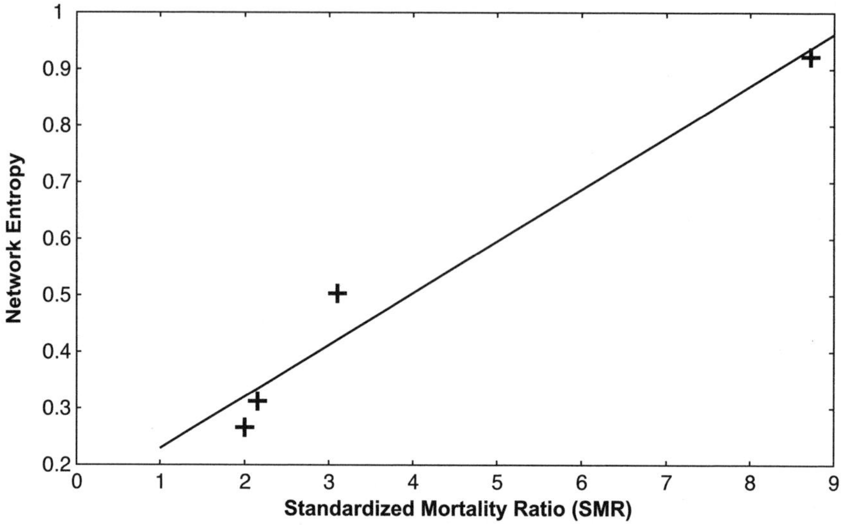

Figure 3. Results of the analysis of SMR as a function of network entropy for directed NDD networks with no nodes removed. The line is a linear fit with $R^2 = 0.96447$.

Table 3. The SMRs from the literature for all the NDD's in the study, and the calculated network entropy values for directed and undirected networks, and for networks where unconnected nodes were removed.

Disease	SMR	Entropy (Directed)	Entropy (Directed, Nodes Removed)	Entropy (Undirected)	Entropy (Undirected, Nodes Removed)
Parkinson's	1.91[30]	0.266	0.336	0.182	0.477
Huntington's	2.15[31]	0.312	0.667	0.295	0.775
Alzheimer's	3.1[32,33]	0.503	1.004	0.683	1.363
ALS	8.72[34,35]	0.922	1.196	1.170	1.446

centrality correctly identified the importance of cytochrome c in Parkinson's dieases[36,37,38], huntingtin in Huntington's disease[39], tau protein in Alzheimer's disease[40], and superoxide dismutase in ALS[41,42]. The betweenness centrality method also identified as important in Huntington's disease the glutamate receptor, NMDA-1; this protein has been recently identified as having an effect on the age of onset of the disease[43].

Table 4. The top three betweenness centrality values for each NDD, along with the names of the associated protein with that value.

	First (B1)	First (B2)	First (B3)
Parkinson's	0.00775 (Cytochrome C)	0.00258 (Caspase 9)	0.00074 (Ubiquitin B)
Huntington's	0.00714 (Huntingtin)	0.00070 (Guanine Nucleotide Binding Protein)	0.00045 (Glutamate Receptor, NMDA-1)
Alzheimer's	0.00051 (Cyclin-dependent Kinase 5, reg. subunit 1 (p35))	0.00051 (Microtubule assoc. protein tau)	0.00043 (Guanine Nucleotide Binding Protein)
ALS	0.34392 (Superoxide dismutase, Cu-Zn Family)	0.23048 (Neural Filament Light Protein)	0.14824 (Apoptosis Regulator BCL-2)

4. Conclusions

By using these network analysis methods to study disease states, it is possible to identify and target important areas for both more research and for drug discovery. In particular, our findings indicate that as with cancer neurodegenerative diseases are far from simple, and indeed a systems biology framework is a useful approach for their analysis. There are further algebraic and topological indices of pathway networks that support the view that greater network complexity is associated with poorer survival probability[14]. Moreover, due to the complexity involved in the initiation and progression of these diseases, a multi-component therapeutic approach is called for, and that no single-agent cures should be expected.

Acknowledgments

J.A.T. acknowledges funding from the Natural Sciences and Engineering Research Council of Canada, the Allard Foundation, Alberta Advanced Education and Technology, the Canadian Breast Cancer Foundation and the Alberta Cancer Foundation. The authors thank Philip Winter for assistance.

References

1. Dreze, M., Monachello, D., Lurin, C., Cusick, M. E., Hill, D. E., Vidal, M. & Braun, P. High-quality binary interactome mapping. *Meth. Enzymol.* **470**, 281–315 (2010).
2. Yu, H., Tardivo, L., Tam, S., Weiner, E., Gebreab, F., Fan, C., Svrzikapa, N., Hirozane-Kishikawa, T., Rietman, E., Yang, X., Sahalie, J., Salehi-Ashtiani, K., Hao, T., Cusick, M. E., Hill, D. E., Roth, F. P., Braun, P. & Vidal, M. Next-generation sequencing to generate interactome datasets. *Nat. Methods* **8**, 478–480 (2011).
3. Chatr-Aryamontri, A., Breitkreutz, B.-J., Heinicke, S., Boucher, L., Winter, A., Stark, C., Nixon, J., Ramage, L., Kolas, N., O'Donnell, L., Reguly, T., Breitkreutz, A., Sellam, A., Chen, D., Chang, C., Rust, J., Livstone, M., Oughtred, R., Dolinski, K. & Tyers, M. The BioGRID interaction database: 2013 update. *Nucleic Acids Res.* **41**, D816–823 (2013).
4. Kandasamy, K., Mohan, S. S., Raju, R., Keerthikumar, S., Kumar, G. S. S., Venugopal, A. K., Telikicherla, D., Navarro, J. D., Mathivanan, S., Pecquet, C., Gollapudi, S. K., Tattikota, S. G., Mohan, S., Padhukasahasram, H., Subbannayya, Y., Goel, R., Jacob, H. K. C., Zhong, J., Sekhar, R., Nanjappa, V., Balakrishnan, L., Subbaiah, R., Ramachandra, Y. L., Rahiman, B. A., Prasad, T. S. K., Lin, J.-X., Houtman, J. C. D., Desiderio, S., Renauld, J.-C., Constantinescu, S. N., Ohara, O., Hirano, T., Kubo, M., Singh, S., Khatri, P., Draghici, S., Bader, G. D., Sander, C., Leonard, W. J. & Pandey, A. NetPath: a public resource of curated signal transduction pathways. *Genome Biol.* **11**, R3 (2010).
5. Kanehisa, M. & Goto, S. KEGG: kyoto encyclopedia of genes and genomes. *Nucleic Acids Res.* **28**, 27–30 (2000).
6. Kanehisa, M., Goto, S., Kawashima, S., Okuno, Y. & Hattori, M. The KEGG resource for deciphering the genome. *Nucleic Acids Res.* **32**, D277–280 (2004).
7. Kanehisa, M., Goto, S., Hattori, M., Aoki-Kinoshita, K. F., Itoh, M., Kawashima, S., Katayama, T., Araki, M. & Hirakawa, M. From genomics to chemical genomics: new developments in KEGG. *Nucleic Acids Res.* **34**, D354–357 (2006).
8. Kanehisa, M., Goto, S., Furumichi, M., Tanabe, M. & Hirakawa, M. KEGG for representation and analysis of molecular networks involving diseases and drugs. *Nucleic Acids Res.* **38**, D355–360 (2010).
9. Kanehisa, M., Goto, S., Sato, Y., Furumichi, M. & Tanabe, M. KEGG for integration and interpretation of large-scale molecular data sets. *Nucleic Acids Res.* **40**, D109–114 (2012).
10. KEGG: Kyoto Encyclopedia of Genes and Genomes. http://www.genome.jp/kegg/
11. De Las Rivas, J. & Fontanillo, C. Protein-protein interactions essentials: key concepts to building and analyzing interactome networks. *PLoS Comput. Biol.* **6**, e1000807 (2010).

12. Breitkreutz, D., Hlatky, L., Rietman, E. & Tuszynski, J. A. Molecular signaling network complexity is correlated with cancer patient survivability. *Proc. Natl. Acad. Sci. U.S.A.* 109, 9209–9212 (2012).
13. SEER Web Site. http://seer.cancer.gov/
14. Hinow, P., Rietman, E. A. & Tuszynski, J. A. Algebraic and topological indices of molecular pathway networks in human cancers. Submitted to *Proc. Natl. Acad. Sci. U.S.A.* (2013).
15. Wang, B., Tang, H., Guo, C. & Xiu, Z. Entropy optimization of scale-free networks' robustness to random failures. *Physica A: Statistical Mechanics and its Applications* **363**, 591–596 (2006).
16. Newman, M. E. J. *Networks: an introduction.* (Oxford University Press: Oxford, 2010).
17. Kranenburg, O. The KRAS oncogene: past, present, and future. *Biochim. Biophys. Acta* **1756**, 81–82 (2005).
18. Knox, C., Law, V., Jewison, T., Liu, P., Ly, S., Frolkis, A., Pon, A., Banco, K., Mak, C., Djoumbou, Y., Eisner, R., Guo, A. C. & Wishart, D. S. DrugBank 3.0: a comprehensive resource for 'omics' research on drugs. *Nucleic Acids Res.* **39**, D1035–1041 (2011).
19. HUGO Gene Nomenclature Committee Home Page — HUGO Gene Nomenclature Committee. http://www.genenames.org/
20. Brookmeyer, R., Johnson, E., Ziegler-Graham, K. & Arrighi, H. M. Forecasting the global burden of Alzheimer's disease. *Alzheimer's & Dementia* **3**, 186–191 (2007).
21. De Lau, L. M. L. & Breteler, M. M. B. Epidemiology of Parkinson's disease. *Lancet Neurol.* **5**, 525–535 (2006).
22. Ropper, A. H., Adams, R. D., Victor, M. & Brown, R. H. Degenerative diseases of the nervous system. *Adams and Victor's principles of neurology* 1058–1070 (McGraw-Hill: New York, 2010).
23. Sharon, I., Sharon, R., Wilkens, J. P. & Ersan, T. Huntington Disease Dementia. *Medscape* (2012). http://emedicine.medscape.com/article/289706-overview
24. Hebert, L. E., Scherr, P. A., Bienias, J. L., Bennett, D. A. & Evans, D. A. Alzheimer disease in the US population: prevalence estimates using the 2000 census. *Arch. Neurol.* **60**, 1119–1122 (2003).
25. Forstl, H. & Kurz, A. Clinical features of Alzheimer's disease. *Eur. Arch. Psychiatry Clin. Neurosci.* **249**, 288–290 (1999).
26. Jankovic, J. Parkinson's disease: clinical features and diagnosis. *J. Neurol. Neurosurg. Psychiatr.* **79**, 368–376 (2008).
27. Jasmin, L., Dugdale, D. C. & Zieve, D. Amyotrophic lateral sclerosis. *MedlinePlus Medical Encyclopedia* (2012). http://www.nlm.nih.gov/medlineplus/ency/article/000688.htm
28. Huntington. http://www.huntingtonsociety.ca/english/index.asp
29. Gentleman, R. C., Carey, V. J., Bates, D. M., Bolstad, B., Dettling, M., Dudoit, S., Ellis, B., Gautier, L., Ge, Y., Gentry, J., Hornik, K., Hothorn, T., Huber, W., Iacus, S., Irizarry, R., Leisch, F., Li, C., Maechler, M., Rossini, A. J., Sawitzki, G., Smith, C., Smyth, G., Tierney, L., Yang, J. Y. H. & Zhang,

J. Bioconductor: open software development for computational biology and bioinformatics. *Genome Biol.* **5**, R80 (2004).
30. Ishihara, L. S., Cheesbrough, A., Brayne, C. & Schrag, A. Estimated life expectancy of Parkinson's patients compared with the UK population. *J. Neurol. Neurosurg. Psychiatry* **78**, 1304–1309 (2007).
31. Foroud, T., Gray, J., Ivashina, J. & Conneally, P. M. Differences in duration of Huntington's disease based on age at onset. *J. Neurol. Neurosurg. Psychiatr.* **66**, 52–56 (1999).
32. Ganguli, M., Dodge, H. H., Shen, C., Pandav, R. S. & DeKosky, S. T. Alzheimer disease and mortality: a 15-year epidemiological study. *Arch. Neurol.* **62**, 779–784 (2005).
33. Burns, A., Lewis, G., Jacoby, R. & Levy, R. Factors affecting survival in Alzheimer's disease. *Psychol. Med.* **21**, 363–370 (1991).
34. Louwerse, E. S., Visser, C. E., Bossuyt, P. M. & Weverling, G. J. Amyotrophic lateral sclerosis: mortality risk during the course of the disease and prognostic factors. The Netherlands ALS Consortium. *J. Neurol. Sci.* **152** Suppl 1, S10–17 (1997).
35. Chio, A., Logroscino, G., Hardiman, O., Swingler, R., Mitchell, D., Beghi, E., Traynor, B. G. & Eurals Consortium Prognostic factors in ALS: A critical review. *Amyotroph. Lateral Scler.* **10**, 310–323 (2009).
36. Mounsey, R. B. & Teismann, P. Mitochondrial Dysfunction in Parkinson's Disease: Pathogenesis and Neuroprotection. *Parkinson's Disease* **2011**, 617472 (2011).
37. Sherer, T. B., Betarbet, R., Stout, A. K., Lund, S., Baptista, M., Panov, A. V., Cookson, M. R. & Greenamyre, J. T. An in vitro model of Parkinson's disease: linking mitochondrial impairment to altered alpha-synuclein metabolism and oxidative damage. *J. Neurosci.* **22**, 7006–7015 (2002).
38. Hashimoto, M., Takeda, A., Hsu, L. J., Takenouchi, T. & Masliah, E. Role of cytochrome c as a stimulator of alpha-synuclein aggregation in Lewy body disease. *J. Biol. Chem.* **274**, 28849–28852 (1999).
39. Walker, F. O. Huntington's disease. *Lancet* **369**, 218–228 (2007).
40. Garcia, M. L. & Cleveland, D. W. Going new places using an old MAP: tau, microtubules and human neurodegenerative disease. *Curr. Opin. Cell Biol.* **13**, 41–48 (2001).
41. Deng, H. X., Hentati, A., Tainer, J. A., Iqbal, Z., Cayabyab, A., Hung, W. Y., Getzoff, E. D., Hu, P., Herzfeldt, B. & Roos, R. P. Amyotrophic lateral sclerosis and structural defects in Cu,Zn superoxide dismutase. *Science* **261**, 1047–1051 (1993).
42. Al-Chalabi, A. & Leigh, P. N. Recent advances in amyotrophic lateral sclerosis. *Curr. Opin. Neurol.* **13**, 397–405 (2000).
43. Diguet, E., Fernagut, P.-O., Normand, E., Centelles, L., Mulle, C. & Tison, F. Experimental basis for the putative role of GluR6/kainate glutamate receptor subunit in Huntington's disease natural history. *Neurobiol. Dis.* **15**, 667–675 (2004).

MODELLING MALARIA DYNAMICS IN TEMPERATE REGIONS WITH LONG TERM INCUBATION PERIOD

KYEONGAH NAH & GERGELY RÖST

Bolyai Institute
University of Szeged
Aradi vértanúk tere 1, H–6720 Szeged, Hungary
E-mail: knah@math.u-szeged.hu, rost@math.u-szeged.hu

YONGKUK KIM

Department of Mathematics
Kyungpook National University
702–701, Daegu, Republic of Korea
E-mail: yongkuk@knu.ac.kr

The incubation period of malaria can vary depending on the species of parasite or the geographic regions. In particular, in endemic areas of temperate climate, the incubation period of *Plasmodium vivax* shows bimodal distribution of short and long term incubation periods. In this paper, we compare two transmission models for *P. vivax* malaria, where we model the long term incubation period using ordinary differential equations or delay differential equations. We show that, while the qualitative behaviors of the two models are similar, the ODE model overestimates the basic reproduction number and also the level of endemicity, compared to the DDE model. However, when we incorporate seasonality, the interplay of the time delay and the periodicity results that in some situations the DDE model predicts higher prevalence of malaria.

1. Introduction

Malaria is a mosquito-borne infectious disease caused by protozoan parasites of the genus *Plasmodium*. While feeding on humans, infected female mosquitoes inject parasites into the bloodstream, which infect liver cells. Parasites stay in liver cells until they are released back into the bloodstream and ready to be spread to another mosquito. The time between the mosquito bite and the release of the parasites from the liver is the incubation period [1]. The incubation period can vary depending on the species of the parasite or the geographic regions. In particular, the incubation period of *Plasmodium vivax* – the malaria inducing parasite species most

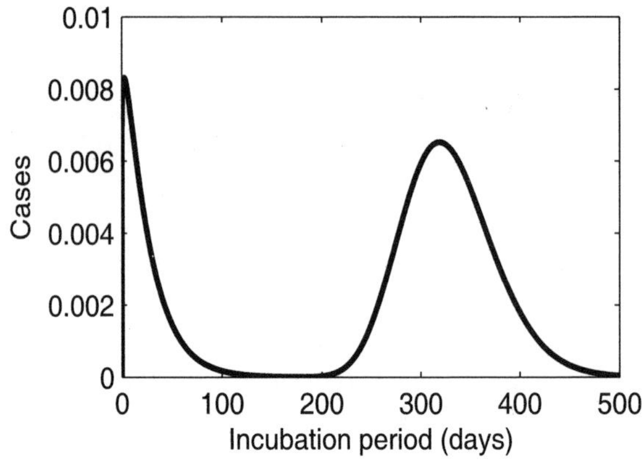

Figure 1. Estimation of the probability density function of *P. vivax* incubation time in Korea [8,15].

prevalent in temperate zones – shows a bimodal distribution, with clearly distinct short term and long term incubation periods.

The classical mathematical models for the dynamics of malaria transmission are based on differential equations[1,2], following the works of Ross and Macdonald. The simplest models describe the incubation period by exponential distribution, thus using ordinary differential equations in which the latent compartment decays exponentially in the absence of inflow from the susceptible compartment. Meanwhile, the delayed Ross-Macdonald model considers that the incubation periods of hosts and vectors have fixed lengths, thus including constant delay terms in the model equations. Most of the previous malaria models adapted either exponential distribution or Dirac–delta distribution[3,4,5,6,7].

The research of Nah *et al.*[7] used two separated compartments for the latent class – distinguishing short and long incubation times – in order to express the distribution of incubation times of *P. vivax* observed in Korea. That model assumes exponential distribution for both the short and the long incubation times, with different mean values. However, based on the empirical estimations of *P. vivax* incubation time in Korea[8,9], it is natural to use discrete delay for the long term incubation period, as a much better approximation of the empirical observation (see Fig. 1) than the exponential distribution assumption. The cumulative distribution functions are depicted in Fig. 2, and one can see that assuming fixed length for the long term incubation period gives a distribution that is much closer to

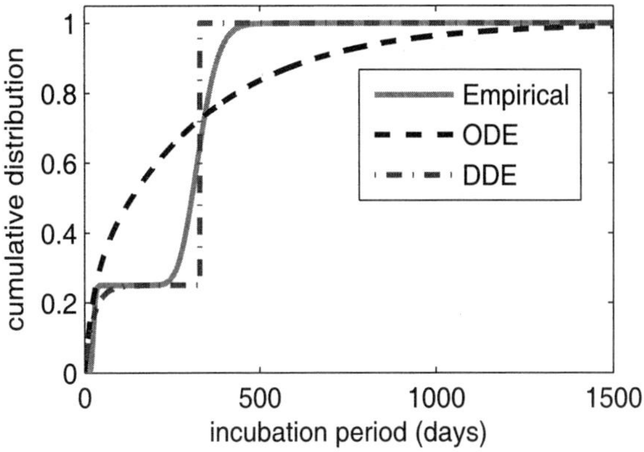

Figure 2. Cumulative distribution functions of incubation time. The Kantorovich distance between the distributions of the incubation time using DDE model and Empirical observation was 31.4, being much smaller than the distance between ODE model and Empirical observation, which is 154.8.

the empirical distribution in the most common probability metrics (such as the Kantorovich metric or the Lévy metric[a]), than the exponentially distributed long term incubation period.

In this paper, we introduce two models for *P. vivax* transmission dynamics where both short and long incubation times are present. In both cases, we separate the exposed individuals into two distinct compartments, depending on the length of their incubation period (short term or long term). In the first model, we assume exponential distribution for the long term incubation period, thus resulting a system of ordinary differential equations (ODE). In the second model, we assume fixed length for the long term incubation period, obtaining a system of delay differential equations (DDE). Our goal is to compare the two models by means of mathematical analysis, to investigate the qualitative and quantitative differences between the two models, and to discuss the implications of these two approaches. Since a contact rate between mosquitoes and humans has strong seasonality in temperate regions where *P. vivax* is endemic, we also study the disease dynamics given by those two models in a periodic environment.

[a]If F_X and F_Y are the distribution functions of random variables X and Y, the Kantorovich distance is defined by $d_K(X,Y) := \int_{-\infty}^{\infty} |F_X(x) - F_Y(x)|dx$, and the Lévy distance is $d_L(X,Y) := \inf\{\epsilon : F_Y(x-\epsilon) - \epsilon \leq F_X(x) \leq F_Y(x+\epsilon) + \epsilon\}$ [10].

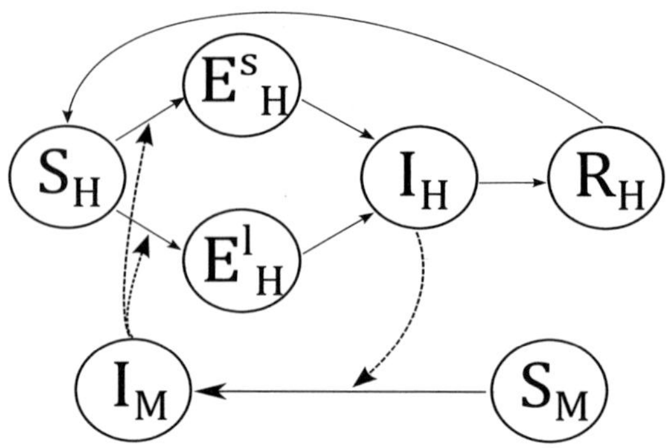

Figure 3. *P. vivax* transmission model.

2. Model description

To describe the transmission of *P. vivax* malaria, we assume SEIRS disease dynamics for the human and SI for the mosquito population (Fig. 3). Exposed humans are divided into two classes by having short term or long term incubation periods. If a susceptible human (S_H) is successfully infected by a mosquito (I_M), then this individual goes through short incubation period (E^s_H) with probability p, or long incubation period (E^l_H) with probability $1-p$; then becomes infectious (I_H) after this incubation time and be able to infect susceptible mosquitoes (S_M). Recovered humans are in the class R_H, and return to S_H after their immunity wanes.

The cross-infection between mosquitoes and humans is described by the terms $abI_M\frac{S_H}{H}$ and $acS_M\frac{I_H}{H}$, where a is the per capita biting rate of mosquitoes with b, c transmission efficiency, and H is the human population size. Assuming constant human and mosquito populations (H and M), our system can be rescaled by introducing the new variables

$$s_{_H} = \frac{S_H}{H}, \quad e^s_{_H} = \frac{E^s_H}{H}, \quad e^l_{_H} = \frac{E^l_H}{H}, \quad i_{_H} = \frac{I_H}{H}, \quad r_{_H} = \frac{R_H}{H},$$

$$s_{_M} = \frac{S_M}{M} \quad \text{and} \quad i_{_M} = \frac{I_M}{M}.$$

The cross-infection terms become $abmi_{_M}s_{_H}$ and $acs_{_M}i_{_H}$, where $m = \frac{M}{H}$.

We consider two models: the short term incubation time has an exponential distribution in both models, but the distributions of long term incubation time are different. The first model assumes exponential distribution for long incubation time, as in previous research. In the second model, individuals going through long term incubation time have the same length of incubation time, i.e. long term incubation time has Dirac–delta distribution. The second model better describes the observed distribution of long term incubation time than the first one (see Fig. 1 and 2), however, the second model is more difficult to analyze mathematically.

2.1. *ODE model with exponentially distributed incubation periods*

First we consider the system

$$\begin{cases} \dfrac{ds_H}{dt} = \xi - abms_H i_M + \omega r_H - \xi s_H, \\ \dfrac{de^s_H}{dt} = pabms_H i_M - d_s e^s_H - \xi e^s_H, \\ \dfrac{de^l_H}{dt} = (1-p)abms_H i_M - d_l e^l_H - \xi e^l_H, \\ \dfrac{di_H}{dt} = d_s e^s_H + d_l e^l_H - ri_H - \xi i_H, \\ \dfrac{dr_H}{dt} = ri_H - \omega r_H - \xi r_H, \\ \dfrac{ds_M}{dt} = \mu - acs_M i_H - \mu s_M, \\ \dfrac{di_M}{dt} = acs_M i_H - \mu i_M. \end{cases} \tag{1}$$

For the explanation of the parameters we refer to Table 1. Model (1) is modified from the model of Nah *et al.* [7], and assumes that individuals leave the exposed compartments at constant rate. The feasible domain $D = \{(s_H, e^s_H, e^l_H, i_H, r_H, s_H, i_M) \in \mathbb{R}^7_+ | s_H + e^s_H + e^l_H + i_H + r_H = 1, s_M + i_M = 1\}$ is clearly invariant.

The term abm describes the successful contacts infectious mosquitoes have with humans per unit time, $\frac{1}{\mu}$ is the length of the infectious period of mosquitoes. Since $p\frac{d_s}{d_s+\xi} + (1-p)\frac{d_l}{d_l+\xi}$ is the probability that an infected human survives the exposed state and becomes infectious, ac is the number of valid contacts infectious humans have with mosquitoes per unit time and $\frac{1}{r+\xi}$ is the length of the infectious period of a human; we can define the basic reproduction number R_o for the ODE model by

$$R_o = \sqrt{\frac{a^2bcm}{(r+\xi)\mu}\left(p\frac{d_s}{d_s+\xi} + (1-p)\frac{d_l}{d_l+\xi}\right)}, \tag{2}$$

where we adapted the convention of taking the square root as reproduction requires two epidemiological generations. We show that R_o works as a threshold for the existence and stability of equilibria of system (1).

Table 1. Description of model parameters.

parameter	description
ξ (μ)	human (mosquito) birth/death rate
b (c)	transmission efficiency from infected mosquito (human) to human (mosquito)
a	biting rate of mosquitoes
m	proportion of mosquito population to human population
d_s (d_l)	rate of progression from the short (long) term exposed state to the infectious state
r	recovery rate
ω	rate of loss of immunity
p	probability of exposed humans going through short term incubation periods
τ	length of long-term incubation period

Lemma 2.1. *The disease free equilibrium (DFE)* $(1,0,0,0,0,1,0)$ *of system* (1) *always exists. An endemic equilibrium (EE) exists if and only if* $R_o > 1$ *and it is given by the following relations:*

$$i_H^* = \frac{R_o^2 - 1}{\frac{ac}{\mu} + K_o R_o^2}, \quad e_H^{s*} = \frac{\frac{p}{d_s+\xi}(r+\xi)}{p\frac{d_s}{d_s+\xi} + (1-p)\frac{d_l}{d_l+\xi}} i_H^*, \quad e_H^{l*} = \frac{1-p}{d_l+\xi}\frac{d_s+\xi}{p} e_H^{s*},$$

$$r_H^* = \frac{r}{\omega+\xi} i_H^*, \qquad i_M^* = \frac{\frac{ac}{\mu} i_H^*}{1 + \frac{ac}{\mu} i_H^*}, \qquad s_H^* = 1 - e_H^{s*} - e_H^{l*} - i_H^* - r_H^*$$

and $s_H^* = 1 - i_M^*$, *where* $K_o = \frac{\frac{p}{d_s+\xi} + \frac{1-p}{d_l+\xi}}{p\frac{d_s}{d_s+\xi} + (1-p)\frac{d_l}{d_l+\xi}}(r+\xi) + 1 + \frac{r}{\omega+\xi}$.

Proof. See Appendix A. □

Theorem 2.1. *The DFE of system* (1) *is locally asymptotically stable if* $R_o < 1$ *and is unstable if* $R_o > 1$. *The EE is locally asymptotically stable whenever exists, i.e. if* $R_o > 1$.

Proof. See Appendix C. □

2.2. *DDE model with fixed length incubation period*

For the ODE model in section 2.1, exponential distribution was assumed for the long incubation period, with mean $1/d_l$. In this section, we introduce a DDE model assuming every individual has the same length of long incubation period, τ. For the sake of comparison with the ODE model, we set $\tau = 1/d_l$. The model reads as

$$\begin{cases} \dfrac{ds_H}{dt} = \xi - abms_H i_M + \omega r_H - \xi s_H, \\ \dfrac{de^s_H}{dt} = pabms_H i_M - d_s e^s_H - \xi e^s_H, \\ \dfrac{de^l_H}{dt} = (1-p)abms_H i_M - (1-p)abms_H(t-\tau)i_M(t-\tau)e^{-\xi\tau} - \xi e^l_H, \\ \dfrac{di_H}{dt} = d_s e^s_H + (1-p)abms_H(t-\tau)i_M(t-\tau)e^{-\xi\tau} - ri_H - \xi i_H, \\ \dfrac{dr_H}{dt} = ri_H - \omega r_H - \xi r_H, \\ \dfrac{ds_M}{dt} = \mu - acs_M i_H - \mu s_M, \\ \dfrac{di_M}{dt} = acs_M i_H - \mu i_M. \end{cases} \tag{3}$$

To guarantee that solutions remain in the feasible domain, compared to the ODE model here we need the additional condition that the initial functions satisfy $e^l_H(0) \geq (1-p)abm \int_{-\tau}^{0} s_H(u)i_M(u)e^{-\xi u}du$.

The basic reproduction number R_d of the DDE model is given by

$$R_d = \sqrt{\frac{a^2bcm}{\mu(r+\xi)}\left((1-p)e^{-\xi\tau} + p\frac{d_s}{d_s+\xi}\right)}, \tag{4}$$

being defined in the same manner as R_o. Comparing with Eq. (2), the term $(1-p)e^{-\xi\tau} + p\frac{d_s}{d_s+\xi}$ is the only different part, which is the probability that a human will survive the exposed state to become infectious. R_d is a stability threshold of system (3).

Lemma 2.2. *The disease free equilibrium (DFE)* $(1,0,0,0,0,1,0)$ *of system* (3) *always exists. An endemic equilibrium (EE) exists if and only if* $R_d > 1$ *and it is given by the following relations:*

$$i^*_H = \frac{R_d^2 - 1}{\frac{ac}{\mu} + K_d R_d^2}, \quad e^{s*}_H = \frac{p\xi}{(1-p)(d_s+\xi)e^{-\xi\tau} + pd_s}\frac{r+\xi}{\xi}i^*_H,$$

$$e^{l*}_H = \frac{(1-p)(d_s+\xi)(1-e^{-\xi\tau})}{p\xi}e^{s*}_H, \quad r^*_H = \frac{r}{\omega+\xi}i^*_H, \quad i^*_M = \frac{\frac{ac}{\mu}i^*_H}{1+\frac{ac}{\mu}i^*_H},$$

$$s^*_H = 1 - e^{s*}_H - e^{l*}_H - i^*_H - r^*_H \quad \textit{and} \quad s^*_M = 1 - i^*_M, \quad \textit{where}$$

$$K_d = \frac{\frac{p}{d_s+\xi} + (1-p)\frac{1-e^{-\xi\tau}}{\xi}}{p\frac{d_s}{d_s+\xi} + (1-p)e^{-\xi\tau}}(r+\xi) + 1 + \frac{r}{\omega+\xi}.$$

Proof. See Appendix B. □

Theorem 2.2. *The DFE of system* (3) *is locally asymptotically stable if* $R_d < 1$ *and is unstable if* $R_d > 1$. *The EE is locally asymptotically stable whenever exists, i.e. if* $R_d > 1$.

Proof. See Appendix D. □

3. Results

3.1. *Comparison of models*

In Section 2, we have shown that for both models there exists a threshold value determining the existence and stability of equilibria. Now we compare these threshold values and also the endemic equilibria of the two models.

Proposition 3.1. *When all parameters are fixed, the basic reproduction number* R_o *of the ODE model is greater than the basic reproduction number* R_d *of the DDE model. Moreover, when* $R_d > 1$, i_H^* *of the ODE model (denoted by* i_o*) is greater than* i_H^* *of the DDE model (denoted by* i_d*).*

Proof. Comparing (2) and (4), we obtain

$$R_o = \sqrt{\frac{a^2bcm}{(r+\xi)\mu}\left(p\frac{d_s}{d_s+\xi} + (1-p)\frac{1}{1+\xi\tau}\right)}$$
$$> \sqrt{\frac{a^2bcm}{(r+\xi)\mu}\left(p\frac{d_s}{d_s+\xi} + (1-p)\frac{1}{e^{\xi\tau}}\right)} = R_d,$$

because $e^{\xi\tau} > 1 + \xi\tau$. To compare the equilibria, consider

$$i_o - i_d = \frac{(R_o^2 - 1)\left(\frac{ac}{\mu} + K_d R_d^2\right) - (R_d^2 - 1)\left(\frac{ac}{\mu} + K_o R_o^2\right)}{\left(\frac{ac}{\mu} + K_o R_o^2\right)\left(\frac{ac}{\mu} + K_d R_d^2\right)}$$
$$= \frac{\frac{ac}{\mu}(R_o^2 - R_d^2) + (K_d - K_o)R_o^2 R_d^2}{\left(\frac{ac}{\mu} + K_o R_o^2\right)\left(\frac{ac}{\mu} + K_d R_d^2\right)}.$$

Since $R_o > R_d$, it is sufficient to show that $K_d > K_o$. Since $e^{\xi\tau} > 1 + \xi\tau$, $e^{-\xi\tau} < \frac{1}{1+\xi\tau}$. Recall that $\tau = 1/d_l$, hence

$$e^{-\xi\tau} < \frac{d_l}{d_l + \xi}. \tag{5}$$

Table 2. Baseline parameter values for simulations.

parameter	value	ref.
ξ	0.00004	human life span
μ	0.2 [0.1,0.24]	[11]
b	0.5	[6]
c	0.5	[6]
a	0.3 [0.25,0.5]	[12]
d_s	0.04	[8,9]
d_l	0.003	[8,9]
r	0.07 [0.005,0.5]	[13,12,14]
p	0.25	[15]
ω	1/365	
P	365	
L	365/2	
a_s	0.3	

Thus, $1 - e^{-\xi\tau} > 1 - \frac{1}{1+\xi\tau} = \frac{\xi}{d_l+\xi}$, implying

$$\frac{1-e^{-\xi\tau}}{\xi} > \frac{1}{d_l+\xi}. \tag{6}$$

By (5) and (6), we find

$$\frac{\frac{p}{d_s+\xi} + (1-p)\frac{1-e^{-\xi\tau}}{\xi}}{p\frac{d_s}{d_s+\xi} + (1-p)e^{-\xi\tau}} > \frac{\frac{p}{d_s+\xi} + \frac{1-p}{d_l+\xi}}{p\frac{d_s}{d_s+\xi} + (1-p)\frac{d_l}{d_l+\xi}}$$

which is equivalent to $K_d > K_o$. □

In conclusion, the ODE model gives a larger basic reproduction number than the DDE model, because of the higher probability of surviving the incubation period. Fig. 4 and Fig. 5 show numerical solutions. Fig. 4 shows the case $R_o > 1$ and $R_d > 1$, when $i_H(t)$ converges to the endemic equilibrium for both models. Fig. 5 shows a particular case when $R_o > 1$ but $R_d < 1$. Despite that all parameters are the same, here $i_H(t)$ of ODE model converges to the endemic equilibrium, and $i_H(t)$ of DDE model converges to disease free equilibrium, thus the two models provide very different predictions. To investigate the robustness of the basic reproduction number with respect to the long term incubation time, in Fig. 6 we compared

$$\frac{\partial R_o}{\partial \tau} = \frac{-1}{2R_o}\frac{a^2bcm}{\mu(r+\xi)}\frac{(1-p)\xi}{(1+\xi\tau)^2} \quad \text{and} \quad \frac{\partial R_d}{\partial \tau} = \frac{-1}{2R_d}\frac{a^2bcm}{\mu(r+\xi)}(1-p)\xi e^{-\xi\tau}.$$

The magnitude of $\frac{\partial R_o}{\partial \tau}$ is smaller than $\frac{\partial R_d}{\partial \tau}$ when τ is in the reasonable range, so the basic reproduction number is more sensitive to the long term incubation time in the DDE model.

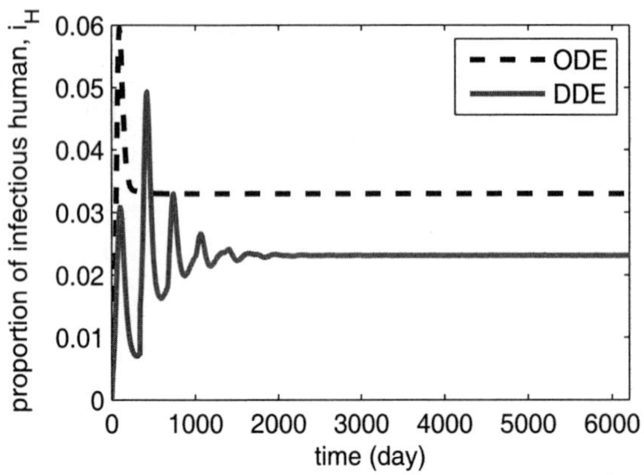

Figure 4. $R_o > 1$ and $R_d > 1$. For both models, $i_H(t)$ converges to endemic equilibrium with $i_o > i_d$. To clearly show the difference of i_o and i_d, we set $\xi = 0.004$ and $m = 10$. Other parameter values are as indicated in Table 2. Initial condition for ODE model is $(s_H, e^s_H, e^l_H, i_H, r_H, s_M, i_M)(0) = (1, 0, 0, 0, 0, 0.09, 0.01)$. For sake of convenience, initial condition of DDE model is set to be $s_H(t) = 1$ and $s_M(t) = 1$ for $t < 0$, introducing infectious mosquito at $t = 0$, $(s_H, e^s_H, e^l_H, i_H, r_H, s_M, i_M)(0) = (1, 0, 0, 0, 0, 0.09, 0.01)$.

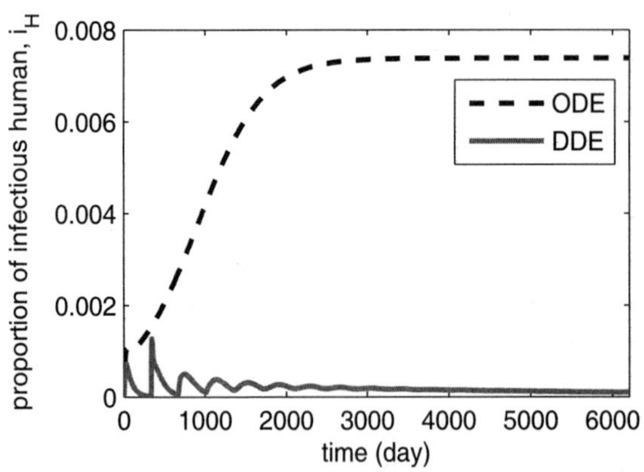

Figure 5. $R_o > 1$ and $R_d < 1$. For the ODE model, $i_H(t)$ converges to EE, while it converges to DFE for the DDE model. To compare with Fig. 4, we used parameter value $m = 1.5$. Other parameter values and initial condition are same as the one in Fig. 4.

3.2. *Role of the mosquito population*

In Fig. 7, the infectious human component of the endemic equilibrium is plotted for various mosquito populations. As expected from Proposition

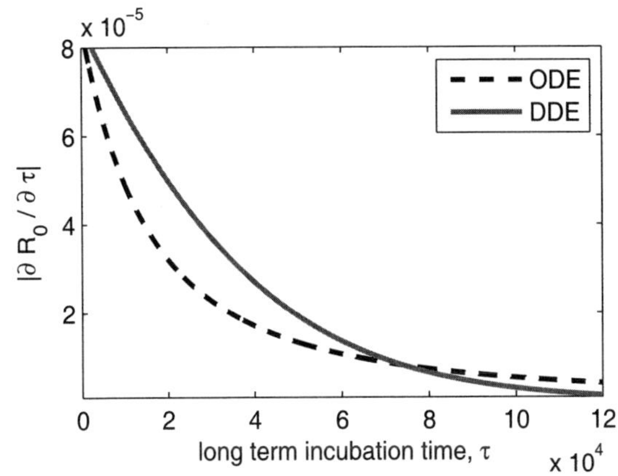

Figure 6. Sensitivity of the basic reproduction number to the long term incubation time. For τ being in reasonable range, $|\frac{\partial R_o}{\partial \tau}| < |\frac{\partial R_d}{\partial \tau}|$. Parameter values are as in Table 2 with $m = 20$.

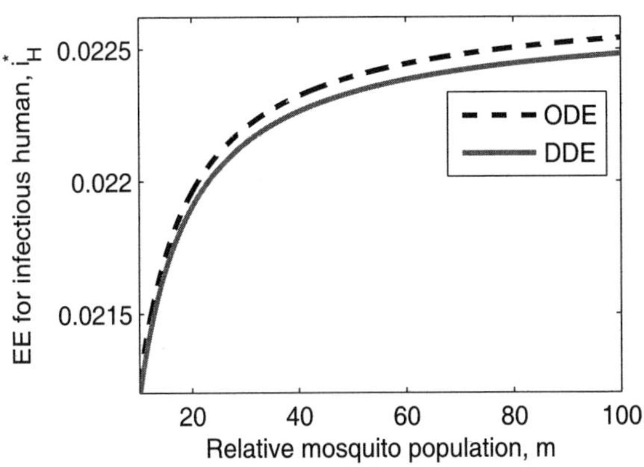

Figure 7. Relation of m and i_H^*. Change of i_H^* is more drastic at smaller mosquito population.

3.1, the infectious human equilibrium for the ODE model is greater than for the DDE model. Moreover, we can see that a small change in the mosquito population affects the level of endemicity more significantly when the mosquito population is relatively small.

3.3. *Comparison with seasonality*

In temperate regions, mosquito populations show huge seasonal variation, and so the transmission of *P. vivax* malaria is seasonal as well. To account

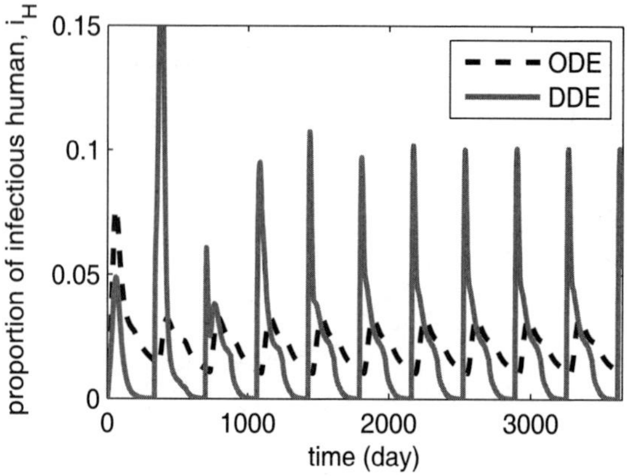

Figure 8. Dynamics of the infectious human population proportion with periodic biting rate. In both models, $i_H(t)$ converges to a periodic attractor. While the DDE model has a higher peak of infection, its annual average is smaller than that of the ODE model. Parameter values are as in Table 2 with $m = 20$. Initial condition is the same as the one used for Fig. 4.

for seasonality in a simplified way, a year is divided into a mosquito season, during which the parasite is transmitted via the mosquitoes, and an off-season, during which no new infection occurs. Let L be the length for a mosquito season and P be the natural period (one year). For the sake of simplicity, we incorporate temporal variation into the biting rate, thus replacing the constant a by

$$a(t) = \begin{cases} a_s & kP \leq t < kP + L, \\ 0 & kP + L \leq t < (k+1)P, \end{cases}$$

where k is an integer and $L \leq P$.

Numerical simulations are shown in Fig. 8, 9, 10 (compare to Fig. 4, 5). With such a periodic biting rate, $i_H(t)$ converges to a periodic attractor, instead of a steady state. Generally, the DDE model shows larger oscillations and predicts higher peaks and lower yearly bottoms of malaria prevalence. In contrast to the non-seasonal case, in some situations the DDE model even has higher annual average of infectious humans (Fig. 9) than its ODE counterpart. In addition, in some cases the DDE model predicts the persistence of the disease even though the infection dies out for the ODE model with the same parameter values (Fig. 10).

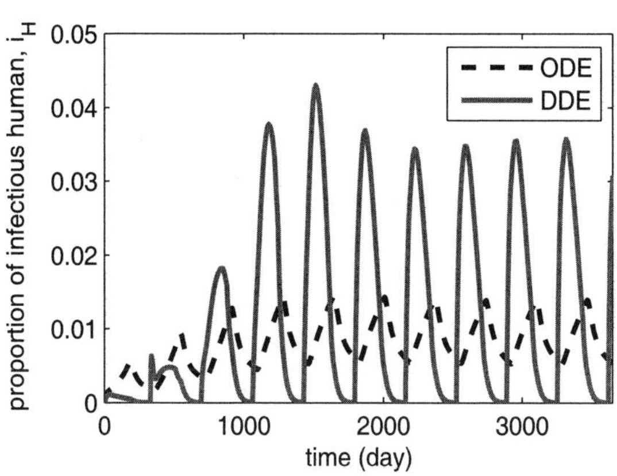

Figure 9. To compare with Fig. 9, we used the parameter value $m = 2$. Other parameter values and initial condition are the same as in Fig. 8. The DDE model has a higher peak of infection, just as in Fig. 9, however, its annual average is also greater than that of the ODE model.

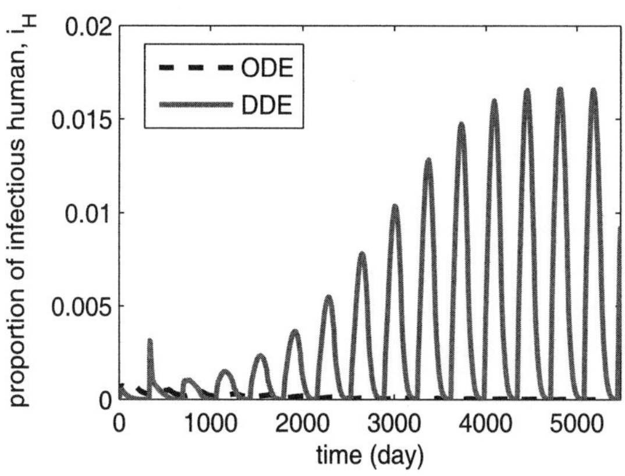

Figure 10. For comparison, we use $m = 1$. Here $i_H(t)$ dies out in the ODE model, however, it converges to a periodic attractor in the DDE model.

4. Conclusion

The exact mechanism governing the development of malaria parasites from dormancy to activation is not known [16]. Motivated by the empirical estimations of the incubation times in Korea, in this work we compared two models having different distribution of long term incubation time,

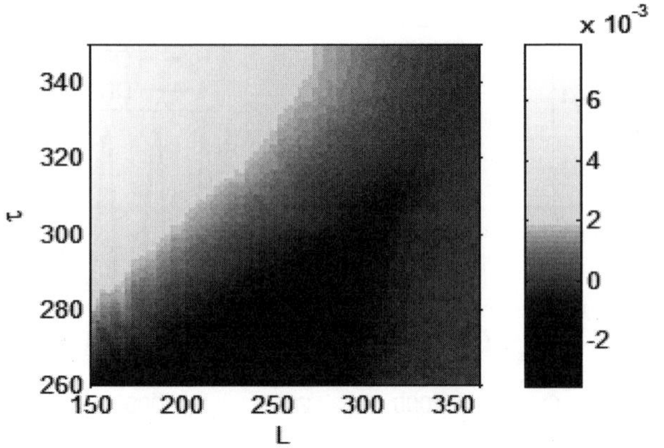

Figure 11. Differences between the yearly average number of infections for the ODE and the DDE model, as a function of length of incubation period τ and mosquito season length L. Black (white) areas represent a parameter region where the DDE (ODE) model predicts higher prevalence of malaria.

resulting an ODE and a DDE system. For both models we identified the basic reproduction number as a threshold value determining the stability of the disease free equilibrium and the existence of the endemic equilibrium. Having all parameters fixed, the exponential distribution assumption for the long term incubation (ODE model) predicts a greater basic reproduction number and higher level of endemicity than the constant length assumption (DDE model). Also, the DDE model generates more oscillatory behavior even without seasonal coefficients. As it was pointed out[17], if the inter-annual cycle of malaria is generated not only by seasonal change, but also by internal mechanisms, the baseline autonomous model should be able to produce oscillations itself. Given that the distribution generated by the constant length assumption approximates much better the empirically estimated distribution, than the exponential distribution assumption, we consider the DDE as a better model for the description of *P. vivax* malaria transmission dynamics, thus previous ODE-based *P. vivax* models are likely overestimate the basic reproduction number.

However, when seasonality is also included in the model, things become more complicated. It is known that periodic delay differential equations can produce unexpected behavior, such as resonances[18], and their dynamics is not completely understood even in the scalar case[19]. In the present situation, the time periodic DDE results larger annual oscillations than the ODE (i.e. higher peaks and lower bottoms); as expected in the case of time delays. However, due to the complicated interplay of the time delay and periodicity, in some situations the DDE model predicts a higher number of infections throughout a year than the ODE, in contrast with the autonomous case. In other cases, the DDE predicts lower average prevalence (as in the autonomous case), depending on the particular choice of parameters. To illustrate this striking behavior, we prepared Fig. 11, where we plotted these domains on the $L-\tau$ parameter plane (length of mosquito season and long incubation period). The results stress the importance of future work incorporating both delay and seasonality into *P. vivax* models in temperate regions.

Acknowledgement. Research was supported by the European Union and co-funded by the European Social Fund under the project "Telemedicine-focused research activities on the field of Mathematics, Informatics and Medical sciences" of project number TÁMOP-4.2.2.A-11/1/KONV-2012-0073 and European Research Council Starting Investigator Grant No. 259559. Yongkuk Kim was supported by Basic Science Research Program through the National Research Foundation of Korea(NRF) funded by the Ministry of Education, Science and Technology(2012R1A1A2003398).

Appendix A. Proof of Lemma 2.1

To find equilibria, we set the LHS of system (1) to zero, and obtain

$$\begin{aligned}
0 &= pabms^*_H i^*_M - (d_s+\xi)e^{s*}_H, && \text{(A.1)}\\
0 &= (1-p)abms^*_H i^*_M - (d_l+\xi)e^{l*}_H, && \text{(A.2)}\\
0 &= d_s e^{s*}_H + d_l e^{l*}_H - (r+\xi)i^*_H, && \text{(A.3)}\\
0 &= ri^*_H - \omega r^*_H - \xi r^*_H, && \text{(A.4)}\\
0 &= acs^*_M i^*_H - \mu i^*_M. && \text{(A.5)}
\end{aligned}$$

If either $i^*_H = 0$ or $i^*_M = 0$, we have a DFE $(s^*_H, e^{s*}_H, e^{l*}_H, i^*_H, r^*_H, s^*_M, i^*_M) = (1,0,0,0,0,1,0)$. Consider the case $i^*_H > 0$ and $i^*_M > 0$. Adding three equations, (A.1) multiplied by $\frac{d_s}{\xi+d_s}$, (A.2) multiplied by $\frac{d_l}{\xi+d_l}$, and (A.3),

$$abm\left(p\frac{d_s}{\xi+d_s} + (1-p)\frac{d_l}{\xi+d_l}\right)s^*_H i^*_M = (r+\xi)i^*_H. \tag{A.6}$$

By (A.5),

$$acs^*_M i^*_H = \mu i^*_M. \tag{A.7}$$

Multiplying each sides of (A.6) and (A.7), and dividing by $(r + xi)\mu i^*_H i^*_M$,

$$R_0^2 s^*_H s^*_M = 1. \tag{A.8}$$

Multiplying $\frac{ac}{(r+\xi)\mu}$ to (A.6) gives

$$R_o^2 s^*_H (1 - s^*_M) = \frac{ac}{\mu} i^*_H. \tag{A.9}$$

By (A.8) and (A.9), we get

$$s^*_H = \frac{1}{R_o^2} + \frac{ac}{R_o^2 \mu} i^*_H. \tag{A.10}$$

Comparing (A.1) and (A.6), we have

$$e^{s*}_H = abms^*_H i^*_M \frac{p}{d_s + \xi} = \frac{\frac{p}{\xi + d_s}(r + \xi)}{p\frac{d_s}{\xi + d_s} + (1 - p)\frac{d_l}{\xi + d_l}} i^*_H. \tag{A.11}$$

Comparing (A.2) and (A.6), we have

$$e^{l*}_H = abms^*_H i^*_M \frac{1 - p}{d_l + \xi} = \frac{\frac{1-p}{\xi + d_l}(r + \xi)}{p\frac{d_s}{\xi + d_s} + (1 - p)\frac{d_l}{\xi + d_l}} i^*_H. \tag{A.12}$$

By (A.4),

$$r^*_H = \frac{r}{\omega + \xi} i^*_H. \tag{A.13}$$

By (A.11), (A.12) and (A.13),

$$\begin{aligned} s^*_H &= 1 - e^{s*}_H - e^{l*}_H - i^*_H - r^*_H \\ &= 1 - \frac{\left(\frac{p}{\xi + d_s} + \frac{1-p}{\xi + d_l}\right)(r + \xi)}{p\frac{d_s}{\xi + d_s} + (1 - p)\frac{d_l}{\xi + d_l}} i^*_H - i^*_H - \frac{r}{\omega + \xi} i^*_H \\ &= 1 - K_o i^*_H. \end{aligned} \tag{A.14}$$

By (A.10) and (A.14), we get $i^*_H = \frac{R_o^2 - 1}{\frac{ac}{\mu} + K_o R_o^2}$, which exists when $R_o > 1$.

Appendix B. Proof of Lemma 2.2

We put LHS of the system (3) to be zero and obtain the following equations

$$0 = pabms^*_H i^*_M - (d_s + \xi)e^{s*}_H, \tag{B.1}$$

$$0 = (1 - p)abms^*_H i^*_M - (1 - p)abms^*_H i^*_M e^{-\xi\tau} - \xi e^{l*}_H, \tag{B.2}$$

$$0 = d_s e^{s*}_H + (1 - p)abms^*_H i^*_M e^{-\xi\tau} - (r + \xi)i^*_H, \tag{B.3}$$

$$0 = ri^*_H - (\omega + \xi)r^*_H, \tag{B.4}$$

$$0 = acs^*_M i^*_H - \mu i^*_M. \tag{B.5}$$

If either $i^*_H = 0$ or $i^*_M = 0$, we have a DFE $(s^*_H, e^{s*}_H, e^{l*}_H, i^*_H, r^*_H, s^*_M, i^*_M) = (1,0,0,0,0,1,0)$. Consider the case $i^*_H > 0$ and $i^*_M > 0$. Adding two equations, (B.1) multiplied by $\frac{d_s}{\xi+d_s}$, and (B.3), one has

$$\left((1-p)e^{-\xi\tau} + p\frac{d_s}{\xi+d_s}\right) abms^*_H i^*_M = (r+\xi)i^*_H. \tag{B.6}$$

By (B.5),

$$acs^*_M i^*_H = \mu i^*_M. \tag{B.7}$$

Multiplying both sides of (B.6) and (B.7), and dividing by $(r+\xi)\mu i^*_H i^*_M$,

$$R_d^2 s^*_H s^*_M = 1. \tag{B.8}$$

Meanwhile, multiplying $\frac{ac}{(r+\xi)\mu}$ to (B.6),

$$R_d^2 s^*_H (1 - s^*_M) = \frac{ac}{\mu} i^*_H. \tag{B.9}$$

By (B.8) and (B.9), we get

$$s^*_H = \frac{1}{R_d^2} + \frac{ac}{R_d^2 \mu} i^*_H. \tag{B.10}$$

Comparing (B.1) and (B.6), we have

$$e^{s*}_H = abms^*_H i^*_M \frac{p}{d_s+\xi} = \frac{\frac{p}{\xi+d_s}(r+\xi)}{(1-p)e^{-\xi\tau} + p\frac{d_s}{\xi+d_s}} i^*_H. \tag{B.11}$$

Comparing (B.2) and (B.6), we have

$$e^{l*}_H = abms^*_H i^*_M \frac{(1-p)(1-e^{-\xi\tau})}{\xi} = \frac{\frac{(1-p)(1-e^{-\xi\tau})}{\xi}(r+\xi)}{(1-p)e^{-\xi\tau} + p\frac{d_s}{\xi+d_s}} i^*_H. \tag{B.12}$$

By (B.4),

$$r^*_H = \frac{r}{\omega+\xi} i^*_H. \tag{B.13}$$

By (B.11), (B.12) and (B.13),

$$\begin{aligned} s^*_H &= 1 - e^{s*}_H - e^{l*}_H - i^*_H - r^*_H \\ &= 1 - \frac{\frac{p}{\xi+d_s} + (1-p)\frac{1-e^{-\xi\tau}}{\xi}}{(1-p)e^{-\xi\tau} + p\frac{d_s}{\xi+d_s}}(r+\xi)i^*_H - i^*_H - \frac{r}{\omega+\xi} i^*_H \\ &= 1 - K_d i^*_H. \end{aligned} \tag{B.14}$$

By (B.10) and (B.14), we get $i^*_H = \frac{R_d^2 - 1}{\frac{ac}{\mu} + K_d R_d^2}$, which exists when $R_d > 1$.

Appendix C. Proof of Theorem 2.1

Consider the linearized system of (1) at an equilibrium:

$$\begin{cases} \dfrac{de^s_H}{dt} = pabms^*_H i_M - pabmi^*_M(e^s_H + e^l_H + i_H + r_H) - (d_s+\xi)e^s_H, \\ \dfrac{de^l_H}{dt} = (1-p)abms^*_H i_M - (1-p)abmi^*_M(e^s_H + e^l_H + i_H + r_H) - (d_l+\xi)e^l_H, \\ \dfrac{di_H}{dt} = d_s e^s_H + d_l e^l_H - (r+\xi)i_H, \\ \dfrac{dr_H}{dt} = ri_H - (\omega+\xi)r_H, \\ \dfrac{di_M}{dt} = acs^*_M i_H - aci^*_H i_M - \mu i_M. \end{cases}$$

The characteristic function $F(\lambda)$ is

$$\begin{vmatrix} \lambda+d_s+\xi+A & A & A & A & -pabms^*_H \\ B & \lambda+d_l+\xi+B & B & B & -(1-p)abms^*_H \\ -d_s & -d_l & \lambda+r+\xi & 0 & 0 \\ 0 & 0 & -r & \lambda+\omega+\xi & 0 \\ 0 & 0 & -acs^*_M & 0 & \lambda+\mu+aci^*_H \end{vmatrix},$$

where $A = pabmi^*_M$, $B=(1-p)abmi^*_M$. After simplification,

$$\begin{aligned} F(\lambda) = \quad & (\lambda+r+\xi)(\lambda+\omega+\xi)(\lambda+\mu+aci^*_H)[(\lambda+d_s+\xi)(\lambda+d_l+\xi) \\ & + abmi^*_M\{p(\lambda+d_s+\xi)+(1-p)(\lambda+d_l+\xi)\}] \\ & + \{-a^2bcms^*_H s^*_M(\lambda+\omega+\xi) + abmi^*_M(\lambda+\omega+\xi)(\lambda+\mu+aci^*_H) \\ & + rabmi^*_M(\lambda+\mu+aci^*_H)\}\{(1-p)d_l(\lambda+d_s+\xi)+pd_s(\lambda+d_l+\xi)\} \\ & + abmi^*_M(d_l-d_s)(\lambda+\mu+aci^*_H)(-1+2p)(\lambda+\omega+\xi)(\lambda+r+\xi). \end{aligned}$$

At the DFE, it reduces to

$$\begin{aligned} F(\lambda) = (\lambda+\omega+\xi)\{ & (\lambda+d_s+\xi)(\lambda+d_l+\xi)(\lambda+r+\xi)(\lambda+\mu) \\ & -(1-p)d_l a^2bcm(\lambda+d_s+\xi) - pd_s a^2bcm(\lambda+d_l+\xi)\}\,. \end{aligned}$$

Assume that $R_o < 1$. Suppose there exists a root of $F(\lambda)=0$ with non-negative real part. Then,

$$\begin{aligned} & \left|\frac{\lambda}{d_s+\xi}+1\right|\left|\frac{\lambda}{d_l+\xi}+1\right|\left|\frac{\lambda}{r+\xi}+1\right|\left|\frac{\lambda}{\mu}+1\right| \\ \le\ & \frac{a^2bcm}{(r+\xi)\mu}\left((1-p)\frac{d_l}{d_l+\xi}\left|\frac{\lambda}{d_s+\xi}+1\right| + p\frac{d_s}{d_s+\xi}\left|\frac{\lambda}{d_l+\xi}+1\right|\right) \\ \le\ & \left|\frac{\lambda}{d_s+\xi}+1\right|\left|\frac{\lambda}{d_l+\xi}+1\right| R_o^2, \end{aligned}$$

which contradicts to $R_o < 1$. Therefore, the roots of $F(\lambda) = 0$ have negative real part, implying that the DFE is locally asymptotically stable if $R_o < 1$.

Now, assume $R_o > 1$. Note that $F(\lambda) = 0$ has at least one real root. Since $F(\lambda) \to \infty$ for real $\lambda \to \infty$ and

$$F(0) = (\omega + \xi)(d_s + \xi)(d_l + \xi)(r + \xi)\mu(1 - R_o^2) < 0,$$

$F(\lambda) = 0$ has a positive real root. Therefore, the DFE is unstable.

Suppose that the EE is not LAS. Then there exists a characteristic root λ with nonnegative real part. For the EE, $s^*_H s^*_M = 1/R_o^2$ and the characteristic equation can be re-written as

$$\begin{aligned}
&\frac{a^2bcm}{R_o^2}(\lambda + \omega + \xi)\{(1-p)d_l(\lambda + d_s + \xi) + pd_s(\lambda + d_l + \xi)\} \\
&= (\lambda + r + \xi)(\lambda + \omega + \xi)(\lambda + \mu + aci^*_H)[(\lambda + d_s + \xi)(\lambda + d_l + \xi) \\
&\quad + abmi^*_M\{p(\lambda + d_s + \xi) + (1-p)(\lambda + d_l + \xi)\}] + abmi^*_M(\lambda + \mu + aci^*_H) \\
&\quad \{(\lambda + \omega + \xi) + r\}\{(1-p)d_l(\lambda + d_s + \xi) + pd_s(\lambda + d_l + \xi)\}.
\end{aligned}$$

Suppose there exists a root of the equation with nonnegative real part. Dividing by $\{(1-p)d_l(\lambda + d_s + \xi) + pd_s(\lambda + d_l + \xi)\}(r + \xi)\mu(\omega + \xi)$ gives

$$\begin{aligned}
&\left|\frac{1}{p\frac{d_s}{d_s+\xi} + (1-p)\frac{d_l}{d_l+\xi}}\right| \left|\frac{\lambda}{\omega + \xi} + 1\right| \\
&= \left|\frac{\lambda}{\mu} + 1 + \frac{aci^*_H}{\mu}\right| \left|\left(\frac{\lambda}{r + \xi} + 1\right)\left(\frac{\lambda}{\omega + \xi} + 1\right)\right. \\
&\quad \times \frac{(\lambda + d_s + \xi)(\lambda + d_l + \xi)}{(1-p)d_l(\lambda + d_s + \xi) + pd_s(\lambda + d_l + \xi)} \\
&\quad + abmi^*_M\left(\frac{\lambda}{r + \xi} + 1\right)\left(\frac{\lambda}{\omega + \xi} + 1\right)\frac{p(\lambda + d_s + \xi) + (1-p)(\lambda + d_l + \xi)}{(1-p)d_l(\lambda + d_s + \xi) + pd_s(\lambda + d_l + \xi)} \\
&\quad \left. + \frac{abmi^*_M}{r + \xi}\left(\frac{\lambda}{\omega + \xi} + 1\right) + \frac{rabmi^*_M}{(r + \xi)(\omega + \xi)}\right| \\
&\geq \left|\frac{\lambda}{\omega + \xi} + 1\right| \left|\frac{(\lambda + d_s + \xi)(\lambda + d_l + \xi)}{(1-p)d_l(\lambda + d_s + \xi) + pd_s(\lambda + d_l + \xi)}\right|.
\end{aligned}$$

This implies

$$\begin{aligned}
p\frac{d_s}{d_s + \xi} + (1-p)\frac{d_l}{d_l + \xi} &\leq \left|\frac{(1-p)d_l(\lambda + d_s + \xi) + pd_s(\lambda + d_l + \xi)}{(\lambda + d_s + \xi)(\lambda + d_l + \xi)}\right| \\
&= \left|(1-p)\frac{d_l}{d_l + \xi}\frac{1}{\frac{\lambda}{d_l+\xi} + 1} + p\frac{d_s}{d_s + \xi}\frac{1}{\frac{\lambda}{d_s+\xi} + 1}\right| \\
&< (1-p)\frac{d_l}{d_l + \xi}\frac{1}{\left|\frac{\lambda}{d_s+\xi} + 1\right|} + p\frac{d_s}{d_s + \xi}\frac{1}{\left|\frac{\lambda}{d_l+\xi} + 1\right|},
\end{aligned}$$

which is a contradiction, as one can check that 0 is not a root of $F(\lambda)$. Hence, the EE is LAS.

Appendix D. Proof of Theorem 2.2

Consider the linearized system of (3) at an equilibrium:

$$\begin{cases} \dfrac{de^s_H}{dt} = -\, pabmi^*_M(e^s_H(t) + e^l_H(t) + i_H(t) + r_H(t)) + pabms^*_H i_M(t) \\ \qquad -\,(d_s+\xi)e^s_H(t), \\ \dfrac{de^l_H}{dt} = -\,(1-p)abmi^*_M(e^s_H(t) + e^l_H(t) + i_H(t) + r_H(t)) + (1-p)abms^*_H i_M(t) \\ \qquad +\,(1-p)abmi^*_M(e^s_H(t-\tau) + e^l_H(t-\tau) + i_H(t-\tau) + r_H(t-\tau))e^{-\xi\tau} \\ \qquad -\,(1-p)abms^*_H i_M(t-\tau)e^{-\xi\tau} - \xi e^l_H(t), \\ \dfrac{di_H}{dt} = -\,(1-p)abmi^*_M(e^s_H(t-\tau) + e^l_H(t-\tau) + i_H(t-\tau) + r_H(t-\tau))e^{-\xi\tau} \\ \qquad +\, d_s e^s_H(t) + (1-p)abms^*_H i_M(t-\tau)e^{-\xi\tau} - ri_H(t) - \xi i_H(t), \\ \dfrac{dr_H}{dt} = ri_H(t) - \omega r_H(t) - \xi r_H(t), \\ \dfrac{di_M}{dt} = acs^*_M i_H(t) - aci^*_H i_M(t) - \mu i_M(t). \end{cases}$$

The characteristic function $F(\lambda)$ is the determinant of the matrix

$$\lambda I + \begin{pmatrix} d_s+\xi+A & A & A & A & -pabms^*_H \\ B & \xi+B & B & B & -(1-p)abms^*_H(1-e^{-\xi\tau-\lambda\tau}) \\ -d_s+C & C & r+\xi+C & C & -(1-p)abms^*_H e^{-\xi\tau-\lambda\tau} \\ 0 & 0 & -r & \omega+\xi & 0 \\ 0 & 0 & -acs^*_M & 0 & \mu+aci^*_H \end{pmatrix},$$

where $A = pabmi^*_M$, $B = (1-p)abmi^*_M(1-e^{-\xi\tau-\lambda\tau})$, and $C = (1-p)abmi^*_M e^{-\xi\tau-\lambda\tau}$. After simplification,

$$\begin{aligned} F(\lambda) = {} & (\lambda+\mu+aci^*_H)(\lambda+\omega+\xi)(\lambda+\xi+abmi^*_M)(\lambda+d_s+\xi)(\lambda+r+\xi) \\ & -\omega rabmi^*_M\{pd_s + (1-p)e^{-\xi\tau-\lambda\tau}(\lambda+d_s+\xi)\}(\lambda+\mu+aci^*_H) \\ & -a^2bcms^*_H s^*_M\{pd_s + (1-p)e^{-\xi\tau-\lambda\tau}(\lambda+d_s+\xi)\}(\lambda+\omega+\xi)(\lambda+\xi). \end{aligned}$$

At the DFE, it reduces to

$$\begin{aligned} F(\lambda) = {} & (\lambda+\omega+\xi)(\lambda+\xi)\{(\lambda+d_s+\xi)(\lambda+r+\xi)(\lambda+\mu) \\ & -(\lambda+d_s+\xi)(1-p)a^2bcme^{-\xi\tau-\lambda\tau} - pa^2bcmd_s\}. \end{aligned}$$

Assume that $R_d < 1$. Suppose there exists a root for $F(\lambda) = 0$ with nonnegative real part. Then,

$$|(\lambda+d_s+\xi)(\lambda+\mu)(\lambda+r+\xi)| = |(\lambda+d_s+\xi)(1-p)a^2bcme^{-\xi\tau-\lambda\tau}+pa^2bcmd_s|.$$

Dividing both sides by $(d_s+\xi)\mu(r+\xi)$ gives

$$\begin{aligned}
&\left|\frac{\lambda}{d_s+\xi}+1\right|\left|\frac{\lambda}{\mu}+1\right|\left|\frac{\lambda}{r+\xi}+1\right| \\
&= \frac{a^2bcm}{\mu(r+\xi)}\left|\left(\frac{\lambda}{d_s+\xi}+1\right)(1-p)e^{-\xi\tau-\lambda\tau}+p\frac{d_s}{d_s+\xi}\right| \\
&\leq \left|\frac{\lambda}{d_s+\xi}+1\right| R_d^2,
\end{aligned}$$

which contradicts to $R_d < 1$. Therefore, every root of $F(\lambda) = 0$ has negative real part, implying the DFE is locally asymptotically stable if $R_d < 1$.

Now assume $R_d > 1$. Note that $F(\lambda) = 0$ has at least one real root. Since $F(\lambda) \to \infty$ for real $\lambda \to \infty$ and

$$F(0) = (d_s+\xi)\mu(r+\xi)(1-R_d^2) < 0,$$

$F(\lambda) = 0$ has a positive real root. Therefore the DFE is unstable.

Suppose that the EE is not LAS. Then there exists a characteristic root λ with nonnegative real part. For the EE, $s_H^* s_M^* = 1/R_d^2$ holds and the characteristic equation can be re-written as

$$\begin{aligned}
&\mu(r+\xi)\frac{pd_s+(1-p)e^{-\xi\tau-\lambda\tau}(\lambda+d_s+\xi)}{(1-p)e^{-\xi\tau}+p\frac{d_s}{d_s+\xi}}(\lambda+\omega+\xi)(\lambda+\xi) \\
&= (\lambda+\mu+aci_H^*)(\lambda+\omega+\xi)(\lambda+\xi+abmi_M^*)(\lambda+d_s+\xi)(\lambda+r+\xi) \\
&\quad -\omega rabmi_M^*\{pd_s+(1-p)e^{-\xi\tau-\lambda\tau}(\lambda+d_s+\xi)\}(\lambda+\mu+aci_H^*).
\end{aligned}$$

Dividing both sides by $\mu(r+\xi)(d_s+\xi)(\omega+\xi)\xi$ gives

$$\begin{aligned}
&\frac{p\frac{d_s}{d_s+\xi}+(1-p)e^{-\xi\tau-\lambda\tau}(\frac{\lambda}{d_s+\xi}+1)}{(1-p)e^{-\xi\tau}+p\frac{d_s}{d_s+\xi}}\left(\frac{\lambda}{\omega+\xi}+1\right)\left(\frac{\lambda}{\xi}+1\right) \\
&= \left(\frac{\lambda}{\mu}+1+\frac{aci_H^*}{\mu}\right)\left[\left(\frac{\lambda}{\omega+\xi}+1\right)\left(\frac{\lambda}{\xi}+1+\frac{abmi_M^*}{\xi}\right)\left(\frac{\lambda}{d_s+\xi}+1\right)\right. \\
&\quad \times\left(\frac{\lambda}{r+\xi}+1\right) \\
&\quad \left. -\frac{\omega}{\omega+\xi}\frac{r}{r+\xi}\frac{abmi_M^*}{\xi}\left\{p\frac{d_s}{d_s+\xi}+(1-p)e^{-\xi\tau-\lambda\tau}\left(\frac{\lambda}{d_s+\xi}+1\right)\right\}\right].
\end{aligned}$$

Then,

$$\left|\frac{\lambda}{\mu}+1+\frac{aci^*_H}{\mu}\right|\left|\frac{\lambda}{\omega+\xi}+1\right|\left|\frac{\lambda}{\xi}+1+\frac{abmi^*_M}{\xi}\right|\left|\frac{\lambda}{d_s+\xi}+1\right|\left|\frac{\lambda}{r+\xi}+1\right|$$

$$= \left|\frac{p\frac{d_s}{d_s+\xi}+(1-p)e^{-\xi\tau-\lambda\tau}(\frac{\lambda}{d_s+\xi}+1)}{(1-p)e^{-\xi\tau}+p\frac{d_s}{d_s+\xi}}\left(\frac{\lambda}{\omega+\xi}+1\right)\left(\frac{\lambda}{\xi}+1\right)\right.$$
$$+\frac{\omega}{\omega+\xi}\frac{r}{r+\xi}\frac{abmi^*_M}{\xi}\left\{p\frac{d_s}{d_s+\xi}+(1-p)e^{-\xi\tau-\lambda\tau}\left(\frac{\lambda}{d_s+\xi}+1\right)\right\}$$
$$\left.\times\left(\frac{\lambda}{\mu}+1+\frac{aci^*_H}{\mu}\right)\right|$$

$$= \left|\frac{p\frac{d_s}{d_s+\xi}+(1-p)e^{-\xi\tau-\lambda\tau}(\frac{\lambda}{d_s+\xi}+1)}{(1-p)e^{-\xi\tau}+p\frac{d_s}{d_s+\xi}}\right|\left|\left(\frac{\lambda}{\omega+\xi}+1\right)\left(\frac{\lambda}{\xi}+1\right)\right.$$
$$\left.+\frac{\omega}{\omega+\xi}\frac{r}{r+\xi}\frac{abmi^*_M}{\xi}\left\{(1-p)e^{-\xi\tau}+p\frac{d_s}{d_s+\xi}\right\}\left(\frac{\lambda}{\mu}+1+\frac{aci^*_H}{\mu}\right)\right|$$

$$\leq \left|\frac{\lambda}{d_s+\xi}+1\right|\left|\left(\frac{\lambda}{\omega+\xi}+1\right)\left(\frac{\lambda}{\xi}+1\right)\right.$$
$$\left.+\frac{\omega}{\omega+\xi}\frac{r}{r+\xi}\frac{abmi^*_M}{\xi}\left\{(1-p)e^{-\xi\tau}+p\frac{d_s}{d_s+\xi}\right\}\left(\frac{\lambda}{\mu}+1+\frac{aci^*_H}{\mu}\right)\right|$$

$$\leq \left|\frac{\lambda}{d_s+\xi}+1\right|\left(\left|\frac{\lambda}{\omega+\xi}+1\right|\left|\frac{\lambda}{\xi}+1\right|+\frac{abmi^*_M}{\xi}\left|\frac{\lambda}{\mu}+1+\frac{aci^*_H}{\mu}\right|\right)$$

$$< \left|\frac{\lambda}{d_s+\xi}+1\right|\left|\frac{\lambda}{\omega+\xi}+1\right|\left|\frac{\lambda}{\xi}+1\right|\left|\frac{\lambda}{\mu}+1+\frac{aci^*_H}{\mu}\right|\left(1+\frac{abmi^*_M}{\xi}\right)$$

which is a contradiction, hence the EE is LAS.

References

1. R. Anderson and R. May, Infectious diseases of humans, *Oxford, UK* (1992).
2. R. Ross, Some quantitative studies in epidemiology, *Nature* **87**,466–467 (1911).
3. R. Aguas, M. U. Ferreira and M. G. M. Gomes, Modeling the effects of relapse in the transmission dynamics of malaria parasites, *Journal of Parasitology Research* (2012).
4. A. Kammanee, N. Kanyamee and I. M. Tang, Basic reproduction number for the transmission of Plasmodium vivax malaria, *Southeast Asian J. Trop. Med. Public Health* **32**, 702–706 (2001).
5. F. Chamchod and J. C. Beier, Modeling Plasmodium vivax: Relapses, treatment, seasonality, and G6PD deficiency, *Journal of Theoretical Biology* **316**, 25–34 (2013)
6. S. Ruan, D. Xiao, J. C. Beier, On the delayed Ross–Macdonald model for malaria transmission, *Bull. Math. Biol.* **70**, 1098 –1114 (2008).

7. K. Nah, Y. Kim and J. M. Lee, The dilution effect of the domestic animal population on the transmission of P. vivax malaria, *Journal of Theoretical Biology* **266**, 299 –306 (2010).
8. K. Nah, I. Choi and Y. Kim, Estimation of the incubation period of P. vivax Malaria in Korea from 2006 to 2008, *J. Korean Data Inf. Sci. Soc.* **21**, 1237 –1242 (2010).
9. H. Nishiura, H. W. Lee, S. H. Cho, W. G. Lee, T. S. In, S. U. Moon, G. T. Chung and T. S. Kim, Estimates of short and long term incubation periods of Plasmodium vivax malaria in the Republic of Korea, *Transactions of the Royal Society of Tropical Medicine and Hygiene* **101**, 338 –343 (2007).
10. A. L. Gibbs and F. E. Su, On choosing and bounding probability metrics, *International Statistical Institute* **70**, 419–435 (2002)
11. H. I. Ree, Studies on Anopheles sinensis, the vector species of vivax malaria in Korea, *The Korean Journal of Parasitology* **43**, 75 –92 (2005).
12. Korea Centers for Disease Control and Prevention, Mathematical modelling on P. vivax malaria transmission and development of its application program (2009), `http://www.cdc.go.kr/CDC/cms/content/68/12268_view.html`.
13. H. W. Lee, H. Nishiura, S. H. Cho, W. G. Lee, T. S. In, S. U. Moon, G. T. Chung and T. S. Kim, Factors affecting delayed diagnosis of Plasmodium vivax malaria in the Republic of Korea, *Res. J. Parasitology* **2**, 72 –75 (2007).
14. R. J. Maude, W. Pontavornpinyo, S. Saralamba, R. Aguas, S. Yeung, A. M. Dondorp, N. P. J. Day, N. J. White and L. J. White, The last man standing is the most resistant: eliminating artemisinin-resistant malaria in Cambodia, *Malaria Journal* **8**, 31 (2009).
15. N. A. Tiburskaja and O. S. Vrublevskaja, The course of infection caused by the North Korean strain of Plasmodium vivax, *WHO MAL.* **895**, 1 –19 (1977).
16. L. Hulden and L. Hulden, Activation of the hypnozoite: a part of Plasmodium vivax life cycle and survival, *Malaria Journal* **10**, 90 (2011).
17. M. Martcheva and O. Prosper. Unstable dynamics of vector-borne diseases: Modeling through delay-differential equations. *Dynamic Models of Infectious Diseases*, Springer New York, 43 –75 (2013).
18. G. Röst, Bifurcation of the Time-One Maps of Delay Differential Equations at Points of Resonance, *Funct. Differ. Eq.*, **13**(3-4), 585–602 (2006)
19. R. Qesmi and J. Wu, Periodic systems of delay differential equations,preprint

A SIMULATION OF THE U.S. INFLUENZA OUTBREAK IN 2009-2010 USING A PATCH SIR MODEL BASED ON AIRPORT TRANSPORTATION DATA

D. I. WALLACE, M. CHEN

Dartmouth College
Hanover, NH, 03755, USA

We simulate the progression of the novel H1N1 virus outbreak in Spring 2009 using a patch SIR model for six U.S. regions. A sensitivity analysis shows that the timing of peak prevalence is greatly affected by epidemiological parameters but little affected by migration rates. The method presented here has the advantage of using publicly available transportation data for only a few large regions and epidemiological parameters that may be estimated from the first few cases and the households in which they occur. Comparison with data shows the method to have predictive value for timing of epidemics.

Keywords: SIR models, patch models, regular graphs, epidemiology, influenza, H1N1

1. Introduction

In March of 2009 a new strain of H1N1 influenza was detected in Southern California [1]. Within months cases were reported in distant sites across the U.S. Some, but not all, cases were traced to foreign travel. A compelling statistical argument has been made for linking regional outbreaks with frequency of inbound flights from Mexico [2]. However, large regional airline hubs serve all parts of the U.S. as well, so disease may well have been transmitted from within the U.S. once it arrived. This hypothesis is the one modeled and analyzed here. This paper is not an attempt to fit a model precisely to data. Rather, it is a test of how well one might predict the progression of disease using only information available at the start of an epidemic.

The classic model of an infectious disease is given by three coupled ordinary differential equations for susceptible, infectious and recovered individuals (SIR models)[3]. This model is appropriate for large homogeneous well-mixed populations. A recent survey article points out the limitations of the SIR model in capturing spacial dynamics, pointing out that agent based models represent these dynamics but have their own computational limitations [4]. Patch models make it possible to investigate spacial phenomena at an intermediate level, by representing regions of a certain size (the patches) as homogeneously mixed while linking well separated regions via migration. In either case, the accuracy of SIR models depends completely on the accuracy of the data used to estimate initial prevalence, transmission, and recovery rates for any given disease.

Global airline networks have been studied in an effort to model large scale disease propagation. One study, of the worldwide air network of 3,880 airports, indicates that properties of the statistical distribution of travel are thought to influence rates of disease propagation [5]. It would be useful to know to what extent a data from a few major airports capture the dynamics of disease transmission and, in particular, the timing and intensity of disease prevalence. In this study we use publicly available airport usage data to construct a patch SIR model for six U.S. regions with major airports and compare the results of the model with publicly available data from the 2009 influenza outbreak.

2. Methods

Models for diseases that spread from city to city consider populations that are separated into patches, each of which is described by an SIR model, with migration between patches given as a linear term. Equations 1-3 below give the general form for the patch SIR model used here.

$$S_i' = (a-b)(S_i)(1 - S_i/k_i) - B\frac{S_i I_i}{(S_i + I_i + R_i)} - \sum_{j \sim i} m_{(i,j)} S_i + \sum_{j \sim i} m_{(j,i)} S_j \tag{1}$$

$$I_i' = B\frac{S_i I_i}{(S_i + I_i + R_i)} - (v + b + d)I_i - \sum_{j \sim i} m_{(i,j)} I_i + \sum_{j \sim i} m_{(j,i)} I_j \tag{2}$$

$$R_i' = vI_i - bR_i - \sum_{j \sim i} m_{(i,j)} R_i + \sum_{j \sim i} m_{(j,i)} R_j \tag{3}$$

Here the growth rate of the susceptible population, as a function of S_i only, is given by a logistic term. b is the natural death rate, d is the death rate of the disease only, B is the transmission coefficient, $m_{(i,j)}$ is the relative rate of travel from vertex i to vertex j, assumed to be equal for susceptible, infectious and recovered populations at vertex i. Sums denoted by $j \sim i$ indicate vertices adjacent to the ith vertex. Note that making birth a function of only susceptible individuals is not the most general form that could be taken, but it is reasonable for short term diseases. In a previous paper [6] we show that stability of the disease free equilibrium for symmetric versions of such models depends only on local parameters, *i. e.* not migration rates.

Six major U.S. areas connected by large airports (CA, CO, TX, GA, IL, NY) were used as a basis for parametrizing the model in equations 1-3. For this simulation, the growth rate was set to $(a-b)(1-S_i/k_i)-dS_i$. The birth rate, a, and death rate, b, were taken as the U.S. average[78] and k_i was chosen to give an equilibrium population at the disease free equilibrium corresponding to the actual population of each region[9]. The transmission and death rates (B and d respectively) from this particular strain can be estimated from commonly known statistics reported in the newspaper as 1.5 new cases per three days [10], which correlate roughly with those in the literature. We used .5 as the transmission rate, estimated as a reproduction rate of 1.5 spread over a three day duration. Actual likelihood of transmission was reported to vary between 1 (for households with two members) and approximately .2 (for households with six members) [11]. A review of multiple swine flu studies puts the reproduction number for the 2009 outbreak between 1.4 and 1.6 new infections per infected individual, but does not say over how many days this occurs [12]. A World Health Organization study gives a reproduction rate of 1.58 and confirms the same three day period based on data from the early outbreak in Mexico [2]. The recovery rate can be estimated from data available at the Center for Disease Control and corresponds to a disease of 14 days duration [13].

The graph between patches was the complete graph on six vertices, with migration rates estimated from airport data[14]. To model migration rates we assumed that the traffic to and from destinations other than the six modeled was negligible.

Migration rates for each state included were two forms: migration from state 1 to state 2, *e. g.* ($m_{CA,NY}$) and total migration away from a specific state *e. g.* (m_{CA}).

Let q_{NY} be the total average daily traffic through New York airports (and define q similarly for all six regions). Let Σ_{NY} be the sum of average daily traffic through the five remaining airports in the model (and define Σ similarly for all six regions).

The calculations for the migration rates from one state to the other are given by Equation 4.

$$m_{CA,NY} = q_{CA}q_{NY}/\Sigma_{CA} \tag{4}$$

This expression weights traffic from CA to NY according to the amount of traffic in NY. Notice that summing all the outgoing traffic from CA returns the total traffic in CA. The total migration away from a specific state is just the sum of the outgoing rates to all other states. The expression for New York is given by Equation 5.

$$m_{NY} = m_{NY,CO} + m_{NY,CA} + m_{NY,TX} + m_{NY,GA} + m_{NY,IL} \tag{5}$$

A simple model was created to test the migration rates with no disease present to verify that populations of the various regions remained approximately accurate at equilibrium. The model was then run with disease introduced in California and Texas. A second version of the model was run in which birth rates depended on the entire population of the region rather than on just susceptible individuals, in order to see if this adjustment to the model made any significant difference to the results. (It did not.)

The simulation was then compared with the patient data from clinics in each area (Regions 2,4,5,6,8,9) that report to the Center for Disease Control [15]. The regions for which FluView data was used included regions 2 (New York, New Jersey), 4 (Kentucky, Tennessee, North Carolina, South Carolina, Georgia, Florida, Alabama, Mississippi), 5 (Minnesota, Wisconsin, Michigan, Illinois, Indiana, Ohio), 6 (New Mexico, Texas, Oklahoma, Louisiana, Arkansas), 8 (Utah, Colorado, Wyoming, Montana, South Dakota, North Dakota), and 9 (California, Nevada, Arizona), which correspond to NY, GA, IL, TX, CO, and CA, respectively. The regions used in this study did not include just the city in which the airport was located, but the surrounding area as well. Population data corresponding with the regions around each airport was taken from a 2010 report on tuberculosis incidence [16]. One assumption of the model was that the heavily populated regions near accounted for most of the disease prevalence.

A sensitivity analysis was conducted on parameters and initial conditions with respect to the error observed between peak case load (for each state and for the US total) in the model versus data. In a subsequent simulation, transmission rates were modified for each state for a best match with data (for the first version of the model). Table 1 shows parameters used for the original model. Not shown are initial conditions for infected individuals (2 in California, 1 in Texas, 0 elsewhere) or recovered individuals (0 for all regions).

Table 1. Parameters for the model

Notation	Value	Units
$k_{NY}, S_{NY}(0)$	19151072	individuals in NY area
$k_{CO}, S_{CO}(0)$	2599235	individuals in CO area
$k_{CA}, S_{CA}(0)$,	17293449	individuals in CA area
$k_{TX}, S_{TX}(0)$	12564909	individuals in TX area
$k_{GA}, S_{GA}(0)$	5540092	individuals in GA area
$k_{IL}, S_{IL}(0)$	9622245	individuals in IL area
$m_{NY,CO}$	$5.93 * 10^{-5}$	daily migration from NY to CO as a percent of regional population
$m_{NY,CA}$	0.000133771	daily migration from NY to CA
$m_{NY,TX}$	0.00012474	daily migration from NY to TX
$m_{NY,GA}$	0.000104113	daily migration from NY to GA
$m_{NY,IL}$	9.67E-05	daily migration from NY to IL
$m_{CO,NY}$	$5.30 * 10^{-5}$	daily migration from CO to NY
$m_{CO,CA}$	$5.86 * 10^{-5}$	daily migration from CO to CA
$m_{CO,TX}$	$5.46 * 10^{-5}$	daily migration from CO to TX
$m_{CO,GA}$	$4.56 * 10^{-5}$	daily migration from CO to GA
$m_{CO,IL}$	4.23E-05	daily migration from CO to IL
$m_{CA,NY}$	0.000137123	daily migration from CA to NY
$m_{CA,CO}$	$6.72 * 10^{-5}$	daily migration from CA to CO
$m_{CA,TX}$	0.000141255	daily migration from CA to TX
$m_{CA,GA}$	0.000117897	daily migration from CA to GA
$m_{CA,IL}$	0.000109452	daily migration from CA to IL
$m_{TX,NY}$	0.000125624	daily migration from TX to NY
$m_{TX,CO}$	$6.16 * 10^{-5}$	daily migration from TX to CO
$m_{TX,CA}$	0.000138777	daily migration from TX to CA
$m_{TX,GA}$	0.00010801	daily migration from TX to GA
$m_{TX,IL}$	0.000100273	daily migration from TX to IL
$m_{GA,NY}$	0.000100813	daily migration from GA to NY
$m_{GA,CO}$	$4.94 * 10^{-5}5$	daily migration from GA to CO
$m_{GA,CA}$	0.000111369	daily migration from GA to CA
$m_{GA,TX}$	0.00010385	daily migration from GA to TX
$m_{GA,IL}$	$8.05 * 10^{-5}$	daily migration from GA to IL
$m_{IL,NY}$	$9.23 * 10^{-5}$	daily migration from IL to NY
$m_{IL,CO}$	$4.52 * 10^{-5}$	daily migration from IL to CO
$m_{IL,CA}$	0.000101972	daily migration from IL to CA
$m_{IL,TX}$	$9.51 * 10^{-5}$	daily migration from IL to TX
$m_{IL,GA}$	$7.94 * 10^{-5}$	daily migration from IL to GA
m_{NY}	0.000518612	daily migration away from NY
m_{CO}	0.000254107	daily migration away from CO
m_{CA}	0.000572914	daily migration away from CA
m_{TX}	0.000534237	daily migration away from TX
m_{GA}	0.000445897	daily migration away from GA
m_{IL}	0.000413958	daily migration away from IL
a	0.0135	natural birth rate percent per day
b	0.008036	natural death rate percent per day
B	0.5	transmission rate individuals per day
d	0.01043575	death rate due to disease percent per day
v	0.07142857	recovery rate percent per day

3. Results

Figure 1 shows the infected populations in all cities and the total. The first two peaks are for the California and Texas regions, while subsequent peaks are for all other regions.

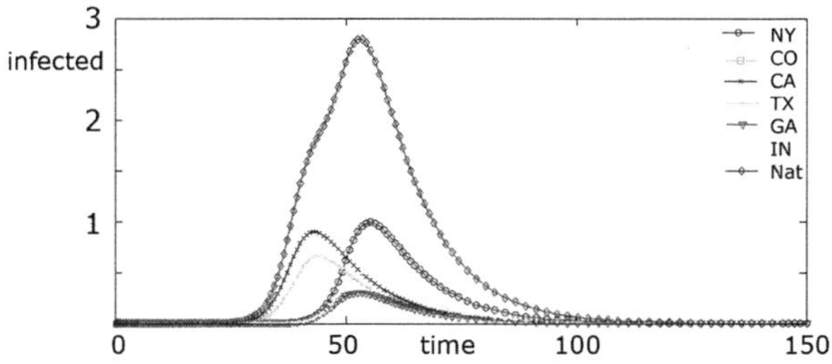

Figure 1. Model output for default parameters. Left scale is to be multiplied by 10^7.

3.1. *Disease prevalence*

The Center for Disease Control reports a total of 43,677 laboratory-confirmed cases of influenza (H1N1) 2009 in the United States through July 2009 [17]. They estimate the under reporting of influenza as 79 actual infections per reported case. Their statistical model estimates 1.8 million to 5.7 million cases occurred. Our model predicts about 15 times that prevalence. Our model is only for heavily populated regions around the six major airports, so it seems like the model overestimates the number of cases. When broken down by region and reported cases we see consistently large overestimates of the peak prevalence, as in Table 2, which gives model calculations and data comparison for constant transmission rate of .5.

Table 2. Peak prevalence results for constant transmission rate

Region	Peak Value (Model)	Peak Value (Data)	79 * Peak Value (Data)
NY	9,981,985	275	21,725
CO	1,442,394	2,134	168,586
CA	9,042,460	1,192	94,168
TX	6,614,867	652	51,508
GA	3,010,005	815	64,385
IL	5,106,028	1,763	139,277
National	28,000,000	10,050	793,950

However the model may just produce a distorted distribution. We can also compare the total number of cases in each region to the integrated infected compartment. If we assume a 14 day duration of disease and use the first fifteen weeks of data for each region to estimate total cases during this period we have the comparison described in Table 3, which gives model calculations and data comparison for constant transmission rate of .5. Total cases for the first 15 weeks of outbreak in each region are multiplied by 79 as recommended. Integrated infected compartments for each region are divided by a 14 day average duration of illness, consistent with the recovery rate. Our model predicts values about 15 times as large as the statistically adjusted data for the total number of cases and most of the individual regions.

Table 3. Total prevalence results compared

Region	79 * Total cases (Data)	Integrated Value (Model)/14
NY	236,052	14,975,307
CO	103,964	2,192,060
CA	474,869	13,758,223
TX	278,870	10,087,391
GA	514,053	4,551,770
IL	643,455	7,691,961
National	3,077,919	53,256,713

3.2. *Timing of peak prevalence*

The number of days until the model reached peak prevalence was compared to the peak in the corresponding data set for all regions and the total. For the default transmission rate of .5 errors were within a couple of weeks for most regions. The exceptionally poor matches were for California and Texas. The data for these states had two separate peaks– an early one and a late one. The model placed a peak between the two. Adjusting the transmission rate for a single state could often bring the model and data into agreement, as is shown for the Colorado region in Figure 2.

Figure 3 shows the results of altering transmission rates one at a time for each region. In this figure, California's two peaks are plotted separately, and these show the largest disagreement with the model, which produces a peak in the middle. Only the first peak of the Texas data was used for comparison.

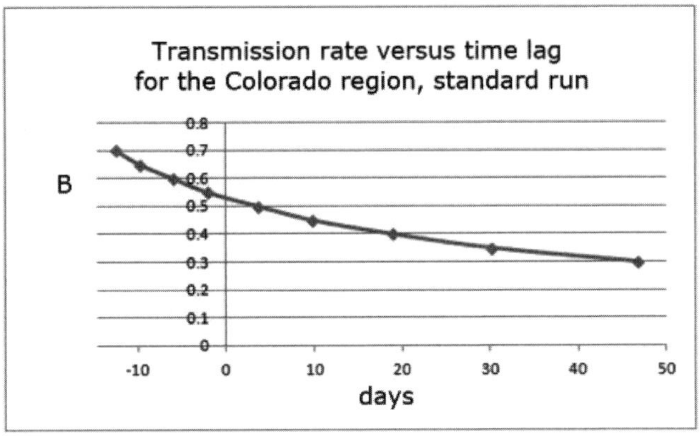

Figure 2. Time lag (peak of model - peak of data) versus transmission rates for Colorado

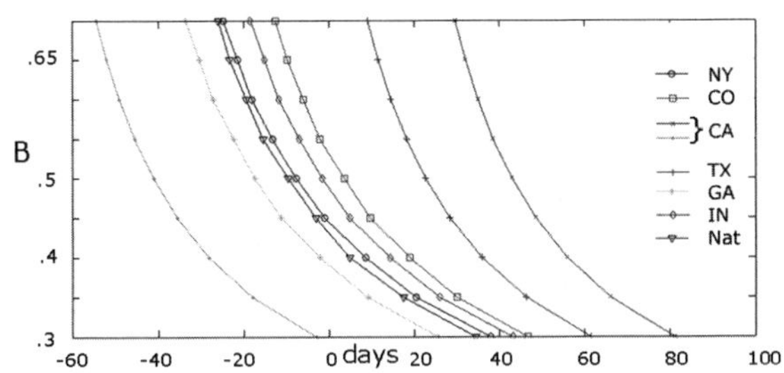

Figure 3. Time lag versus transmission rates for all regions

3.3. *Sensitivity of parameters*

A sensitivity analysis was carried out on all parameters. Specifically, we were interested in the error between the time to peak prevalence in the model and the data, for each region. Figure 4 shows, for Colorado, the result of varying each parameter up and down 10%. The result in Figure 4 is typical of all the regions and also the total. Disease related parameters have, in general, a much larger effect than migration rates, with the transmission rate having the largest effect of all. Table 4 summarizes the results for all regions with parameters listed in descending order of effect on time lag between data peak and model peak.

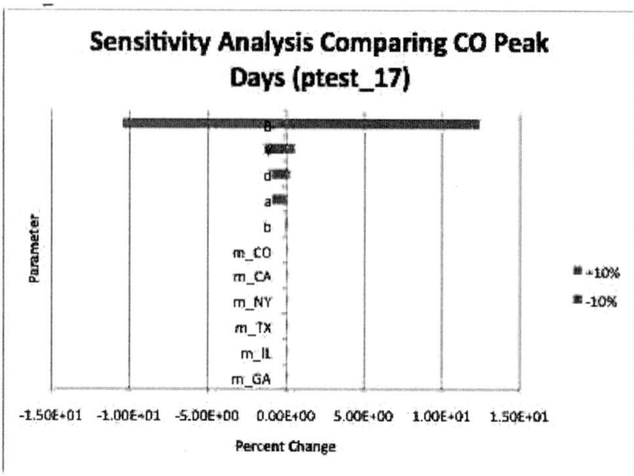

Figure 4. As each parameter is varied from its default value by +/- 10%, the time to peak prevalence varies. The transmission rate is seen to have by far the largest effect; migration rates have almost no effect.

Table 4. Sensitivity of error in peak time prediction

Region	Top 5	Negligible effect
NY	B, v, d, a, b	$m_{CO}, m_{CA}, m_{TX}, m_{NY}, m_{IL}, m_{GA}$
CO	B, v, d, a, b	$m_{CO}, m_{CA}, m_{NY}, m_{TX}, m_{IL}, m_{GA}$
CA	B, v, d, b, m_{CA}	$m_{NY}, m_{CO}, m_{TX}, m_{IL}, a, m_{GA}$
TX	B, v, d, b, m_{CA}	$m_{NY}, m_{CO}, m_{TX}, m_{IL}, a, m_{GA}$
GA	B, v, d, b, a	$m_{NY}, m_{CA}, m_{CO}, m_{TX}, m_{IL}, m_{GA}$
IL	B, v, d, b, a	$m_{NY}, m_{CA}, m_{CO}, m_{IL}, m_{TX}, m_{GA}$
NAT	B, v, m_{NY}, a, d	$b, m_{CO}, m_{CA}, m_{IL}, m_{TX}, m_{GA}$

4. Conclusions

4.1. *Estimates of disease prevalence*

The data we used gives the percent of those patients who, having presented with flu-like symptoms, tested positive for influenza[15]. Early cases were detected in California and Texas[18]. These were taken as initial conditions in the model, with disease incidence at all other locations taken to be zero. The FluView application has a disclaimer that explains that the data is from both the U.S. World Health Organization (WHO) Collaborating Laboratories and the National Respiratory and Enteric Virus Surveillance System (NREVSS). The disclaimer also mentions that the methods/testing practices varied from region to region. One last thing the disclaimer states is that WHO collaborating laboratories report on the influenza A subtypes of H1 or H3 while the majority of NREVSS laboratories do not report them.

There are approximately 145 participating laboratories in the U.S. How we chose to aggregate this data influences how our model compares to it. Perhaps if we had included different regions in the vicinity of each airport the local estimates would have been closer to the CDC estimate of 79 times the reported cases. The model prediction for the total is also greater than the CDC estimate for the U.S. of 1.8 to 5.7 million cases [17] for April to July of 2009.

In addition, altering the transmission rate would also change the predicted prevalence. Our model does not take into account behavioral changes of individuals which may take place when an outbreak is known and publicized, as data for the resulting change in transmission rates was not available. These changes may also be modeled explicitly if relevant parameters are known [19]. Doing so should lower the estimated number of cases in each region.

The estimates of prevalence we obtain from the integral of infected individuals over the time period under consideration (and divided by duration of the disease) were consistently higher than CDC estimates, indicating that a patch SIR model would have to be adjusted in order to predict the disease burden in advance.

4.2. *Timing of peak prevalence*

With two notable exceptions the timing of peak prevalence in our model was within a week or two of peak number of cases reported in the data. With transmission rates within a range of .4 to .55 we have good agreement on timing with data from GA, NY, CO, IL and the whole U.S., as shown in Figure 3. Strangely, the simulations for states where the disease starts (TX, CA) have the worst agreement with data. In both cases the data show double peaks. Texas has an early peak followed by a period of low case reports and then a second peak, which may be a second strain of virus. It is possible that, due to overland travel from Mexico, the original number of cases was under-reported in Texas, which could certainly alter the timing of the peak in that state. California also shows two peaks, but this is probably because the disease started in the San Diego region and later reached Northern California. We suspect that if we had treated the California region as two separate patches we would have had better agreement with the data. Nonetheless, the model produced results within a week or two of the peaks in the data set for most regions, indicating that patch models based on airport migration data could be a useful tool in predicting the timing of disease spread.

The timing also suggests that only a few cases arriving from abroad have the potential to spread through migration within the country. It is unnecessary to suppose that every outbreak was initiated through travel from Mexico. Regions assumed to acquire the disease by migration from Texas or California (or each other) had peak prevalence substantially later than those two states, both in the model and in the data. Had infected visitors from abroad been the source of infection it is unlikely that the timing would be so consistent.

4.3. *General utility of this method*

Although the transportation data is only an average, our sensitivity analysis shows that small variations in it do not have a noticeable effect on timing of peak prevalence. Of all the parameters in this model, the transmission rate has the largest effect. Unfortunately, this parameter can be difficult to estimate and can vary regionally and over time as people increase their awareness of a possible epidemic. How well a patch SIR model will predict the timing of an epidemic is highly dependent on getting an accurate estimate of the transmission rate and other disease-related parameters. However, even a rough estimate of the timing of an epidemic, in advance, is very useful. The method presented here has the advantage of using publicly available transportation data for only a few large regions and epidemiological parameters that may be estimated from the first few cases and the households in which they occur. Comparison with data shows the method to have predictive value.

References

1. Novel Swine-Origin Influenza A (H1N1) Virus Investigation Team, *N Engl J Med* **360** 2605-2615 (2009)
2. C. Fraser *et al*, *Science* **324** 1557 (2009)
3. R. M. Anderson and R. M. May, *Nature* **280**, 361 (1979).
4. M. J. Keeling and L. Danon, Mathematical modelling of infectious diseases *British Medical Bulletin***92** (2009)
5. V. Colizza, A. Barrat, M. Barthelémy, and A. Vespignani *PNAS* **103 7** 20152020 (2006)
6. Wallace, Hu-Wang, Chen *Symmetries and Groups in Contemporary Physics*

(Nankai Series in Pure, Applied Mathematics and Theoretical Physics), World Scientific, in press (2013)
7. Center for Disease Control and Prevention (Atlanta 2009), [Data File]. Retrieved May 3, 2012 from`http://www.cdc.gov/nchs/births.htm`
8. Center for Disease Control and Prevention (Atlanta 2009), [Data File]. Retrieved May 3, 2012 from `http://www.cdc.gov/nchs/deaths.htm`
9. Center for Disease Control and Prevention (Atlanta 2010). Retrieved May 3, 2012 from `http://www.cdc.gov/tb/statistics/reports/2010/pdf/report2010.pdf`
10. A. Gardner (ABC News, New York 2009). Retrieved May 3, 2012 from `http://abcnews.go.com/Health/Healthday/story?id=8439860`
11. S. Cauchemez, *et al*, *N Engl J Med* **361** 2619-2627 (2009)
12. B. J. Coburn, B. G. Wagner and S. Blower, *BMC Medicine* **7** **30** (2009)
13. Center for Disease Control and Prevention (Atlanta 2010) Retrieved May 3, 2012 from `http://www.cdc.gov/h1n1flu/qa.htm`
14. Airports Council International, North America (Montreal 2009), [Data File] Retrieved May 3, 2012 from `http://aci-na.org/content/airport-traffic-reports`
15. Center for Disease Control and Prevention (Atlanta, 2010). [Data Set]. Retrieved Oct. 3, 2012 from `http://gis.cdc.gov/grasp/fluview/fluportaldashboard.html`
16. Center for Disease Control and Prevention (Atlanta 2010).[Data Set]. Retrieved Oct. 3, 2012 from the Division of Tuberculosis Eliminations online ordering system at `http://www.cdc.gov/tb/`.
17. C. Reed, F. J. Angulo, D. L. Swerdlow, M. Lipsitch, M. I. Meltzer, D. Jernigan, *et al*, *Emerg Infect Dis* (2009) Retrieved Jul. 31, 2013 from `http://wwwnc.cdc.gov/eid/article/15/12/09-1413.htm`
18. Center for Disease Control and Prevention (Atlanta 2009). *MMWR* **58** **15**, 400-402. Retrieved Oct. 10, 2012 from `http://www.cdc.gov/mmwr/preview/mmwrhtml/mm5815a5.htm`
19. S. Funk, Marcel Salathé and V. A. A. Jansen, *J. R. Soc. Interface* **7** 12471256 (2010)

MODELLING DIRECTLY TRANSMITTED INFECTIONS CONSIDERING AGE-STRUCTURED CONTACT RATE AND VACCINATION*

H. M. YANG† and C. H. DEZOTTI

DMA, IMECC, UNICAMP; Praça Sérgio Buarque de Holanda, 651
CEP: 13083-859, Campinas, SP, Brazil; E-mail: hyunyang@ime.unicamp.br

We analyzed an infectious disease transmission model taking into account age-structured vaccination and contact rates. To show that the bifurcation from trivial to non-trivial solution is really determined by the reproduction number (obtained as the spectral radius of a linear operator), we proved the uniqueness of the non-trivial solution. We also determined the stability of the trivial solution based on the basic reproduction number, which can be evaluated in the upper and lower bounds. We applied the results to assess the basic reproduction number and the threshold of the vaccination rate considering the vaccination scheme against rubella introduced in the São Paulo State, Brazil.

1. Introduction

Directly transmitted childhood infections, like rubella and measles, have been used as good examples for the application of mathematical models. The models are formulated basically by taking into account the force of infection depending on the contact rate, which is related to the pattern of contacts among susceptible and infectious individuals. Additionally, vaccination is introduced in the model aiming to assess the controlling efforts.

When constant contact rate is used, the basic reproduction number R_0 and the lower value (threshold) for the vaccination rate above which the disease can be considered eradicated[1] are easily obtained. However, when considering age-structured contact rate, the calculation of R_0 and the threshold of the vaccination rate are not an easy task.[2–3]

In this paper, we present briefly an age-structured model[4] in section 2. This model is analyzed to show that the non-trivial solution is unique

*This work was supported by FAPESP (grant 09/15098-0) – Projeto temático
†CAPES (grant 10525/13-0)

(section 3), which guarantees that R_0 is related to the bifurcation value. We present the upper and lower bounds of R_0 for age-dependent contact rate (section 4). The stability of the trivial solution is presented in section 5. Discussion and conclusion are presented in section 6.

2. Model

Briefly, the model presented in a previous paper[4] divided a community into four groups: $X(a,t)$, $H(a,t)$, $Y(a,t)$ and $Z(a,t)$ which are, respectively, susceptible, exposed, infectious and immune individuals, distributed according to age a at time t. The exposed individuals are moved to the infectious class at a constant rate σ, and enters to the immune class at a rate γ. All individuals are under a constant mortality rate μ. The dynamics of the disease transmission is given by

$$\begin{cases} \frac{\partial}{\partial t} X(a,t) + \frac{\partial}{\partial a} X(a,t) = -\left[\lambda(a,t) + \nu(a) + \mu\right] X(a,t) \\ \frac{\partial}{\partial t} H(a,t) + \frac{\partial}{\partial a} H(a,t) = \lambda(a,t) X(a,t) - (\mu+\sigma) H(a,t) \\ \frac{\partial}{\partial t} Y(a,t) + \frac{\partial}{\partial a} Y(a,t) = \sigma H(a,t) - (\mu+\gamma) Y(a,t), \end{cases} \quad (1)$$

where $Z(a,t)$ is decoupled from the system, and $\nu(a)$ is the age-dependent vaccination rate. The age-specific force of infection at time t is $\lambda(a,t) = \int_0^L \beta(a,a') Y(a',t)\, da'$, where $\beta(a,a')$ is the age-structured contact rate, that is, the contact among susceptible individuals of age a with infectious individuals of age a', and L is the maximum age attainable by the human population.

The boundary conditions of system (1) are $X(0,t) = X_b$, the rate of newborns, and $H(0,t) = Y(0,t) = 0$. The initial conditions are assumed to be the solutions of (1) at the steady state, letting $\nu = 0$.

In a previous paper[4] we characterized the reproduction number R_ν (when $\nu = 0$, R_0 is the basic reproduction number) by applying the spectral radius theory. We showed that the bifurcation from the trivial to the non-trivial solution occurs when the spectral radius assumes unity value. In another words, we stated that when $R_0 < 1$ the disease is not able to establish in a disease free community, and when $R_0 > 1$, the disease can be settle at an endemic level. In this paper, we present further results obtained from the model given by system of equations (1).

3. Uniqueness of the non-trivial solution

We study the conditions regarding to the operator T (defined below) in order to have a unique non-trivial solution, which can be attained by successive approximations if some conditions are satisfied. In a previous paper,[4] we showed the existence of the non-trivial solution depending on the spectral radius $r(T'(0))$.

Theorem 3.1. ***(Existence Theorem)*** *Let us consider the operator $T : C[0,L] \to C[0,L]$ described by the equation*

$$Tu(a) = \int_0^L B(a,\zeta) M(\zeta, u(\zeta), \nu(\zeta)) u(\zeta) d\zeta, \qquad (2)$$

where $M(\zeta, u(\zeta), \nu(\zeta)) = e^{-\int_0^\zeta u(s)ds} \times e^{-\int_0^\zeta \nu(s)ds}$, and the kernel is

$$B(a,\zeta) = \sigma X_b e^{-N(\zeta)} \int_\zeta^L e^{-\sigma(s-\zeta)} e^{\gamma s} \left[\int_s^L \beta(a,a') e^{-(\mu+\gamma)a'} da' \right] ds, \qquad (3)$$

with X_b being newborn rate. If $r(T'(0)) \leq 1$, the only solution of equation

$$\lambda(a) = \int_0^L B(a,\zeta) M(\zeta, \lambda(\zeta), \nu(\zeta)) \lambda(\zeta) d\zeta$$

is the trivial solution. Otherwise, if $r(T'(0)) > 1$, then there is at least one non-trivial positive solution for this equation.

Proof: This Theorem was proven in Dezotti and Yang.[4] ■

Our goal is to prove that the non-trivial solution is unique.

Definition 3.1. The operator $A : E \to E$, where E is a Banach space with cone K, is **concave** if there exists a non-zero element $u_0 \in K$ such that for an arbitrary non-zero $x \in K$, the inequalities

$$\alpha(x) u_0 \leq Ax \leq \beta(x) u_0,$$

where α and β are positive, are valid, and if for every $x \in K$ such that

$$\alpha_1(x) u_0 \leq x \leq \beta_1(x) u_0,$$

with $\alpha_1, \beta_1 > 0$, the following relation, with $0 \leq t \leq 1$, is valid:

$$A(tx) \geq tA(x).$$

Definition 3.2. The operator $A : E \to E$, where E is a Banach space with cone K, is **$\mathbf{u_0}$-monotone** if it is concave; for every $x \in K$ such that $x \geq \gamma u_0$ $(\gamma > 0)$ and for $t \neq 0, 1$, then $A(tx) \neq tA(x)$ and it follows from $x \geq y$ and $x \neq y$ that $Ax \geq Ay + \varepsilon_0 u_0$, where $\varepsilon_0 = \varepsilon_0(x,y) > 0$.

Theorem 3.2. *If the operator A is u_0-monotone, then the equation $Ax = \eta x$ does not have two distinct non-zero solutions in the cone for some value of the parameter η.*

Theorem 3.3. *Let the equation $Ax = x$, with a concave monotone operator A, has a unique non-zero solution x^* in the cone K. Let either one of the following conditions be satisfied:*
(a) the cone K is regular, the operator A is continuous.
(b) the cone is normal, the operator A is completely continuous.
Then the successive approximations

$$x_n = Ax_{n-1}, \quad n = 1, 2, \cdots,$$

converge with respect to the norm to x^* independent of the initial approximation $x_0 \in K$, $x_0 \neq 0$.

Let us consider the operator A on Banach space $C[0, L]$ with cone $C[0, L]^+$, and rewrite (2) as

$$Au = \int_0^L B(a, \zeta)\, e^{-\int_0^\zeta u(s)ds} u(\zeta)\, d\zeta. \tag{4}$$

Defining the function $G(a, s)$ as

$$G(a, s) = \int_s^L \beta(a, a')\, e^{-(\mu+\gamma)a'} da',$$

the kernel (3), with $N(\zeta) = \int_0^\zeta \nu(s)\, ds$, can be rewritten as $B(a, \zeta) = \sigma X_b e^{-N(\zeta)} \int_\zeta^L e^{-\sigma(s-\zeta)} e^{\gamma s} G(a, s)\, ds$.

To evaluate $Au - Av$, we rewrite the equation (4) as

$$Au(a) = B(a, 0) + \int_0^L e^{-\int_0^\zeta u(s)ds} \frac{\partial B}{\partial \zeta}(a, \zeta)\, d\zeta, \tag{5}$$

and obtain

$$Au - Av = B(a, 0) + \int_0^L e^{-\int_0^\zeta u(s)ds} \tfrac{\partial B}{\partial \zeta}(a, \zeta)\, d\zeta - B(a, 0)$$
$$- \int_0^L e^{-\int_0^\zeta v(s)ds} \tfrac{\partial B}{\partial \zeta}(a, \zeta)\, d\zeta = \int_0^L \left[e^{-\int_0^\zeta u(s)ds} - e^{-\int_0^\zeta v(s)ds}\right] \tfrac{\partial B}{\partial \zeta}(a, \zeta)\, d\zeta$$
$$= \int_0^L e^{-\int_0^\zeta v(s)ds} \left[e^{-\int_0^\zeta u(s)ds + \int_0^\zeta v(s)ds} - 1\right] \tfrac{\partial B}{\partial \zeta}(a, \zeta)\, d\zeta$$
$$\int_0^L e^{-\int_0^\zeta v(s)ds} \left\{e^{-\int_0^\zeta [u(s)-v(s)]ds} - 1\right\} \tfrac{\partial B}{\partial \zeta}(a, \zeta)\, d\zeta.$$

If $\frac{\partial B}{\partial \zeta}(a,\zeta) < 0$ and $u > v$, we have $Au > Av$, and the operator A is monotone. By calculating $\frac{\partial B}{\partial \zeta}(a,\zeta)$, we have

$$\frac{\partial B}{\partial \zeta}(a,\zeta) = \sigma X_b \frac{\partial}{\partial \zeta}\left[e^{\sigma\zeta - N(\zeta)} \int_\zeta^L e^{-s(\sigma-\gamma)} G(a,s)\, ds\right] = \sigma X_b \left[\sigma - \frac{d}{d\zeta} N(\zeta)\right] \times$$
$$e^{\sigma\zeta - N(\zeta)} \int_\zeta^L e^{-s(\sigma-\gamma)} G(a,s)\, ds - e^{\sigma\zeta - N(\zeta)} e^{-\zeta(\sigma-\gamma)} G(a,\zeta) = e^{\gamma\zeta} G(a,\zeta) -$$
$$\sigma X_b e^{-N(\zeta)} \frac{d}{d\zeta} N(\zeta)\, e^{\sigma\zeta} \int_\zeta^L e^{-s(\sigma-\gamma)} G(a,s)\, ds - \sigma e^{\sigma\zeta} \int_\zeta^L e^{-s(\sigma-\gamma)} G(a,s)\, ds.$$

The condition to have $\frac{\partial B}{\partial \zeta}(a,\zeta) < 0$ is

$$e^{\gamma\zeta} G(a,\zeta) - \sigma \int_\zeta^L e^{-\sigma(s-\zeta)} e^{\gamma s} G(a,s)\, ds > 0.$$

However, observe that

$$\int_\zeta^L e^{-\sigma(s-\zeta)} e^{\gamma\zeta} G(a,\zeta)\, ds = e^{\gamma\zeta} G(a,\zeta) \left(\frac{e^{-\sigma(s-\zeta)}}{-\sigma}\right)\Bigg|_{s=\zeta}^L$$
$$= e^{\gamma\zeta} G(a,\zeta) \left[\frac{1}{\sigma} - \frac{e^{-\sigma(L-\zeta)}}{\sigma}\right],$$

and we can write

$$e^{\gamma\zeta} G(a,\zeta) = \sigma \int_\zeta^L e^{-\sigma(s-\zeta)} e^{\gamma\zeta} G(a,\zeta)\, ds + e^{\gamma\zeta - \sigma(L-\zeta)} G(a,\zeta),$$

such a way that

$$e^{\gamma\zeta} G(a,\zeta) - \sigma \int_\zeta^L e^{-\sigma(s-\zeta)} e^{\gamma s} G(a,s)\, ds =$$
$$\sigma \int_\zeta^L e^{-\sigma(s-\zeta)} e^{\gamma\zeta} G(a,\zeta)\, ds + e^{\gamma\zeta - \sigma(L-\zeta)} G(a,\zeta) - \sigma \int_\zeta^L e^{-\sigma(s-\zeta)} e^{\gamma s} G(a,s)\, ds =$$
$$e^{\gamma\zeta - \sigma(L-\zeta)} G(a,\zeta) + \sigma \int_\zeta^L e^{-\sigma(s-\zeta)} \left[e^{\gamma\zeta} G(a,\zeta) - e^{\gamma s} G(a,s)\right] ds.$$

If $e^{\gamma s} G(a,s)$ is decreasing in s for every a, then we have $\frac{\partial B}{\partial \zeta}(a,\zeta) < 0$.

To prove the next Theorem, we need the following Lemma.

Lemma 3.1. *T is completely continuous positive operator.*

Proof: This Lemma was proven in Dezotti and Yang.[4] ■

Theorem 3.4. *If the function* $e^{\gamma s}G(a,s)$ *is decreasing in* s *for each* a*, then the operator* A *is* u_0*-monotone, where* $u_0 \equiv 1$*, and completely continuous.*

Proof: (i) A is positive. This is straightforwardly checked, once A is the operator (2) rewritten.
(ii) for an arbitrary non-zero $u \in K$, the inequalities

$$\alpha(u)\, u_0 \leq Au \leq \beta(u)\, u_0$$

hold, where α and β are positive. Since

$$Au(a) = \int_0^L B(a,\zeta)\, e^{-\int_0^\zeta u(s)ds} u(\zeta)\, d\zeta = \int_0^L B(a,\zeta) \frac{d}{d\zeta}\left[-e^{-\int_0^\zeta u(s)ds}\right] d\zeta,$$

if we consider the First Mean Value Theorem (see Bartle[5]), we have a $\zeta^* \in [0,L]$ such that

$$Au(a) = B(a,\zeta^*) \int_0^L \frac{d}{d\zeta}\left[-e^{-\int_0^\zeta u(s)ds}\right] d\zeta = B(a,\zeta^*)\left[1 - e^{-\int_0^L u(s)ds}\right].$$

If we consider $b = \inf_{a,\zeta \in [0,L]} B(a,\zeta)$ and $B = \sup_{a,\zeta \in [0,L]} B(a,\zeta)$, we have

$$\alpha(u)\, u_0 \leq Au \leq \beta(u)\, u_0,$$

where $u_0 \equiv 1$, $\alpha(u) = b\left[1 - e^{-\int_0^L u(s)ds}\right]$ and $\beta(u) = B\left[1 - e^{-\int_0^L u(s)ds}\right]$.
If $b = 0$ then there exist a' and ζ' such that $B(a',\zeta') = 0$, and we already demonstrated that this is not possible.
(iii) for every $u \in K$ such that $\alpha_1(u)\, u_0 \leq u \leq \beta_1(u)\, u_0$, with $\alpha_1, \beta_1 > 0$, we have the relation $A(tu) > tAu$, with $0 < t < 1$. Since

$$\begin{aligned} A(tu) - tAu &= \int_0^L B(a,\zeta)\, e^{-\int_0^\zeta tu(s)ds} tu(\zeta)\, d\zeta - t\int_0^L B(a,\zeta)\, e^{-\int_0^\zeta u(s)ds} \\ &\times u(\zeta)\, d\zeta = t\int_0^L B(a,\zeta)\left[e^{-\int_0^\zeta tu(s)ds} - e^{-\int_0^\zeta u(s)ds}\right] u(\zeta)\, d\zeta \\ &= t\int_0^L B(a,\zeta)\, e^{-\int_0^\zeta u(s)ds}\left[e^{-\int_0^\zeta tu(s)ds + \int_0^\zeta u(s)ds} - 1\right] u(\zeta)\, d\zeta \\ &= t\int_0^L B(a,\zeta)\, e^{-\int_0^\zeta u(s)ds}\left[e^{(1-t)\int_0^\zeta u(s)ds} - 1\right] u(\zeta)\, d\zeta, \end{aligned}$$

we have

$$A(tu) - tAu > 0.$$

(iv) for $u \geq v$ and $u \neq v$, we have that $Au \geq Av + \varepsilon_0 u_0$, where $\varepsilon_0 = \varepsilon_0(u, v) > 0$. Using the equation (5), we have that

$$Au - Av = \int_0^L e^{-\int_0^\zeta v(s)ds} \left\{ e^{-\int_0^\zeta [u(s)-v(s)]ds} - 1 \right\} \frac{\partial B}{\partial \zeta}(a, \zeta)\, d\zeta.$$

Since $e^{\gamma s} G(a, s)$ is decreasing in s for every a, then $\frac{\partial B'}{\partial \zeta}(a, \zeta) < 0$ and

$$Au - Av = \int_0^L e^{-\int_0^\zeta v(s)ds} \left\{ e^{-\int_0^\zeta [u(s)-v(s)]ds} - 1 \right\} \frac{\partial B}{\partial \zeta}(a, \zeta)\, d\zeta > 0.$$

(v) A is completely continuous. The operator A is the operator T, given by (2), which was rewritten. So this affirmation was verified in Lemma 3.1. ■

Theorem 3.5. *Let us consider the function*

$$D(a, s) = e^{\gamma s} G(a, s)$$

decreasing in s for each a and the operator $T : C[0, L] \to C[0, L]$ defined by equation (2). If $r(T'(0)) > 1$, then the equation

$$\lambda(a) = \int_0^L B(a, \zeta)\, e^{-\Lambda(\zeta)} \lambda(\zeta)\, d\zeta,$$

where $\Lambda(\zeta) = \int_0^\zeta \lambda(s)\, ds$, has a unique non-zero solution which can be attained by successive approximations,

$$\lambda_n = T\lambda_{n-1}, n = 1, 2, \cdots,$$

and is independent of the initial approximation $\lambda_0 \in K$, $\lambda_0 \neq 0$.

Proof: The operator T is completely continuous and u_0-monotone and $K = C[0, L]^+$ is a normal cone. Using **Existence Theorem**, Theorem 3.2 and Theorem 3.3 we have the desired result. ■

4. Upper and lower bounds for $\mathbf{R_0}$

The main result of this section rests upon the possibility of estimating the spectral radius using inequalities based on the operator $T'(0)$ on a suitable element of K. The upper and lower bounds of the reproduction number are obtained based on the following theorems found in Krasnosel'kii[6] and Zabreyko.[7]

Theorem 4.1. *Let the linear positive operator A be completely continuous. Let the relation $A^p u_0 \geq \alpha u_0$, with $\alpha > 0$, be satisfied for some non-zero element u such that $-u_0 \notin K$, and $u_0 = v - w$, where $v, w \in K$ and p is some natural number. Then the operator has at least one characteristic vector $u^* \in K$,*

$$Au^* = \lambda u^*,$$

where the positive characteristic number λ satisfies the inequality

$$\lambda \geq (\alpha)^{\frac{1}{p}}.$$

Theorem 4.2. *Let the positively completely continuous linear operator A satisfies the inequality*

$$A^p v_0 \leq \beta v_0,$$

where v_0 is a quasi-interior element of the cone K. Then

$$r(A) \leq (\beta)^{\frac{1}{m}},$$

where $r(A)$ is the spectral radius of A.

To prove the next Theorem, we need the following Lemma.

Lemma 4.1. *T is Fréchet differentiable at the point $0 \in C[0, L]$ in the directions of the cone $C[0, L]^+$ and*

$$T'(0)\, h(a) = \int_0^L B(a, \zeta)\, M(\zeta, 0, \nu(\zeta))\, h(\zeta)\, d\zeta. \tag{6}$$

Furthermore $T'(0)$ is strongly positive completely continuous operator.

Proof: This Lemma was proven in Dezotti and Yang.[4] ■

Theorem 4.3. *Let us consider the linear operator on Banach space $C[0, L]$ with cone $C[0, L]^+$ given by the equation*

$$T'(0)\, h(a) = \int_0^L B(a, \zeta)\, h(\zeta)\, d\zeta,$$

where $B(a, \zeta)$ is given by equation (2). Then

$$R_\nu^s = \inf_{a \in [0,L]} \int_0^L B(a, \zeta)\, d\zeta \leq r(T'(0)) \leq \sup_{a \in [0,L]} \int_0^L B(a, \zeta)\, d\zeta = R_\nu^g.$$

Proof: It is enough to consider $m = p = 1$ and $u_0 = v_0 \equiv 1$ in Theorem 4.1 and Theorem 4.2. ■

5. Stability of the trivial solution

Let $x(a,t)$, $h(a,t)$ and $y(a,t)$ be small perturbations from the equilibrium[4] $(X_\infty(a), H_\infty(a), Y_\infty(a))$, that is,

$$\begin{cases} X(a,t) = X_\infty(a) + x(a,t) \\ H(a,t) = H_\infty(a) + h(a,t) \\ Y(a,t) = Y_\infty(a) + y(a,t), \end{cases} \tag{7}$$

giving rise a small perturbation on the force of infection

$$\lambda(a,t) = \lambda_\infty(a) + l(a,t),$$

where $\lambda_\infty(a)$ is the age-specific force of infection in the steady state, and the perturbed force of infection is $l(a,t) = \int_0^L \beta(a,a') y(a',t)\, da'$. If we substitute (7) into (1), and retain only the first order terms, we obtain

$$\begin{cases} \frac{\partial}{\partial t} x(a,t) + \frac{\partial}{\partial a} x(a,t) = -l(a,t) X_\infty(a) - [\lambda_\infty(a) + \nu(a) + \mu] x(a,t) \\ \frac{\partial}{\partial t} h(a,t) + \frac{\partial}{\partial a} h(a,t) = l(a,t) X_\infty(a) + \lambda_\infty(a) x(a,t) - (\mu+\sigma) h(a,t) \\ \frac{\partial}{\partial t} y(a,t) + \frac{\partial}{\partial a} y(a,t) = \sigma h(a,t) - (\mu+\gamma) y(a,t). \end{cases}$$

The boundary conditions become $x(0,t) = h(0,t) = y(0,t) = 0$.

For each variable $x(a,t)$, $h(a,t)$ and $y(a,t)$ we consider solution in the form $v(a,t) = v(a) e^{\omega t}$, where ω is a complex number, and the system of differential equations becomes

$$\begin{cases} \frac{d}{da} x(a) = -l(a) X_\infty(a) - [\lambda_\infty(a) + \nu(a) + \mu + \omega] x(a) \\ \frac{d}{da} h(a) = l(a) X_\infty(a) + \lambda_\infty(a) x(a) - (\mu+\sigma+\omega) h(a) \\ \frac{d}{da} y(a) = \sigma h(a) - (\mu+\gamma+\omega) y(a), \end{cases} \tag{8}$$

with $l(a) = \int_0^L \beta(a,a') y(a')\, da'$, and the initial conditions are $x(0) = h(0) = y(0) = 0$.

Solving the system of differential equations (8), we deduce that $l(a)$ must satisfy the following integral equation

$$l(a) = \int_0^L \int_0^{a'} \int_0^b \sigma X_b \beta(a,a') e^{-(\mu+\gamma+\omega)a'} e^{(\gamma-\sigma)b} e^{-\Lambda^*(\varsigma) - N(\varsigma) + \sigma\varsigma} \times \left[l(\varsigma) e^{\omega\varsigma} - \lambda_\infty(\varsigma) \int_0^\varsigma l(s) e^{\omega s} ds \right] d\varsigma\, db\, da'.$$

In the case of the trivial equilibrium, $\lambda_\infty \equiv 0$, this equation has the form

$$l(a) = \int_0^L \int_0^{a'} \int_0^b \sigma X_b \beta(a,a') e^{-(\mu+\gamma+\omega)a'} e^{(\gamma-\sigma)b} e^{-N(\varsigma)+\sigma\varsigma} l(\varsigma) e^{\omega\varsigma} d\varsigma\, db\, da',$$

and interchanging the order of the integrations, we have

$$l(a) = \int_0^L S(a, \zeta, \nu(\zeta), \omega)\, l(\zeta)\, d\zeta, \tag{9}$$

where

$$S(a, \zeta, \nu(\zeta), \omega) = e^{-N(\zeta)} \int_\zeta^L \left[\int_b^L \sigma X_b \beta(a, a') e^{-\mu a'} e^{(\zeta - a')\omega} e^{(b-a')\gamma} da' \right] e^{(\zeta - b)\sigma} db. \tag{10}$$

Letting $\omega = 0$ in the equations (9) and (10), we obtain

$$l(a) = \int_0^L S(a, \zeta, \nu(\zeta), 0)\, l(\zeta)\, d\zeta = T'(0)\, l(a),$$

and

$$S(a, \zeta, \nu(\zeta), 0) = M(\zeta, 0, \nu(\zeta))\, B(a, \zeta),$$

where $T'(0)$ is the operator given by equation (6). We are interested to determine if the trivial equilibrium is either stable or unstable, which is equivalent to analyze if the real part of ω is either negative or non-negative. Let us consider the case where ω is a real number, and also positive, then $S(a, \zeta, \nu(\zeta), \omega)$ is strictly monotonically decreasing function of ω.

Let us consider the linear operator R_ω on Banach space $C[0, L]$ with the usual norm, $\|u\| = \sup_{a \in [0,L]} |u(a)|$, and cone $K = C[0, L]^+$, defined by

$$R_\omega l(a) = \int_0^L S(a, \zeta, \nu(\zeta), \omega)\, l(\zeta)\, d\zeta. \tag{11}$$

The following Theorems and Lemmas are used in the demonstration of the **Stability Theorem** stated below.

Theorem 5.1. *(**Ascoli's Theorem,** Kreyszig[8]) A bounded equicontinuous sequence $(x_n)_n$ in $C[0, L]$ has a subsequence which converges in the norm on $C[0, L]$ (a sequence $(y_n)_n$ in $C[0, L]$ is said to be **equicontinuous** if for every $\varepsilon > 0$ there is a $\delta > 0$, depending only on ε, such that for all y_n and all $a, a' \in [0, L]$ satisfying $|a - a'| < \delta$, we have $|y_n(a) - y_n(a')| < \varepsilon$).*

Theorem 5.2. *(**Compactness criterion,** Kreyszig[8]) Let $S : Y \to Z$ be a linear operator where Y and Z are normed spaces. Then S is compact* *operator if and only if it maps every bounded sequence $(y_n)_n$ in Y onto a sequence in Z which has a convergent subsequence.*

Theorem 5.3. *(Deimling[9]) Let be X a Banach space, $K \subset X$ a solid cone, that is, $int(K) \neq \emptyset$, and $T : X \to X$ strongly positive compact linear operator. Then:*
(i) $r(T) > 0$, $r(T)$ is a simple eigenvalue with eigenvector $v \in int(K)$ and there is not eigenvalue with positive eigenvector.
(ii) if λ is an eigenvalue and $\lambda \neq r(T)$ then $|\lambda| < r(T)$.
(iii) if $S : X \to X$ is bounded linear operator and $Sx \geq Tx$ on K, then $r(S) \geq r(T)$, while $r(S) > r(T)$ if $Sx > Tx$ for $x \in K, x > 0$.

Lemma 5.1. *R_ω is strongly positive compact continuous linear operator on $C[0,L]$.*

Proof: (i) $R_\omega u$ is linear in u. It is checked straightforwardly.
(ii) $R_\omega u$ is continuous in u. Suppose that $l_1, l_2 \in C[0,L]$, where $\|l_1 - l_2\| = \sup\limits_{a\in[0,L]} |l_1(a) - l_2(a)|$. We desire to calculate $\|R_\omega l_1 - R_\omega l_2\| = \sup\limits_{a\in[0,L]} |R_\omega l_1(a) - R_\omega l_2(a)|$. Given $a \in [0,L]$, we have

$$|R_\omega l_1(a) - R_\omega l_2(a)| = \int_0^L S(a,\zeta,\nu(\zeta),\omega)\, l_1(\zeta)\, d\zeta - \int_0^L S(a,\zeta,\nu(\zeta),\omega)\, l_2(\zeta)\, d\zeta \leq \int_0^L |S(a,\zeta,\nu(\zeta),\omega)|\, |l_1(\zeta) - l_2(\zeta)|\, d\zeta.$$

Let be $m^* \in R$ such that $|S(a,\zeta,\nu(\zeta),\omega)| \leq m^*$ for all $(a,\zeta) \in [0,L] \times [0,L]$. So, given $\varepsilon > 0$, let us consider $\|l_1 - l_2\| \leq \delta = \frac{\varepsilon}{m^* L}$, then

$$|R_\omega l_1(a) - R_\omega l_2(a)| \leq \int_0^L m^* \frac{\varepsilon}{m^* L} d\zeta = \varepsilon,$$

and $\|R_\omega l_1 - R_\omega l_2\| \leq \varepsilon$.
(iii) R_ω is compact operator. Since R_ω is a linear operator on $C[0,L]$ we will use Theorem 5.1 and Theorem 5.2 to prove it. Let $(l_n)_n$ be a bounded sequence in $C[0,L]$. We will prove that $(R_\omega l_n)_n$ is a bounded equicontinuous sequence.
Since $(l_n)_n$ is a bounded sequence in $C[0,L]$ there exists $m^{**} > 0$ such that we have $|l_n(a)| \leq m^{**}$ for all $n = 1, 2, \cdots$, and all $a \in [0,L]$. Given $\varepsilon > 0$,

let be $a_1, a_2 \in [0, L]$. For each $n = 1, 2, \cdots$, we have

$$|R_\omega l_n(a_1) - R_\omega l_n(a_2)| = \int_0^L S(a_1, \zeta, \nu(\zeta), \omega) l_n(\zeta) d\zeta -$$
$$\int_0^L S(a_2, \zeta, \nu(\zeta), \omega) l_n(\zeta) d\zeta \leq \int_0^L m^{**} |S(a_1, \zeta, \nu(\zeta), \omega) - S(a_2, \zeta, \nu(\zeta), \omega)| d\zeta.$$

Since S is continuous and $[0, L]$ is compact, S is uniformly continuous. So, there is $\delta_1 > 0$ such that if $|a_1 - a_2| < \delta_1$, then

$$|S(a_1, \zeta, \nu(\zeta), \omega) - S(a_2, \zeta, \nu(\zeta), \omega)| \leq \frac{\varepsilon}{m^{**} L},$$

consequently $|R_\omega l_n(a_1) - R_\omega l_n(a_2)| \leq \varepsilon$, that is, $(R_\omega l_n)_n$ is equicontinuous. That $(R_\omega l_n)_n$ is bounded is checked straightforwardly.
(iv) R_ω is strongly positive. Let be $l \in K = C[0, L]^+$, $l \neq 0$. R_ω will be strongly positive if $R_\omega(l) \in int(K)$, that is, $R_\omega(L) > 0$. Suppose that there exists $a^* \in [0, L]$ such that $R_\omega(l) a^* = 0$, that is,

$$R_\omega l(a^*) = \int_0^L S(a^*, \zeta, \nu(\zeta), \omega) l(\zeta) d\zeta = 0.$$

Since $S(a, \zeta, \nu(\zeta), \omega) l(\zeta)$ is non-negative and continuous, and we are integrating over all the interval, then $S(a^*, \zeta, \nu(\zeta), \omega) l(\zeta) \equiv 0$. Furthermore, as $l \neq 0$, there exists ζ^* such that $l(\zeta^*) \neq 0$. So $S(a^*, \zeta^*, \nu(\zeta^*), \omega) = 0$, or

$$e^{-N(\zeta^*)} \int_{\zeta^*}^L \int_b^L \sigma X_b \beta(a^*, a') e^{-\mu a'} e^{(\zeta^* - a')\omega} e^{(b - a')\gamma} e^{(\zeta^* - b)\sigma} db da' = 0.$$

Then $\beta(a^*, a') = 0$ for all $a' \in [\zeta^*, L]$, which is not true (recall[4] that $\beta(a, a') > 0$ for all $a, a' \in [0, L]$, except for $a = a' = 0$ where $\beta(a, a') = 0$). Thus, there is not a such a^*.
If $l \in C[0, L]$ then $R_\omega l \in C[0, L]$. It is enough to observe that $R_\omega l$ is a composition of continuous functions. ■

Lemma 5.2. $r(R_{\omega_2}) < r(R_{\omega_1})$ *for every* $\omega_1 < \omega_2$. *Particularly* $r(R_\omega) < r(T'(0))$ *for every positive real number* ω.

Proof: We demonstrate that $S(a, \zeta, \nu(\zeta), \omega_1) > S(a, \zeta, \nu(\zeta), \omega_2)$ if $\omega_1 < \omega_2$, in order to show that $S(a, \zeta, \nu(\zeta), \omega)$ is strictly monotonically decreasing function of ω.

Suppose $\omega_1 < \omega_2$. Since $f(\omega) = e^{-\omega}$ is strictly monotone decreasing function of ω, and $S(a, \zeta, \nu(\zeta), \omega)$ is given by the equation (10), we have, from $e^{(\zeta - a')\omega_1} > e^{(\zeta - a')\omega_2}$, that

$$e^{-N(\zeta)} \int_{\zeta}^{L} \left[\int_{b}^{L} \sigma X_b \beta(a, a') e^{-\mu a'} e^{(\zeta - a')\omega_1} e^{(b-a')\gamma} da' \right] e^{(\zeta - b)\sigma} db >$$
$$e^{-N(\zeta)} \int_{\zeta}^{L} \left[\int_{b}^{L} \sigma X_b \beta(a, a') e^{-\mu a'} e^{(\zeta - a')\omega_2} e^{(b-a')\gamma} da' \right] e^{(\zeta - b)\sigma} db,$$

or, $S(a, \zeta, \nu(\zeta), \omega_1) > S(a, \zeta, \nu(\zeta), \omega_2)$. Then, by definition of R_ω follows that $R_{\omega_1} > R_{\omega_2}$. Now, using Theorem 5.3 (iii) for $T = R_{\omega_2}$ and $S = R_{\omega_1}$ we have that $r(R_{\omega_1}) > r(R_{\omega_2})$. ■

In next Lemma we will use the following theorem.

Theorem 5.4. *(Zabreyko[10]) Let E_1 and E_2 be Banach spaces with cones K_1 and K_2, respectively. Consider $A : E_1 \to E_1$ being a linear operator, $B : E_2 \to E_2$ a positive linear operator and $\varphi : E_1 \to E_2$ an operator satisfying following conditions:*
(i) $\varphi(u + v) \le \varphi(u) + \varphi(v), \forall u, v \in E_1$;
(ii) if $\|\varphi(u_n)\| \to 0$, where $n \to \infty$, then $|u_n| \to 0$, where $n \to \infty$;
(iii) $\varphi(Au) \le B\varphi(u)$, $\forall u \in K_1$.
Then $r(A) \le r(B)$.

Lemma 5.3. *Let us consider the operator R_ω defined by the equation (11), where $\omega \in [0, +\infty)$. Then $r(R_\omega) \to 0$ when $\omega \to +\infty$.*

Proof: Using the Theorem 5.4 for $E_1 = C[0, L]$, $E_2 = R^n$, $K_1 = C[0, L]^+$, $K_2 = \{(\zeta_1, \cdots, \zeta_n); \zeta_i \ge 0, i = 1, ..., n\}$, $A = R_\omega$, $\varphi(u) = \left(\left\|u\,|_{[a_0, a_1]}\right\|, \cdots, \left\|u\,|_{[a_{n-1}, a_n]}\right\|\right)^t$ and $B = S(\omega) = (S_{ij}(\omega))_{1 \le i,j \le n}$, where $S_{ij}(\omega) = \sup_{a_{i-1} \le a \le a_i} \int_{a_{j-1}}^{a_j} S(a, \zeta, \nu(\zeta), \omega) d\zeta$, we have that $r(R_\omega) \le r(S(\omega))$. Using the Lebesgue Dominated Convergence Theorem (see Bartle[11]) we have that $S(a, \zeta, \nu(\zeta), \omega) \to 0$ when $\omega \to +\infty$, particularly, $S_{ij}(\omega) \to 0$ when $\omega \to +\infty$. Since that $r(S(\omega)) = \max_{\lambda \in \sigma(S(\omega))} |\lambda|$, we have $r(S(\omega)) \to 0$ when $\omega \to +\infty$ (see Appendix 6.1). So, we have the desired result. ■

Lemma 5.4. *$r(R_\omega)$ is a continuous function in ω.*

Proof: Remembering that the expression of $R_\omega l(a)$, given by equation (11), is strongly positive compact continuous linear operator defined on solid cone $K = C\,[0, L]^{+}$, let us consider the function

$$\begin{cases} r : [0, +\infty) & \to & R \\ \quad \omega & \longmapsto & r(\omega) = r(R_\omega), \end{cases}$$

which is defined on the whole interval $[0, +\infty)$.
(i) We show that $r(\omega)$ is left continuous for every $\omega > 0$. Let us consider an increasing sequence $(\omega_n)_n$ in $[0, +\infty)$ such that $\omega_n \to \omega$ when $n \to +\infty$. Since $r(\cdot)$ is decreasing function of ω, we have

$$\omega_n \leq \omega_{n+1} \leq \omega \implies r(\omega_n) \geq r(\omega_n + 1) \geq r(\omega).$$

Let us call $r_n = r(\omega_n)$ and $r = r(\omega)$. Since $(r_n)_n$ is a bounded decreasing sequence there exists r^* such that

$$r^* = \lim_{n\to\infty} r_n = \inf\{r_n; n = 1, 2, \cdots\} \geq r$$

(see Bartle[5]). For each n let $l_n \in C\,[0, L]^{+}$ be such that $\|l_n\| = 1$, and

$$R_{\omega_n} l_n = r_n l_n.$$

If we can find $l^* \in C\,[0, L]^{+}$ such that $R_{\varpi} l^* = r^* l^*$, that is, r^* is an eigenvalue of R_ω, then

$$r^* = |r^*| \leq r(R_\omega) = r,$$

and we obtain the desired result.
Since R_ω is compact and $(l_n)_n$ is bounded sequence, we can assume that $(R_\omega l_n)_n$ is convergent. Suppose l^* such that $R_\omega l_n \to l^*$ when $n \to \infty$. From the fact that

$$\|R_{\omega_n} l_n - R_\omega l_n\| \leq \|R_{\omega_n} - R_\omega\| \, \|l_n\| = \|R_{\omega_n} - R_\omega\|,$$

we have $\|R_{\omega_n} l_n - R_\omega l_n\| \to 0$ when $n \to \infty$. Since

$$\begin{aligned} &\|R_{\omega_n} l_n - l^*\| \leq \|R_{\omega_n} l_n - l^* + R_\omega l_n - R_\omega l_n\| \leq \\ &\|R_{\omega_n} l_n - R_\omega l_n\| + \|R_\omega l_n - l^*\|, \end{aligned}$$

then $\|R_{\omega_n} l_n - l^*\| \to 0$ when $n \to \infty$. So, $R_{\omega_n} l_n \to l^*$ when $n \to \infty$. Furthermore

$$\lim_{n\to\infty} l_n = \lim_{n\to\infty} \frac{1}{r_n}(r_n l_n) = \lim_{n\to\infty} \frac{1}{r_n} R_{\omega_n} l_n = \frac{1}{r^*} l^*,$$

thus

$$R_\omega(l^*) = R_\omega\left(r^* \lim_{n\to\infty} l_n\right) = r^* \lim_{n\to\infty} R_\omega(l_n) = r^* l^*.$$

(ii) Now, we show that $r(\omega)$ is right continuous for every $\omega \geq 0$. Let $(\omega_n)_n$ be a decreasing sequence such that $\omega_n \to \omega$ where $n \to \infty$. Then

$$\omega \leq \omega_{n+1} \leq \omega_n \implies r \geq r_{n+1} \geq r_n.$$

Being $(r_n)_n$ a bounded increasing sequence, there exists $r^{**} \in \mathbf{R}$ such that

$$r^{**} = \lim_{n\to\infty} r_n = \sup\{r_n; n = 1, 2, \cdots\} \leq r$$

(see Bartle[5]). Suppose that $0 < r^{**} < r$. Consider the resolvent operator of R_ω defined on a domain $\mathcal{D}$ of $\mathbf{C}$,

$$\mathfrak{R} : \lambda \in \mathcal{D} \longmapsto \mathfrak{R}(\lambda) = (R_\omega - \lambda Id)^{-1},$$

and for each $n = 1, 2, \cdots$, and the resolvent operator of R_{ϖ_n} defined on a domain $\mathcal{D}_n$ of $\mathbf{C}$,

$$\mathfrak{R}_n : \lambda \in \mathcal{D}_n \longmapsto \mathfrak{R}_n(\lambda) = (R_{\omega_n} - \lambda Id)^{-1},$$

which singularities are their eigenvalues, respectively. Since R_ω is a strongly positive compact linear operator we have that r is a isolated singularity which is a simple polo of $\mathfrak{R}$. By the same reason, each r_n is an isolated singularity which is a simple polo of $\mathfrak{R}_n$.
Since $\lim_{n\to\infty} r_n = r^{**} < r$ and

$$r_n = r(R_{\omega_n}) = \sup\{|\lambda|; \lambda \text{ is a eigenvalue of } R_{\omega_n}\},$$

there exists a neighborhood of r, $B_\varepsilon(r) = \{\lambda \in \mathbf{C} \cap \mathcal{D}; |\lambda - r| < \varepsilon\}$, such that we can assume $\mathfrak{R}_n$ is holomorphic into $B_\varepsilon(r)$ and over its boundary $\partial(B_\varepsilon(r)) = \{\lambda \in \mathbf{C} \cap \mathcal{D}; |\lambda - r| = \varepsilon\}$.
Consider ℓ being the simple closed continuous curve described by $\partial(B_\varepsilon(r))$, which shall be considered oriented in the positive sense. Since $\mathfrak{R}_n$ is holomorphic inside and over ℓ, we have

$$\frac{1}{2\pi i} \oint_\ell \mathfrak{R}_n(\lambda)\, d\lambda = 0.$$

Since r is a simple polo of $\mathfrak{R}$ and, the only singularity of $\mathfrak{R}$ inside ℓ, then

$$\frac{1}{2\pi i} \oint_\ell \mathfrak{R}(\lambda)\, d\lambda = \text{residue of } \mathfrak{R} \text{ in } r \neq 0.$$

It follows that

$$0 = \lim_{n\to\infty} \frac{1}{2\pi i} \oint_\ell \mathfrak{R}_n(\lambda)\, d\lambda = \frac{1}{2\pi i} \oint_\ell \lim_{n\to\infty} \mathfrak{R}_n(\lambda)\, d\lambda = \frac{1}{2\pi i} \oint_\ell \mathfrak{R}(\lambda)\, d\lambda \neq 0,$$

(see Dunford and Schwartz[12]) which is a contradiction. So $r^{**} = r$. ■

Theorem 5.5. ***(Stability Theorem)*** *If $r(T'(0)) \leq 1$, then the trivial equilibrium is locally stable. If $r(T'(0)) > 1$, then the trivial equilibrium is locally unstable.*

Proof: If $r(T'(0)) \leq 1$, by Lemma 5.2 we have that $r(R_\omega) < r(T'(0)) \leq 1$, and being R_ω strongly positive completely continuous linear, by Theorem 5.3, its spectral radius is an eigenvalue, and the only solution of equation (9) is the trivial one.
Suppose that $r(T'(0)) > 1$. Being $r(R_\omega)$ decreasing function of ω (see Lemma 5.2) and continuous in ω (see Lemma 5.4), and since $r(R_\omega) \to 0$ when $\omega \to +\infty$ (see Lemma 5.3), there exists $\omega^* > 0$ such that $r(R_{\omega^*}) = 1$, and the corresponding eigenfunction of $r(R_{\omega^*})$ generates the instability in the trivial equilibrium. ■

6. Discussion and conclusion

We showed the uniqueness of the non-trivial solution whenever the spectral radius (from operator $T'(0)$) assumes value greater than 1. Moreover, if the function $D(a,s) = e^{\gamma s} G(a,s)$ is decreasing in s for each a, the non-trivial solution can be obtained by the recursive method. Additionally, complex mathematical calculations are involved to obtain the spectral radius, which was the reason to evaluate the upper and lower bounds for the basic reproduction number R_0. These bounds are more easily calculated. Local stability of the trivial solution was performed taking into account the operator $T'(0)$: If $r(T'(0)) \leq 1$, the trivial solution is locally stable; otherwise, unstable.

We present only two kinds of contact rates in order to illustrate the calculation of R_0. Detailed calculations can be found in Dezotti.[13] We consider that vaccine is administrated in the age interval $[a_1, a_2]$, that is,

$$\nu(a) = \nu\theta(a - a_1)\,\theta(a_2 - a), \tag{12}$$

where ν is a constant vaccination rate, and the Heaviside function is $\theta(x) = 1$, if $x \geq 0$; otherwise, 0.

Let us consider an effective contact concentrated at the same ages, that is, $\beta(a, a') = \beta\delta(a - a')$, where the Dirac delta function is $\delta(x) = \infty$, if $x = 0$; otherwise, 0. In this case, we obtain non-linear homogeneous Volterra integral equation of second type,[14] $\lambda(a) = \int_0^a B(a,\zeta)\, e^{-\int_0^\zeta \lambda(s)ds} \lambda(\zeta)\, d\zeta$, which has a unique solution $\lambda \equiv 0$,[13] where the kernel is

$$B(a,\zeta) = \begin{cases} \frac{\beta\sigma X_b}{\gamma-\sigma} e^{-\mu a}\left[e^{-\sigma(a-\zeta)} - e^{-\gamma(a-\zeta)}\right], & if \ \ \zeta \le a \\ 0, & if \ \ a < \zeta. \end{cases}$$

Hence, the ages of contact must be relaxed to occur the transmission.

Let us consider $\beta(a,a') = \beta e^{-c_1 a} e^{-c_2 a'}$. In this case, the upper and lower bounds are equal, and the reproduction number, letting $L \to \infty$, is

$$R_\nu = R_0 \left\{ 1 - \frac{\nu e^{-(\mu+c_1+c_2)a_1}}{\mu+\nu+c_1+c_2} \left[1 - e^{-(\mu+\nu+c_1+c_2)(a_2-a_1)}\right] \right\},$$

where the basic reproduction number is

$$R_0 = \frac{\beta\sigma X_b}{(\mu+c_1+c_2)(\mu+\gamma+c_2)(\mu+\sigma+c_2)}.$$

The proof that $e^{\gamma s}G(a,s)$ is decreasing in s for each a is omitted.[13] When $\beta(a,a') = f(a)\,g(a')$, separable, it can be shown that the upper (R_ν^g) and lower (R_ν^s) bounds for the reproduction number are equal.[13] By letting $c_1 = 0$ and $c_2 = 0$, we have a constant contact rate $\beta(a,a') = \beta$. In this case, the reproduction number, letting $L \to \infty$, is

$$R_\nu = R_0 \left\{ 1 - \frac{\nu}{\mu+\nu} e^{-\mu a_1} \left[1 - e^{-(\mu+\nu)(a_2-a_1)}\right] \right\}, \tag{13}$$

where the basic reproduction number is

$$R_0 = \frac{\beta\sigma X_b}{\mu(\mu+\gamma)(\mu+\sigma)}.$$

In this case, we have

$$G(a,s) = \int_s^L \beta(a,a')\, e^{-(\mu+\gamma)a'} da' = \int_s^L \beta e^{-(\mu+\gamma)a'} da' = \frac{\beta e^{-(\mu+\gamma)s}}{\mu+\gamma} - \frac{\beta e^{-(\mu+\gamma)L}}{\mu+\gamma}$$

such a way that

$$e^{\gamma s}G(a,s) = e^{\gamma s}\left[\frac{\beta e^{-(\mu+\gamma)s}}{\mu+\gamma} - \frac{\beta e^{-(\mu+\gamma)L}}{\mu+\gamma}\right] = \frac{\beta}{\mu+\gamma}\left[e^{-\mu s-\gamma s+\gamma s} - e^{-\mu L-\gamma L+\gamma s}\right]$$
$$= \frac{\beta e^{-\mu s}}{\mu+\gamma}\left[1 - e^{-\gamma(L-s)-\mu(L-s)}\right] = \frac{\beta e^{-\mu s}}{\mu+\gamma}\left[1 - e^{-(\gamma+\mu)(L-s)}\right],$$

which is decreasing in s for every a. Hence, from Theorem 3.5, the non-trivial solution can be attained by successive approximations when $R_0 > 1$.

Vaccination strategy against rubella[15–17] carried on in São Paulo State, Brazil, in 1992, was adopted based on a model considering constant contact rate $\beta(a,a') = \beta$ and vaccination rate given by (12). In that occasion we used equation (13) as the reproduction number without proof, which was now proven by spectral radius. The vaccination scheme introduced in Brazil against rubella controlled the infection (at very low incidence, practically as the eradication situation). Hence, if the vaccination scheme could be maintained continuously, we can assure that the trivial (eradication) solution is stable according to our results.

In a future work we will analyze the contact rate of the form

$$\beta(a,a') = \beta_0 \times \frac{b_3}{b_2\Gamma(b_1+1)} \times \frac{\left(\frac{a}{b_2}\right)^{b_1} e^{-\frac{a}{b_2}}}{2-e^{-b_3 a}} \times e^{-b_3\left|a-a'\right|},$$

where $\Gamma(x)$ is the gamma function. With respect to the parameters, β_0 (dimension of *time*) is the period of exposure encompassing the infectivity of virus; b_1 is the average number of potentially infective contacts; b_2 (dimension of *time*) is the togetherness period; and b_3 is the infective contact rate (dimension of $time^{-1}$). This kind of contact rate was studied numerically elsewhere,[18–19] because it does not satisfy $\beta(a,a') > 0$ for all $a, a' \in [0, L]$, except for $a = a' = 0$ where $\beta(a,a') = 0$, which is a condition for Lemma 5.1. One of the main results in classical epidemiology is the displacement of the average age of the acquisition of the first infection to elder ages whenever vaccination is introduced.[1] However, model considering age-structured contact rate showed that this is true only for vaccination of lower aged children.[19]

Acknowledgments

We thank Prof. David Greenhalgh for comments and suggestions that contributed to improve this paper.

Appendix A. $r(S(\omega)) \to 0$ when $\omega \to +\infty$

Let be $B = S(\omega) = (S_{ij}(\omega))_{1\le i,j\le n}$, where $S(a,\zeta,\nu(\zeta),\omega)$ is given by equation (10), and $S_{ij}(\omega) = \sup_{a_{i-1}\le a\le a_i} \int_{a_{j-1}}^{a_j} S(a,\zeta,\nu(\zeta),\omega)\,d\zeta$. If $\omega_1 \le \omega_2$, then $S(a,\zeta,\nu(\zeta),\omega_1) \ge S(a,\zeta,\nu(\zeta),\omega_2)$, and $\int_{a_{j-1}}^{aj} S(a,\zeta,\nu(\zeta),\omega_1)\,d\zeta \ge \int_{a_{j-1}}^{aj} S(a,\zeta,\nu(\zeta),\omega_2)\,d\zeta$. Observe that for each $j = 1,\cdots,n$, the function $a \in [a_{i-1},a_i] \longmapsto \int_{a_{j-1}}^{aj} S(a,\zeta,\nu(\zeta),\omega)\,d\zeta$ is continuous in a, for all $i = 1,\cdots,n$. Since $[a_{i-1},a_i]$ is compact, there exists $a^* \in [a_{i-1},a_i]$ such that

$$\sup_{a_{i-1}\le a\le a_i} \int_{a_{j-1}}^{a_j} S(a,\zeta,\nu(\zeta),\omega_2)\,d\zeta = \int_{a_{j-1}}^{a_j} S(a^*,\zeta,\nu(\zeta),\omega_2)\,d\zeta.$$

So

$$S_{ij}(\omega_2) = \sup_{a_{i-1}\le a\le a_i} \int_{a_{j-1}}^{a_j} S(a,\zeta,\nu(\zeta),\omega_2)\,d\zeta = \int_{a_{j-1}}^{a_j} S(a^*,\zeta,\nu(\zeta),\omega_2)\,d\zeta \le$$

$$\int_{a_{j-1}}^{a_j} S(a^*,\zeta,\nu(\zeta),\omega_1)\,d\zeta \le \sup_{a_{i-1}\le a\le a_i} \int_{a_{j-1}}^{a_j} S(a,\zeta,\nu(\zeta),\omega_1)\,d\zeta = S_{ij}(\omega_1),$$

and $S(\omega_1) \ge S(\omega_2)$.

Let us consider in Theorem 5.3 $T = S(\omega_2)$ and $S = S(\omega_1)$. Then it follows that $r(S(\omega_2)) \le r(S(\omega_1))$, that is, $r(S(\omega))$ is monotonically decreasing function of ω. Particularly, if $\omega_1 < \omega_2$, then $r(S(\omega_2)) < r(S(\omega_1))$. In fact, we have that $S(\omega) = (S_{ij}(\omega)) \to 0$ when $\omega \to +\infty$.

Furthermore, $r(S(\omega)) = \max\{|\lambda| : \lambda \text{ is eigenvalue of } S(\omega)\}$, that is, λ is a root of the polynomial equation in x,

$$P(x,\omega) = \det[S(\omega) - xId].$$

A matrix $A_{n\times n}$ is **irreducible** if there is no **permutation matrix** P (a permutation matrix is a square matrix such that $Pe_i = e_j$, with $i \to j$ being a permutation of $\{1,\cdots,n\}$ and $\{e_1,\cdots,e_n\}$, the natural basis in $\mathbf{R}^n$) such that

$$P^{-1}AP = \begin{bmatrix} A_1 & 0 \\ B & A_2 \end{bmatrix},$$

where A_1 and A_2 are square matrices. If it is possible, then A is said **reducible**. Particularly, if $A = [a_{ij}]$ is such that $a_{ij} > 0$ for every i,j, then A is irreducible.

Theorem A.1. *(Deimling[9]) Let $A = (a_{ij})$ be an $n \times n$ matrix with $a_{ij} \ge 0$ for all $i,j \le n$, or $A \ge 0$ for shorter. Then*
(a) $r(A)$ is an eingevalue of A with a positive eigenvalue. If $B \ge 0$ and $A - B \ge 0$, then $r(A) \ge r(B)$.
(b) if A is irreducible, then $r(A)$ is a simple eingevalue with an eigenvector in int $(\mathbf{R}^n_+)$.
(c) if $a_{ij} > 0$ for all i,j, then $|\lambda| < r(A)$ for all eigenvalues $\lambda \ne r(A)$.

In our case, $S_{ij}(\omega) > 0$ for every i,j, and using the theorem above, we have that $r(S(\omega))$ is a simple eigenvalue with eigenvector in $int\,(R^n_+)$.

Lemma A.1. *(Greenhalgh[2]) Let $F(k,\eta) = \det(A(k) - \eta Id)$ be a polynomial equation, where $A(k)$ is an $n \times n$ matrix. Suppose that*
(a) $F(k,\eta)$ is continuous in k for k real, right continuous in k at $k = 0$, and analytic in the complex variable η for $\eta \in \mathcal{D}$, $\mathcal{D}$ a domain in $\mathbf{C}$.

(b) $r(A(k))$ *is a simple root of* $F(k,\eta)$.
(c) $r(A(k))$ *is strictly monotone decreasing in* k.
Then $r(A(k))$ *is continuous in* k *for* $k > 0$*, and right continuous at* $k = 0$.

If we consider $\Phi(\omega) = S(\omega)$ and $F(\omega,\eta) = \det[S(\omega) - \eta Id]$, we notice that $r(S(\omega))$ is a continuous function of ω, particularly,

$$\lim_{\varpi\to\infty} r(S(\omega)) = r\left(\lim_{\varpi\to\infty} S(\omega)\right) = r(0) = 0.$$

Being $r(R_\omega) \leq r(S(\omega))$, we have the desired result .

References

1. R. M. Anderson and R. M. May, *Infectious Diseases of Humans: Dynamics and Control*, Oxford, Oxford University Press (1991).
2. D. Greenhalgh, *Math. Biosc.* **100**, 201 (1990).
3. H. Inaba, *J. Math. Biol.* **28**, 411 (1990).
4. C. H. Dezotti and H. M. Yang, *Proceedings of Biomat 2010*, 106 (2011).
5. R. G. Bartle, *The Elements of Real Analysis second edition*, John Wiley & Sons, New York (1976).
6. M. A. Krasnosel'skii, *Positive Solutions of Operator Equations*, P. Noorddhoff ltda. Groningen, The Netherlands (1964).
7. P. P. Zabreyko, M. A. Krasnosel'skii and V. Y. Stecenko, *Math. Notes* **1, 3 and 4**, 306 (1967).
8. E. Kreyszig, *Introductory Functional Analysis with Applications*, John Wiley & Sons, New York (1989).
9. K. Deimling, *Nonlinear Functional Analysis*, Springer-Verlag, Berlin (1985).
10. P. P. Zabreyko, A. I. Koshelev, M. A. Krasnosel'skii, S. G. Mikhlin, L. S. Rakovshchik and V. Y. Stet'senko, *Integral equations-a reference text*, Noordhoff International Publishing Leyden, The Netherlands (1975).
11. R. G. Bartle, *The Elements of Integration*, John Wiley & Sons, New York (1966).
12. N. Dunford and J. T. Schwartz, *Linear Operators Part I: General Theory*, John Wiley & Sons, New York (1957).
13. C. H. Dezotti, *Thesis*, IMECC-UNICAMP (Brazil) (2000).
14. F. G. Tricomi, *Integral equations*, Dover Publications, Inc., New York (1985).
15. E. Massad, M. N. Burattini, R. S. Azevedo Neto, H. M. Yang, F. A. B. Coutinho and D. M. T. Zanetta, *Epidemiol. Infect.* **112 (3)**, 579 (1994).
16. E. Massad, R. S. A. Azevedo Neto, M. N. Burattini, D. M. T. Zanetta, F. A. B. Coutinho, H. M. Yang, J. C. Moraes, C. S. Pannuti, V. A. U. F. Souza, S. S. B. Silveira, C. H. Struchiner, G. W. Oselka, M. C. C. Camargo, T. M. Omoto and S. D. Passos, *Intern. J. Epidemiol.* **24 (4)**, 842 (1995).
17. H. M. Yang, *Appl. Math. Comput.* **122 (1)**, 27 (2001)
18. H. M. Yang, *Math. Compt. Model.* **29 (8)**, 39 (1999).
19. H. M. Yang, *Math. Compt. Model.* **29 (7)**, 11 (1999)

A GENERAL FRAMEWORK FOR AGENT-BASED MODELLING WITH APPLICATIONS TO INFECTIOUS DISEASE DYNAMICS*

MAREK LASKOWSKI
Agent-Based Modelling Laboratory
York University, 4700 Keele St.
Toronto, Ontario
M3J 1P3, Canada
E-mail: mareklaskowski@gmail.com

SEYED M. MOGHADAS
Agent-Based Modelling Laboratory
York University, 4700 Keele St.
Toronto, Ontario
M3J 1P3, Canada
E-mail: moghadas@yorku.ca

Advanced information and communication technologies are increasingly applied to collect heterogeneous data on various aspects of population health, including the emergence of novel diseases and their spread in human populations. Because information evolves rapidly during an emerging disease epidemic, and information on key epidemiological features of the disease may be subject to substantial uncertainty, more advanced methods and computational capabilities are required to guide optimal disease control policies and support decision-making under uncertainty. Agent Based Modelling appears to be an effective method to analyze such information and data and, in principle, capture the true mechanisms underlying disease processes at both the individual and population levels. In practice, the fidelity of agent-based models is only limited by the amount, variety, and richness of data available; by the human resources necessary to implement the model; and by the availability of computational resources for simulation. Here, we propose a general framework for an agent-based modelling computational system to encapsulate important variables and parameters of demographic, health, and infection, and to simulate the infection dynamics. This framework has the potential to be used for decision-making support in real-time event of an emerging disease.

*This work was in part supported by Mitacs and CIHR (Canada).

1. Introduction

Agent Based Modeling and Simulation (ABMS) is an integrative mutli-modelling approach in which complex system macro phenomena arise from interactions between simpler micro models defined in terms of agents embedded within an environment [1]. This can be interpreted as the model of the whole (one or more complex phenomena) that is constructed through a bottom up framework from the constituent components (e.g., Agents and Environment) [2]. The mediation of relationships between agents and the environment is shown in Figure 1.

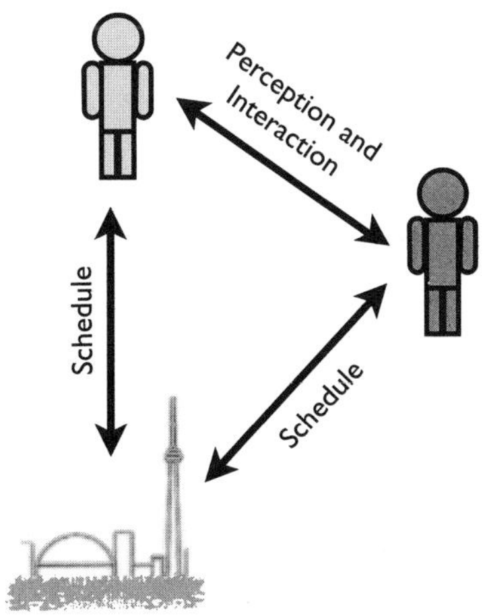

Figure 1. Agents are related to particular locations in the environment as a function of time. Agents can perceive and interact with co-located Agents.

As with any simulation approach, it is most useful in situations where analytic or equation-based methods are intractable. ABMS can in principle incorporate any complex behavior that can be observed experimentally; however, in practice, these models are limited by the time investment required to implement the micro and macro behavior in software, the availability of data to support model assumptions, and the computational time required to carry out the simulations. Therefore, the model design will necessarily require a balance between the desired complexity of the model and the effective use of available resources (e.g., compute resources, human resources, and data).

In this chapter, we propose an agent-based modeling framework that, despite inherent complexity, is capable of linking micro and macro dynamics in a systematic manner that can be computationally well-managed. This framework, when adopted for simulating the spread of a novel infection in a population, has shown enormous potential to uncover behavioural and response patterns that are not encountered in any equation-based dynamical models of disease epidemics [3,4,5,6]. We begin by describing the components of our ABMS, and follow with its structural design and implementation. We discuss new vistas that are currently open for the application of this simulation framework.

2. ABMS Structure

2.1. *Agents and the environment*

In the context of ABMS, an agent can be viewed as a discrete entity that perceives, makes decisions, and takes actions within an environment that includes other agents [7]. The decision process can take into account the perceptions, internal state, and goals [2]. Agents in ABMS are analogous to the eponymous individuals in individual-based modeling within the field of ecology. Often, the terms agent and individual are used interchangeably, although in ABMS there is more of an emphasis on agent adaptive behavior, learning, and intelligence than in individual-based modeling. An ABMS systematically (i.e., by iteration) gives each agent the opportunity to perform a perception-decision-action cycle in a given state of the simulated environment and current simulated time. Agents perception can be thought of as an iteration process through a set of perceivable properties of the environment and other interacting agents. These perceptions are typically limited to some spatial vicinity of the agent. Since agents are discrete encapsulated entities, it is possible that certain properties will be limited to certain agents, or not perceivable by other agents at all.

The agent's decision process can be arbitrarily complex. Often, rule-based behavior is sufficient for describing agents' behavior, especially if learning does not relate to modeling goals. Rules are imputed into the model with expert involvement, utilizing as much data as possible for informing behavior. In cases where learning is important, such as in learning an optimal strategy [8], several Machine Learning approaches have been documented [9,10,11]. It is largely unknown to what degree Machine Learning strategies would reflect human learning, particularly in a complex social context.

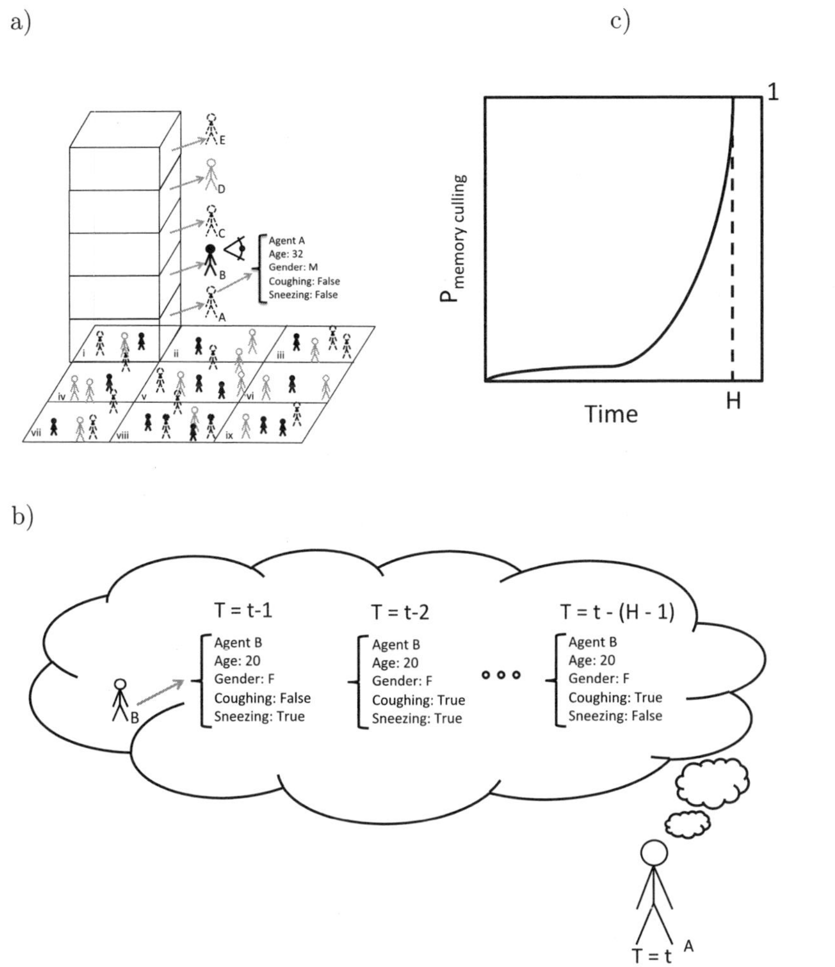

Figure 2. (a) Infectious agent B in the lattice location 'i' is shown currently perceiving the properties of susceptible agent A. Thereafter agent B may probabilistically transmit the disease to agent A. Agent B will perform this for agents C, D and E. Agents A and C, D and E, also perform perceptions, delibration, and action on the current timestep, as do agents in the other lattice locations (ii-ix). (b) Using the appropriate data structures, memory can be implemented that allows agents to store and retrieve memories during the perception-decision-action cycle. Memories (of other agents) can be probabilistically culled until the probability of forgetting a memory approaches 1 at the memory Horizon H, which can be chosen to meet either modeling or computer memory (i.e., RAM) constraints. (c) An example of culling probability function is shown.

Actions can be included as information, in the form of discrete messages being passed between agents. Disease transmission can also be implemented in this way. In general, an agent may change its internal state in response to a message, or even respond with one or more messages of its own. It is recommended that an agent should not be able to modify the encapsulated state of another agent or the environment, rather the first agent would send a message to either the second agent or the environment and let the second entity react to the message (including changing its own encapsulated state). In this way [12], the unrealistic identity between agent actions, and changes in other agents or the environment could be avoided. Figure 2a illustrates an agent during its perception-action-cycle in the context of a population of agents distributed throughout a lattice environment. Figure 2b demonstrates that the agent's memory is fairly straightforward to implement with a function that permits the agent to forget older memories (Figure 2c).

In other complex systems modeling approaches such as static network [13], there is no distinction between agent and the nodes of the network. A network model uses a (dynamic) graph as a formalism to represent some phenomena with an underlying complex network. Space is not necessarily implied in a network model, for example, if the edges represent abstract social relationships. Similarly, in Cellular Automata (CA) [14], there is no distinction between agents and cells that make up the lattice environment. Nevertheless, agent-based models can be implemented using both complex network models [15], and cellular automata [16].

In contrast, in an ABMS, agents are embedded within the environment and in general can move (i.e., change spatial associations) within an environment as shown in Figure 3. In the context of a spatial network, agents can move by changing relationships to the spatial nodes over time.

In an ABMS, the environment can be considered as a spatial network with locations in space corresponding to nodes in the network (Figure 4a). Agents are associated with a particular node or location at a particular time. One or more nodes can serve as origins for Cartesian space in which Agents could move about (Figure 4b). Edges would represent connections or paths traversable by agents between two locations. Other models of the environment will tend to be specializations or variations of this view. An example of a common specialization is a lattice environment (Figure 4c).

The environment may have its own set of properties that could be relevant to some models. Labels, names, or other information and properties that are meaningful to agents can be associated with the environment in the context of a model. The environment may respond to messages from agents, or change with time, and can sometimes generate or absorb agents.

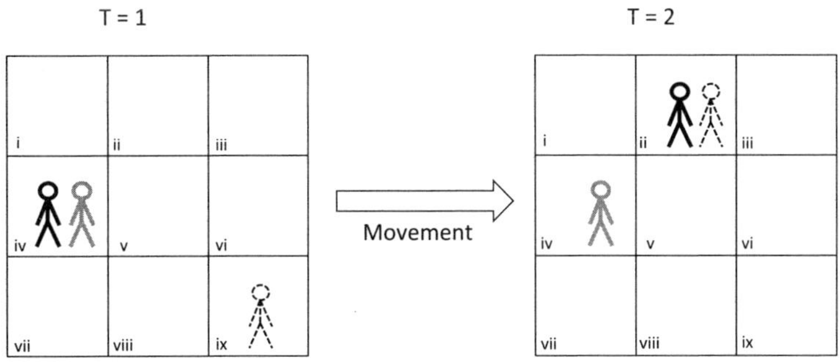

Figure 3. Agents moving in a lattice environment between time T=1 and T=2.

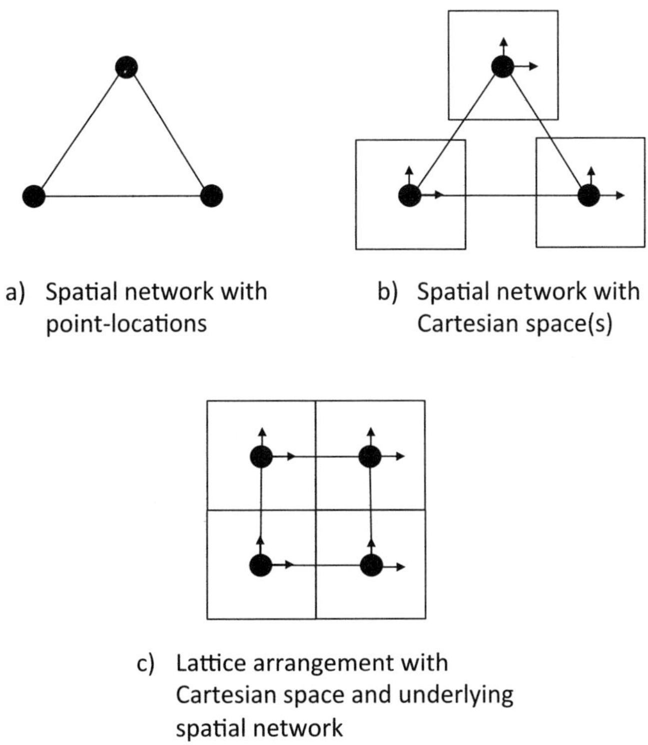

Figure 4. Types of environment in ABMS.

3. The Modeling Process

Beginning with and maintaining the goals of an ABMS in mind throughout, the modeling becomes an iterative process of using data to create sub-models (e.g., agent behaviors, environment), defining how the sub-models interact, calibrating, validating, running the composite model, analyzing results and synthesizing emerging information, and making improvements to the sub-models when required. This process continues until modellers and end-users agree that the model is reasonably valid, that is, whether the model is a "good enough" representation of the target system, and can meet the proposed objectives [7]. The ABMS methodology emphasizes end-user involvement in designing and validating the model.

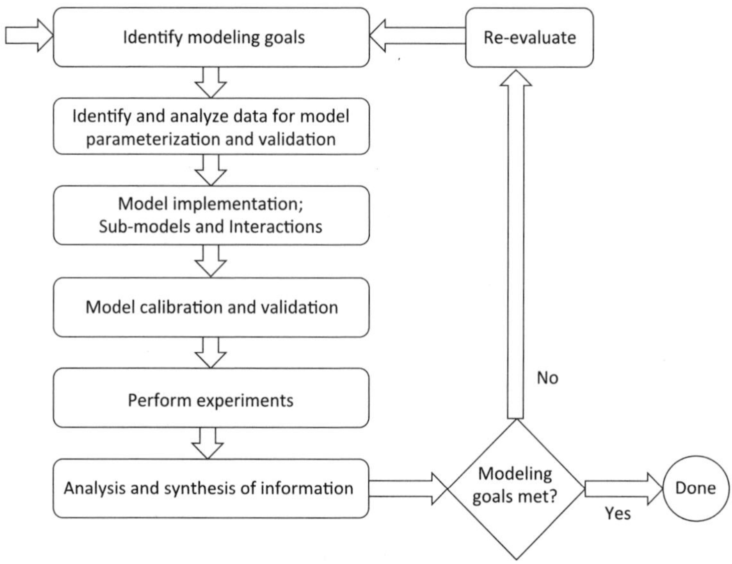

Figure 5. ABMS development cycle. Since it is an iterative process, reuse of resources where possible could improve efficiency. For reuse of software components adherence to software engineering principles is suggested..

3.1. *Model calibration and validation*

An ABMS is a complex adaptive system itself, an interesting property that has implications for model validation and testing. As such, the output is highly sensitive to the input conditions, i.e., the random seeds. Like the prototypical cellular automaton [14], the ABMS beyond some trivial instances cannot be treated analytically, and therefore the simulation computation must be completed to observe the resulting outcomes. This computational irreducibility exacerbates computational requirements that increase with the number of agents, the complexity of interactions between the agents and their perception-decision-action processes. Furthermore, results will tend to exhibit particularly high variance compared to more deterministic models or even static network models with fewer degrees of freedom. Often, trends will become apparent when varying parameters systematically as demonstrated in Figure 6. A Monte Carlo approach to running simulations is recommended [17], each time with different initial conditions (as a result of random seeding), and output is generally averaged over many simulation realizations.

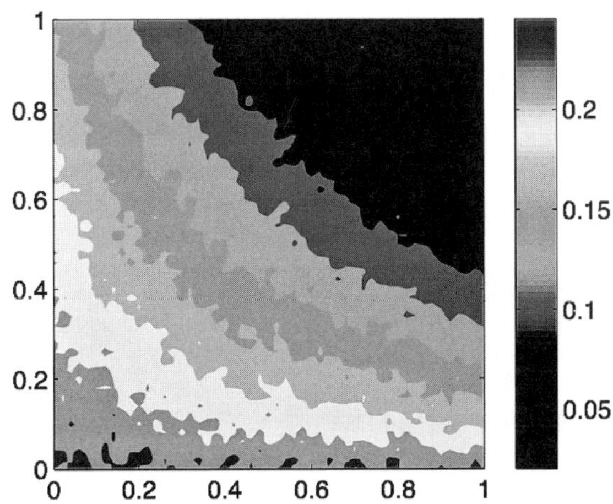

Figure 6. Fraction of infected individuals, indicated by colorbar, as a function of two parameterized (x and y axes) cooperating treatment strategies in an epidemic scenario. Each point in the heatmap is the average of 1000 simulation realizations.

It is often the case that the modelling process will leave at least one unknown parameter or independent variable. If there is no reliable data that is directly useful, then the model can be calibrated using historical data, depending on the nature of data available for the particular phenomenon being modeled, in our case herein, that is the spread of a disease in the population. For example, an emergent property, the average number of secondary cases generated by the first infectious individual introduced into the population can be used to calibrate the model to the specific epidemic scenario [18]. When more historical data are available on the number of cases in particular locations [19], the disease transmission rate can be independently calibrated at each location type. If an emergent property of the model is being compared to data, then the aforementioned Monte Carlo approach should be used to determine the mean value of the emergent property, since there could be considerable variance in estimation of the emergent property. As with other modeling approaches, the number of simulation runs necessary to achieve statistical significance will be unknown until the variance of the output parameters is analyzed. If the distribution of a given model parameter is known, we propose to draw a particular value from the distribution for each model simulation run. For example [20], infectious period for each infectious agent could be drawn (and stored) from a data-generated distribution (Figure 7). A similar approach could be taken for agents' demographic parameters to be drawn from known (e.g., census) distributions. The first infectious individual is also chosen at random at the start of the simulation. Although this parameterization tends to increase the variability in the model, fewer biases are imputed into the model.

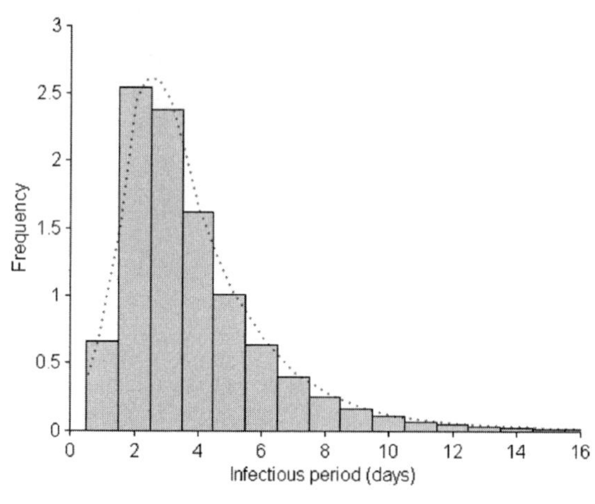

Figure 7. Distributions of the infectious period in days for influenza H1N1 infection caused by the 2009 pandemic virus strain.

A number of approaches to validating computational [21] and agent-based models [7] have been proposed. In addition to designing the model structure, subject expert judgment is useful in interpretation of the model output to establish the so-called face validity [7,21], that is, whether the model outputs are reasonable for particular inputs. Frameworks exist for validating ABMS [22], but they all build upon the very basic concepts of "Plausible

Reasoning Within Mathematics" [23]. In practical sense, the model is used to make projections. If the projections are comparable to historical data (i.e., mimic the patterns of observed phenomena), or reasonable according expert judgment, then the model can be seen as a credible framework. Observing animation of the evolving model state can help communicate the nature of the model to subject expert, and help them validate the output and interaction of sub-models. Close attention must therefore be paid to the emergent behavior arising from sub-model interactions to ensure that any unexpected behavior can be described or eliminated if resulted from computational or software errors or artefacts. An example of such an emergent phenomenon is the changing social network structure due to behavioural responses of individuals during an epidemic (see Figure 8).

Other validation techniques include degeneracy tests and extreme-value analysis; for example, testing boundary conditions such as assuming the probability of transmission is fixed at 0 to ensure that there are no secondary infections (no epidemic) [21]. These techniques can be used to compare simulation outputs with previously validated models, if available, including analytic models. Examples of studies that typify the aforementioned validation techniques can be found in some recent literature [18,19].

4. The Simulator Framework

An agent-based Simulator is a program that is capable of computing the result of one or more agent-based models. The simulator is responsible for driving the behaviors and interactions in the model systematically in time (i.e., scheduling), as demonstrated by mechanical system analogy in Figure 9.

A simulator program (suite) can include other support functionality such as output visualization or analysis modules. Simulators can be generalized to handle a number of different models. Typically, a model implementation will only work with a particular simulator, and the model will need to be re-implemented to work with a different simulator. Often, ABMS are created in an ad-hoc manner leading to simulators that are tightly coupled or highly integrated with the models. A more formal treatment of ABMS is given in Chapter 1 of Urumacher and Weyns [24].

4.1. *Modules*

Since most ABMS are implemented in software, the principles of software engineering should be applied whenever possible. One principle that might

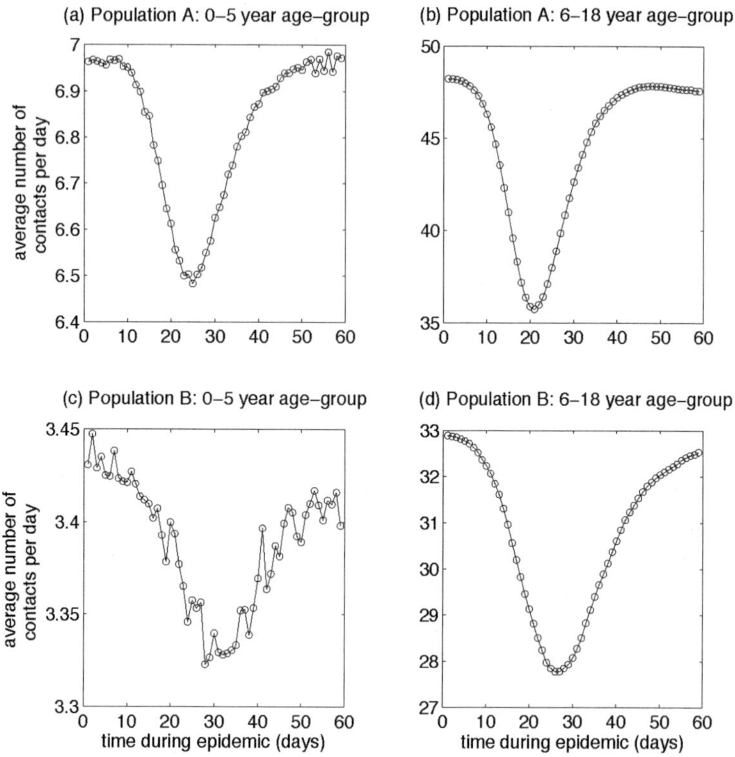

Figure 8. Variability in the number of contacts between agents throughout the course of an epidemic. Populations A and B represent the demographics of a remote/isolated community and an urban setting, respectively. Figures (a) and (c) show variations in contacts that lasted up to one hour for the age group 0-5 years old. Figures (b) and (d) show variations in contacts that lasted up to 6 hours for the age group 6-18 years old.

be of particular importance is modularity. Modular programs, in contrast with monolithic programs, attempt to separate different aspects, or concerns, of the task at hand into discrete modules. The benefits of appropriate modularity include, but are not limited to, making maintenance and re-use easier when considering future changes to the model. In an Object-Oriented language, an ABM can be modularized through the use of classes and interfaces.

Modules communicate or interact through interfaces. These interfaces can be specific to some communication task. In the context of an infectious disease transmission, examples of interface functions in order of increasing

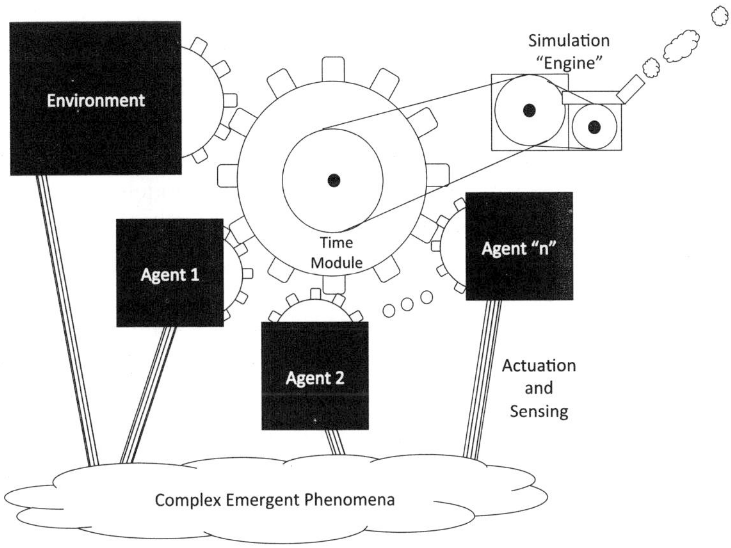

Figure 9. Mechanical analogy for an ABMS system. The Simulation Engine drives the time evolution of the model which in turn drives the environment and agents. The environment and agents interact to form the system level complex emergent phenomena.

generality are transmitInfluenza(); transmitDisease(Disease); transmitDisease (Disease, probability-of-infection); receiveMessage(Message). In the latter interface function, an agent would interpret the message and determine how to react. The general style of message passing interface could be used for any type of communication between agents, from disease transmission to coordinating social behaviors. It also enforces decoupling agents' action and state change from other agents or the environment, since it will be up to the receiver to decide how to change its internal state in response to the message.

4.2. *Time*

Although time can be considered as part of the environment, is discussed separately from environment since it presents an orthogonal set of choices to the modeler. Also for the reason that the concern of modeling time and keeping time is closely related to the task of the agent-based simulator, sometimes these two concerns are kept in the same module.

In practice, ABMS can implement time processes in either *Discrete Event* or *Discrete Time* approaches [25]. The decision to use one of these two approaches should take into account the problem and tools at hand. Examples throughout this article are based on discrete time approach, because they are more simplistic to implement and consider. Both approaches have tradeoffs with respect to challenges of scaling to large numbers of agents in high performance computing environments, and implementation. Neither approach is favorable over the other for all possible models. In either case, a variable that could be considered part of the environment, is used to keep track of the current time in the model, and may be referred to as the virtual clock.

In both discrete time and discrete event styles of simulation, actions or events can appear simultaneously from the perspective of the agents, particularly if perception, action, or communication delay is modelled explicitly. In this context, mechanisms must be built in to resolve, or allow for the agents behaviour to tolerate simultaneous, mutually exclusive actions [24].

4.3. *Discrete time*

In a discrete time simulation, each time step advances the virtual clock by a regular interval, and each agent is given a chance at a perception-decision-action cycle. The time step will typically have some units associated with it such as 5-seconds, 0.5-minute, 1-hour, or 0.5-day. The state of the environment (identifying properties independent from the state of the agents) will also advance accordingly.

In a recent study, we chose a discrete time to model influenza infection spread in a population, where the time advanced in steps of one hour [18]. In the context of influenza infection, other studies have used 4-hour time steps to describe agents' behavior and infection dynamics [26]. If agents' behavior is on a more finely grained scale than daily life patterns, then even finer time-steps such as minutes or seconds are warranted [27]. Listing 1 provides a Pseudocode for a discrete time used to simulate the spread of an infectious disease.

Simulating or executing agents using discrete time is perhaps more natural for modellers who have backgrounds in Multi-Agent Systems (MAS) within Artificial Intelligence that considers ABMS as a successor to cellular automata. In the discrete time approach, each time step is analogous to one perception-decision-action cycle performed in a manner that is apparently concurrent from the perspective of the agents. In an ABMS, the timescales of phenomena may have real world units if there is real world data underpinning the model. Choosing the time steps in discrete time

Listing 1. Pseudocode for a discrete time simulation of an infectious disease spread.

```
initializeEnvironment()
initializeAgents()
while (terminationCriteriaMet() == FALSE) do:
 virtualEnvironment.updateState();
 for each agent in SetOfAllAgents:
    agent.chooseNextLocation();
 end for each
 for each infectiousAgent in ...
    ... SetOfInfectiousAgents:
  for each colocatedAgent in ...
    ... getNearbyAgents(infectiousAgent):
     modelDiseaseSpread(infectiousAgent, ...
    ... colocatedAgent);
  end for each
 end for each
 for each agent in SetOfAllAgents:
   agent.moveToChosenLocation();
 end for each
 for each agent in SetOfAllAgents:
    agent.updateDiseaseState();
 end for each
 updateSimulationTime();
end while do
```

approach is often tied to perception-decision-action cycles of the agent, i.e., the timescales of the agent scheduled activities, or those of the micro dynamics of the system such as the timelines for the spread of a disease in an epidemic episode.

4.4. *Discrete Event*

In a discrete event simulation (DES) [25], time advances in discrete increments with variable magnitude, corresponding to the scheduled time of events, which in an ABMS will be related to perception-decision-action cycles. The discrete event simulator maintains a queue of events ordered increasingly by the time the events are scheduled to occur. In DES, time is not viewed as quantized, although ultimately in any simulation model it is represented in a computer that has finite precision. Often, a floating point number is used to represent the time that an event is scheduled to occur. An event can correspond to an instantaneous agent action. For actions that take measurable amount of time, an event can be used to represent the beginning and the end of such an action. The next event to occur is at the head of the queue, and it is removed when it is said to occur. This can cause other events to be scheduled at some time in the future. Then, the virtual clock "advances" to the time the event for execution is scheduled, as given in Listing 2.

Listing 2. Pseudocode for a Discrete Event Simulation (DES).

```
intializeEventQueue();
initializeEnvironment(); //schedules initial events
initializeAgents(); //schedules any initial events
while (eventQueue.isEmpty() == FALSE AND ...
      ... terminationCriteriaMet() == FALSE) do:
    Event e := eventQueue.removeHead();
    //carries out or simulates Event 'e'
    // and generates new Events
    newEventsList := Execute(e);
    for each event in newEventsList:
        eventQueue.schedule(event);
    end for each
end while do
```

The DES may feel more natural coming from a traditional modelling discipline in computer science or engineering with the view that events are governed by the laws of physics to occur in continuous time (and therefore events are never truly concurrent). It is important to note that the available digital computers have finite precision.

5. Agent-Related Modules

5.1. *Characteristics*

In general, agents will have a set of properties. The values of these properties are associated with a particular agent, and independent from the values of other agents. Examples include demographic variables (e.g., age, gender) that are drawn from distributions (provided by census data) when the agent is created. Some agents' characteristics may affect agents' behavior, and may change over time.

Agents in a model can behave relatively homogenously or heterogenously. Age is an example of a property that may affect agents' schedules, or their behavior during the course of a simulation day. Agent age may change over the time scales relevant to some diseases that have timelines comparable to the agents' life-span such as tuberculosis. Another example of these properties is gender, which also affects employment rates. A more extreme example is membership to the mosquito versus human species. Agents would likely exhibit vastly different behavior and properties based on this class membership. This type of property would not change for the lifetime of the agent, which highlights the issues of mutability, that is, whether a given property can change with time, or in response to some event or stimulus external to the agent.

Properties can consist of binary, integer, floating point, or string variable type. Properties can refer to an association with the environment, for example, a particular discrete location representing an individual's home, workplace or school. Properties can also refer to associations with other agents, for example, familial relationships.

5.2. *Behaviour*

The behaviour modules relate a particular agent to other agents and the environment in time. Behavior can be dependent on agents' properties, and can change with time, either directly through the passage of time or due to alteration of the agents' properties time. The agents' behavioral modules can themselves change the agent properties over time.

Many kinds of possible behaviors can be described by a formalization related to Markov Chains, referred to as the Finite State Automaton (FSA). Agents can share the same behavior, or some agents may have a different FSA describing their behavior. For example, a preschool child may have a FSA describing daily life pattern behavior that is different from FSA for an adult who has work related activities. Generally, agents will each store their own current state, even if they are using an identical FSA to describe their behavior. State transitions or even the entire FSA will therefore depend on particular agent properties.

FSA can be implemented as a set of states, with rules governing transitions between those states, and actions taken on state transitions. Unlike Markov models, transitions are not only probabilistic. There are three types of transition: probabilistic, defined as a probability of an event occurring over time (that follows a Bernoulli process); deterministically time-based, that relates to changes in states after some amount of time elapses; and other-signal-based, such as sending or receiving a message. Figure 10 demonstrates a compartmental model in terms of FSA.

Two kinds of behavior that are key to disease modeling include disease Behavior (i.e., transmission and progression) and Agent Life Pattern (i.e., schedules which relate an agent's position to a location in the environment as a function of time). For the sake of simplicity, these may be represented as intersections of an agent and disease behavioral states [18].

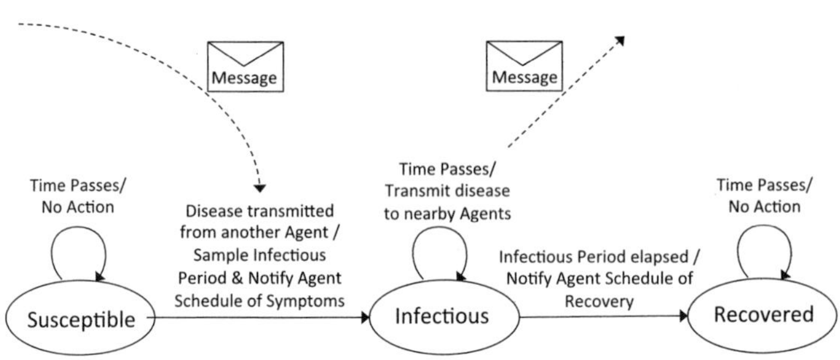

Figure 10. Finite State Automaton implementing a disease model with Susceptible, Infectious, and Recovered compartments. Disease states are represented by ellipsoids, with the name of the state contained within. State transitions are denoted with arrows; the condition for the transition written preceding the slash /, and the action taken following the slash /.

5.3. *Disease transmission and progression module*

A variety of diseases can be modeled formally as a set of FSA. The states of the FSA correspond to disease states in the related compartmental model (Figure 10). In order to implement different behaviors, a different FSA from the set can be assigned to particular agents to describe the behavior of each agent class (as decided by the relevant parameters). Examples of properties where a different FSA could be used to model different behaviors include age, gender, or species of the agent. Some states may cause the agent to pass messages to other agents, possibly depending on the location or properties of the agents.

The disease behavior module can also emit messages to other behavioral modules within the same agent or the environment. State changes can occur as a result of the passage of time (a certain probability per unit time, as in a Markov Chain, or deterministically after the passage of a particular time interval), or in response to messages from other agents, other behavioral modules within the same agent, or the environment. The disease behaviour module can also affect agent properties.

Disease transmission can be implemented using message passing. The infection message can include a probability of infection, which can change with time or location or the characteristics of the agents involved. An example of FSA implementation of a compartmental disease model is presented in Figure 10. Listing 3 provieds a pseudocode for probabilistic disease transmission. Note that in Listing 3 as well as Listing 1, the agent's disease states was updated (from Susceptible to Infected) after all opportunities for disease transmission on a particular time step. Otherwise, without modelling an incubation period, there is a chance that an agent who became infected can transmit the disease to a third agent on the current time step, and so on. This leads to the implausible situation where all the agents could become infected on the same time step.

5.4. *Agent life pattern*

The Agent Life Pattern or Schedule module relates each agent to locations and/or other agents (primarily) as a function of time. Research has shown that a great deal of human behavior is predictable and possible to capture using FSA [28]. Agent schedule is usually dependent on, or at least affected by, time. It can also be affected by messages from within the same agent class (such as coming from the disease behavior module), other agents, or the environment.

As with disease transmission and progression module, agents schedules may be very different from each other depending on the particular properties of the agent, such as age, and they may cause messages to be emitted (in particular states) to other agents, modules within the same agent, or the environment; The Agent Life Pattern module may cause new agents to be created. Figure 11 shows an example of an agent's schedule, conceptualized as an FSA, for the agent who goes to work every day. This schedule features a random choice on the part of the agent, for whom the parameterization is arbitrary and for illustration purposes only.

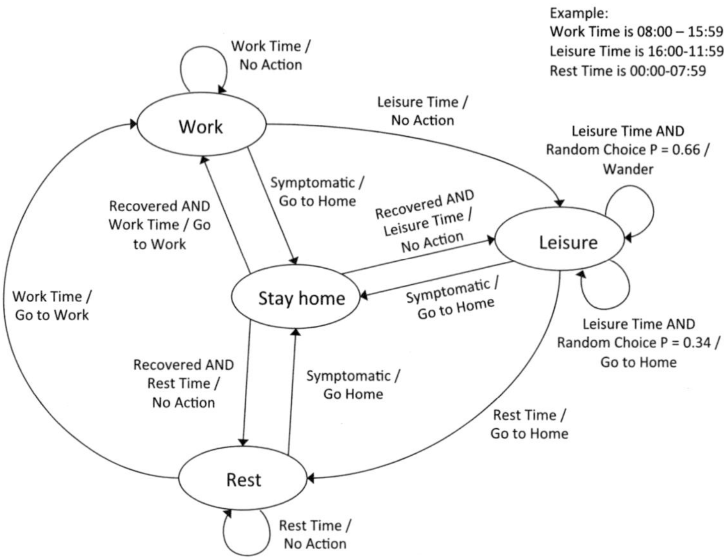

Figure 11. Finite State Automaton implementing a simple agent life pattern schedule for a human agent who goes to work every day. The schedule also accounts for self-isolation when symptomatic infection. Life pattern states are represented by ellipsoids, with the name of the state contained within. State transitions are denoted with arrows; the condition for the transition written preceding the slash "/", and the action taken following the slash "/". The Leisure state uses a random selection mechanism (similar to the one shown in Listing 3). The "Wander" action can be implemented as a Levy Flight.

6. Discussion

ABMS is an explicit mode of data visualization, and a natural framework within which various data sources can be integrated in a meaningful way to provide a clear picture of the system dynamics [29]. In this study, we focused on developing an ABMS framework that can be used to simulate a variety of scenarios pertaining to the dynamics of human diseases.

While the focus of this study has been on the development of the modelling and computational structure of this framework, several important applications of ABMS have already been recognized. For example, the ABMS approach can be used to evaluate the impact of population heterogeneity and demographic factors on disease spread [18]. Furthermore, the risks, benefits, and tradeoffs that are implicit in competing disease intervention strategies can be made explicit through ABMS, such that difficult decisions can be optimized from a societal point of view [20]. Decisions in healthcare and public health depend not only on effectiveness of disease interventions, but also on the availability of health resources in the face of competing priorities. Decision makers need quantitative tools to synthesize available information and emerging data sources in order to understand secular trends in disease occurrence. The ABMS provides such a tool which can be used to measure important deviations from expectations in an accurate manner. More specifically, this tool presents results generated through simulations as possible scenarios, allowing decision makers to assess the likely effect of different health intervention strategies.

Other applications of this tool relate to the understaning of the social phenomena, eological and environmental processes, and evolutionary aspects of living organisims. The types of data aggregated in ABMS for these applications include cellular telecommunications records, census data, travel and transportation surveys, crime statistics, weather and climate data, and laboratory and experimental data. Moreover, the ABMS is capable of integrating highly detailed spatial-temporal information that is being collected by social-networking websites, including capturing the context of users activities in time and space. The complexity and heterogeniety of such databases is only increasing with the wider availability of advanced information and communications technologies, and ABMS provides an important tool for data-use, analysis, and visualization.

References

1. M. Wooldridge, Agent-based software engineering. IEE Proceedings: Software Engineering, Vol. 144, No. 1, 26-37, 1997.
2. L. C. Mostaco-Guidolin LC *et al.*, A software development framework for agent-based infectious disease modelling, In: Biomedical Engineering, Trends, Researches and Technologies, In-Tech, Chapter 33, pp. 641-664, 2011.
3. D. S. Burke *et al.* Individual-based computational modeling of smallpox epidemic control strategies. *Academic Emergency Medicine*, 13:1142–1149, 2006.
4. Y. Meng, R. Davies, K. Hardy, and P. Hawkey. An application of agent-based simulation to the management of hospital-acquired infection. *Journal of Simulation*, 4(1):60–67, 2010.
5. N. Halder, J. K. Kelso, and G.J. Milne. Analysis of the effectiveness of interventions used during the 2009 a/H1N1 influenza pandemic. *BMC public health*, 10:168, 2010.
6. R. Paranjape and A. Sadanand. *Multi-Agent Systems for Healthcare Simulation and Modeling : Applications for System Improvement.* Hershey, New York: Medical Information Science Reference, IGI Global, 2010.
7. B. Bonabeau, Agent-based modeling: methods and techniques for simulating human systems. Proceedings of the National Academy of Sciences, Vol. 99, Supplement 3, 7280-7287, 2002.

8. M. Laskowski A prototype agent based model and machine learning hybrid system for healthcare decision support. International Journal of E-Health and Medical Communications, 2(4), 67-90, 2011.
9. T. Mitchell Machine Learning. Boston, MA: McGraw Hill, 1997.
10. R. Sutton and A. Barto, Reinforcement Learning: An Introduction. Cambridge: MIT Press, 1998.
11. J. Noble and D.W. Franks. Social learning in a multi-agent system. Computing and Informatics, 22(6), 561-574, 2004.
12. C.E. Leiserson, R.L. Rivest, & C. Stein. Introduction to algorithms. T. H. Cormen (Ed.). Cambridge: The MIT press, 2001.
13. M. Newman. Networks: an introduction. Oxford University Press, 2009.
14. S. Wolfram. A new kind of science. Champaign: Wolfram media, 2002.
15. S. Eubank *et al.*, Modelling disease outbreaks in realistic urban social networks. Nature, 429, no. 6988: 180-184, 2004.
16. R. Axtell and J. Epstein. Growing artificial societies: social science from the bottom up. Washington: Brookings Inst. Press, 1996.
17. C. Mooney, ed. Monte carlo simulation. No. 116. Thousand Oaks: Sage, 1997.
18. M. Laskowski, *et al.*, The impact of demographic variables on disease spread: influenza in remote communities. Scientific Reports 1, 2011.
19. D. Burkem *et al.*, Individual-based Computational Modeling of Smallpox Epidemic Control Strategies. Academic Emergency Medicine 13, no. 11: 1142-1149, 2006.
20. M. Laskowski *et al.*, Influenza H3N2 Variant Viruses with Pandemic Potential: Preventing Catastrophe in Remote and Isolated Canadian Communities. Preventive medicine, in press 2013.
21. R. Sargent. Verification and validation of simulation models. In: Proceedings of the 37th conference on Winter simulation, pp. 130-143. Winter Simulation Conference, 2005.
22. M. Niazi, A. Hussain, and M. Kolberg. Verification and validation of agent based simulation using vomas approach. In Proceedings of the third workshop on multi-agent systems and simulation. 2009.
23. G. Pólya. Mathematics and plausible reasoning: Induction and analogy in mathematics. Vol. 1. Princeton University Press, 1990.
24. A. Uhrmacher and D. Weyns, eds. Multi-Agent systems: Simulation and applications. CRC Press, 2010.
25. B. Zeigler, et al. Theory of modeling and simulation. Vol. 19. New York: John Wiley, 1976.
26. K. Carley, *et al.*, BioWar: scalable agent-based model of bioattacks. Systems, Man and Cybernetics, Part A: Systems and Humans, IEEE Transactions on 36, no. 2 : 252-265, 2006.
27. M. Laskowski *et al*,. Models of emergency departments for reducing patient waiting times. PLoS One 4, no. 7, 2009.
28. N. Eagle and A. Pentland. Reality mining: sensing complex social systems. Personal and ubiquitous computing 10, no. 4 : 255-268, 2006.

29. Committee on the Role of Information Technology in Responding to Terrorism, (U.S.) National Research Council (2003). Information Technology for Counterterrorism: Immediate Actions and Future Possibilities.

ANALYSIS OF THE BASIC REPRODUCTION NUMBER FROM THE INITIAL GROWTH PHASE OF THE OUTBREAK IN DISEASES CAUSED BY VECTORS*

R. P. SANCHES, E. MASSAD

University of São Paulo
Medicine School
Avenida Doutor Arnaldo, 455
CEP: 01246-000
São Paulo, SP, Brazil
E-mail: rosangela@dim.fm.usp.br, edmassad@usp.br

The basic reproduction number, R_0, is defined as the expected number of secondary cases of a disease produced by a single infection in a completely susceptible population. The value of R_0 can be estimated in several ways, for example, of the stability analysis of a compartmental model, through the matrix of next generation, etc. In this work we studied the method for estimating R_0 from the initial growth phase of the outbreak. Some authors proposed different methods for estimating the value of R_0 using the initial growth phase of the outbreak without assuming exponential growth of cases, which is suggested in most studies. We used the method proposed by Macdonald and studied by Massad et. al.(2010)[1] and the method proposed by Nishiura(2010)[2]. Massad et. al.(2010)[1] studied the basic reproduction number proposed by Macdonald's, which it was divided in two components:the vector-to-human component (T_{V-H}) and the human-to-vector component (T_{H-V}), R_0 is the product of those two components. Nishiura(2010)[2] presented a correction of the actual reproduction number (R_a), he showed through this correction that the basic reproduction number and actual reproduction number are equal, then he offered a framework for estimating R_0. Our objective is to evaluate which technique best estimates the basic reproduction number applying them to diseases caused by vectors, in this particular case we used data of dengue.

Keywords: basic reproduction number, dengue, mathematical model.

*This work is supported by CAPES.

1. Introduction

A major concern in terms of public health is about the increase in the number of cases of emerging and re-emerging infections like: dengue, malaria, yellow fever, several mosquito-borne encephalitis, leishmaniasis, and Lyme disease [5].

In works about infectious diseases one of the main goals is to evaluate the potential transmission of this diseases. The principal parameter used to evaluate the potential transmission is the basic reproduction number (R_0)[5], which can be used to determine if a control strategy can stop an epidemic or not, and the proportion of a population must be immunized[2].

The basic reproduction number, R_0, is defined as the expected number of secondary infections produced by a single infective person in a completely susceptible population. If $R_0 > 1$ on average each infected individual will to cause more than one infection, so we expect the disease spreads in the community. But if $R_0 < 1$ on average each infected individual will not to cause more than one infection, so we don't expect a epidemic, the disease will die out[6,7,8,9]. R_0 is a measure of how quickly a disease spreads in its initial phase and this measure can predict if a disease will become endemic or it will die out [8].

There are several ways to estimate R_0, for example, from the stability analysis of a compartmental model, through the matrix of next generation, etc[10]. In this work we study the method to estimate R_0 from the initial growth phase of the outbreak in diseases caused by vectors. In this case R_0 is defined as the expected number of people who would be infected from a single person initially infected by a vector[1,5].

The value of R_0 has been used to determine the proportion of the population will be immunize in order to eradicate a disease. According to some works, if a proportion of the population greater than $1 - 1/R_0$ is constantly protected, the disease cannot spread, and it will die out over time [2,11].

Several authors proposed different methods to estimate R_0 from the initial growth phase of the outbreak, they assumed exponential growth, what not always is true, there are cases wherein the initial growth curve has a concave shape unlike what happens in the growth exponencial[1].

In the papers of Massad et. al.(2010)[1] and Nishiura(2010)[2], the value of R_0 was estimated without assuming exponential growth of cases, and they had good results.

Massad et. al. (2010)[1] presents a method for estimating R_0 from the initial growth phase of the outbreak, where is not assumed growth exponential, they found that is possible to exist an outbreak even if $R_0 < 1$, however an outbreak self-limited.

They studied the basic reproduction number proposed by Macdonald (1952)[3], which was divided in two components: the vector-to-human component (T_{V-H}) and the human-to-vector component (T_{H-V}), R_0 is the product of those two components. They noted for $R_0 < 1$ the curve is concave down and for $R_0 > 1$ the curve is convex(exponential). They applied this method to dengue in São Paulo(Brazil) and Londrina(Brazil).

Nishiura(2010)[2] presents a correction of the actual reproduction number (R_a), he shows through this correction that the basic reproduction number and actual reproduction number are equal, then he offers a framework for estimating R_0. He applied this method to the HIV epidemic in European countries.

Dengue is a vector-born disease caused by dengue fever virus, with four serotypes, namely, DENV-1, DENV-2, DENV-3 and DENV-4, that belongs to genus *Flavivirus*, family *Flaviviridae*. Dengue is an urban disease and its virus is kept in a lifecycle that involves humans and mosquitoes *Aedes*[12,13]. The serotype DENV-2 and DENV-3 are probable to contribute the most to disease severity and mortality [13]. Dengue is transmitted to humans through bite of infected female of mosquitoes *Aedes*, mainly by *Aedes aegypti* and *Aedes albopictus*[12,13,14].

The transmission cycle occurs as follows, the female of mosquitoes *Aedes* feeds of the blood an infected person during her/his period of viremia, after an extrinsic incubation period, about 8 to 12 days, the mosquito transmits the virus to other uninfected people[12,13]. The incubation period of virus in humans is in general of 4 to 7 days, although can range between 2 to 15 days [12,13,14]. This disease is not transmitted by contact with a infected people or their secretions, neither by sources of water or food[15].

Infected people present a clinical picture that ranges from asymptomatic infection or mild febrile illness and even lethal disease. However the general symptoms are: high fever, tiredness, muscle pain, etc[13]. Sometimes this disease is confounded with other viral or bacterial diseases [16]. People infected by one serotype acquire lifelong immunity only to this serotype, thus people can be infected by four dengue serotypes during their lifetime[12].

Our aim is to evaluate which technique best estimates the basic reproduction number applying them to diseases caused by vectors without assuming exponential growth of cases, in this particular case we used data of dengue provided by CVE[17].

2. Methods and Materials

In this section we present a detailed description of the methods studied by Massad et. al.(2010)[1] and Nishiura(2010)[2] to estimate the basic reproduction number, without assuming exponential growth from the initial growth phase of the outbreak.

2.1. *Mathematical Model studied by Massad et al. (2010)*

The first method presented in this section was studied by Massad et. al.(2010)[1], in this method the basic reproduction number was estimated using the Ross-Macdonald[3] model for vector-borne diseases, wherein the populations are divided into host population $N_H(t)$ and vector population $N_V(t)$.

The host population is divided into susceptible hosts $S_H(t)$, infected hosts $I_H(t)$ and recovered hosts $R_H(t)$. The vector population $N_V(t)$ is divided into susceptible vectors $S_V(t)$, latent vectors $L_V(t)$ and infected vectors $I_V(t)$.

The following equations describe the dynamics of the disease:

$$\begin{cases} \dfrac{dS_H(t)}{dt} = -abI_V(t)\dfrac{S_H(t)}{N_H(t)} + \mu_H(N_H(t) - S_H(t)), \\ \dfrac{dI_H(t)}{dt} = abI_V(t)\dfrac{S_H(t)}{N_H(t)} - (\mu_H + \gamma)I_H(t), \\ \dfrac{dR_H(t)}{dt} = \gamma I_H(t) - \mu_H R_H(t), \\ \dfrac{dS_V(t)}{dt} = -acS_V(t)\dfrac{I_H(t)}{N_H(t)} + \mu_V(N_V(t) - S_V(t)), \\ \dfrac{dL_V(t)}{dt} = acS_V(t)\dfrac{I_H(t)}{N_H(t)} - \mu_V L_V(t) - acS_V(t-\tau)\dfrac{I_H(t-\tau)}{N_H(t-\tau)}e^{-\mu_V\tau}, \\ \dfrac{dI_V(t)}{dt} = acS_V(t-\tau)\dfrac{I_H(t-\tau)}{N_H(t-\tau)}e^{-\mu_V\tau} - \mu_V I_V(t). \end{cases} \tag{1}$$

The total hosts population and total vectors population is kept constant, (natality replacing mortality), thus $N_H = S_H + I_H + R_H$ and $N_V = S_V + L_V + I_V$.

The biological meaning of the parameters used in the model (1) are given in the Table 1.

Table 1. Biological meaning of the parameters used in the model (1).

Parameter	Biological meaning
a	Biting rate by *A. aegypti.*
γ	Recovery rate for dengue.
τ	Dengue extrinsic incubation period.
μ_H	Natural mortality rate of humans.
c	Probability that a vector to get infected.
b	Probability that a host to get infected.
μ_V	Mortality rate of *A. aegypti.*

Macdonald (1925) demonstrated the existence of a threshold, the basic reproduction number, given by

$$R_0 = m\frac{a^2bce^{-\mu_V\tau}}{(\gamma+\mu_H)\mu_V}, \tag{2}$$

where $m = \frac{N_V}{N_H}$.

Massad et. al.(2010)[1] shows that R_0 can be divided into two components, $T_{(V-H)}$ (vector-host component) that describes the transmission of the infection from the vector to the host, in other words, this component represents the arrival of the infected vectors in a disease-free host population, and $T_{(H-V)}$ (host-vector component) that describes the transmission of the infection from the host to the vector, in other words, this component represents the arrival of the infected hosts in a disease-free vector population. R_0 is the product of those two components,

$$T_{(H-V)} = \frac{N_V a}{N_H(\gamma+\mu_H)}c, \tag{3}$$

and

$$T_{(V-H)} = \frac{a}{\mu_V}be^{-\mu_V\tau}. \tag{4}$$

In this study Massad et. al. (2010)[1] made an analysis of R_0, wherein they used an approximate linearizing method, it was demonstrated that if $R_0 < 1$ the initial growth curve is concave down and tends to zero over time. But if $R_0 > 1$ the initial growth curve has a convex (exponential) shape, and the disease tends to an endemic state over time.

They showed that if $T_{(V-H)} > 1$ and $T_{(H-V)} << 1$ resulting in a value of $R_0 < 1$, which means the introduction of a certain amount of infected vectors can cause an outbreak, the curve presents a concave shape.

2.2. *Method used by Nishiura(2010)*

Nishiura(2010)[2] presents a correction of the actual reproduction number (R_a), which is defined as the product of the average duration of infectiousness and the ratio of incidence to prevalence [18].

Initially, the author calculates the basic reproduction number through a statistical estimate. He assumes that $j(t)$ is the number of new infections at time t, and $A(\tau)$ is the generation secondary cases rate by each infected individual, at time τ, $j(t)$ is given by,

$$j(t) = \int_0^\infty A(\tau) j(t-\tau) d\tau. \tag{5}$$

Hereupon the author uses the definition of R_0 as total number of secondary cases, that a initial case generates during the entire course of infection [20],

$$R_0 = \int_0^\infty A(\tau) d\tau. \tag{6}$$

Nishiura assumes that $j(t)$ follows an exponential growth, after some calculations he concludes,

$$\hat{R_0} = \frac{1}{\int_0^\infty g(\tau) exp(-r\tau) d\tau}, \tag{7}$$

where $g(\tau)$ is defined as frequency of secondary transmissions (generation interval),

$$g(\tau) = \frac{A(\tau)}{R_0}. \tag{8}$$

As previously mentioned R_a is defined as the product of the average duration of infectiousness and the ratio of incidence to prevalence, therefore the prevalence at time t is given by,

$$p(t) = \int_0^\infty \Gamma(\tau) j(t-\tau) d\tau, \tag{9}$$

where $\Gamma(\tau)$ can be understood as probability of being infectious at infection-age τ. Let $\beta(\tau)$ defined as transmission rate at time τ, then $A(\tau) = \beta(\tau)\Gamma(\tau)$ [19]. So R_a is given by

$$R_a = \frac{j(t)}{p(t)} D = \frac{\int_0^\infty \beta(\tau)\Gamma(\tau) j(t-\tau) d\tau}{\int_0^\infty \Gamma(\tau) j(t-\tau) d\tau} D, \tag{10}$$

where D is defined as mean generation interval, Nishiura replaces $A(\tau)$ in the equation (6) and he considers that transmission rate is constant, thus R_0 is given by:

$$R_0 = \int_0^\infty \beta(\tau)\Gamma(\tau)d\tau = \beta D. \tag{11}$$

Therefore, in this case where $\beta(\tau)$ is constant, the basic reproduction number and the actual reproduction number are equal.

However, in the case that $\beta(\tau)$ is not constant Nishiura proposes a correction to concept of R_a.

The first step is to rewrite the equation given in (10),

$$R_a = \frac{j(t)}{\frac{1}{D}\int_0^\infty \Gamma(\tau)j(t-\tau)d\tau}, \tag{12}$$

this equation is understood as follows, the numerator represents the numbers of new infections at time t, and the denominator that was previously understood as the total number of infectious individuals, now it represents the total number of effective contacts. R_a is written as:

$$\grave{R}_a = \frac{j(t)}{\int_0^\infty g(\tau)j(t-\tau)d\tau}. \tag{13}$$

In the next step Nishiura replaces $g(\tau)$ on the right side of the equation (13) by $g(\tau) = \frac{\beta(\tau)\Gamma(\tau)}{R_0}$,

$$\grave{R}_a = \frac{j(t)}{\frac{1}{R_0}\int_0^\infty \beta(\tau)\Gamma(\tau)j(t-\tau)d\tau} = R_0. \tag{14}$$

Then $R_a = R_0$, so the basic reproduction number can be estimated without assume exponential growth of cases.

Nishiura rewrites the equation(13) in discrete-time, because his data are discrete:

$$\hat{R_0} = \frac{j_t}{\sum_{s=0}^\infty g_s j_{t-s}}, \tag{15}$$

where g_s is given by $g_s = G(s+1) - G(s)$, and $G(s)$ is the cumulative function of generation interval of length s, then he proposes the following estimation, the first step is to reverse of both sides of equation (15),

$$\frac{1}{R_0} = \frac{\sum_{s=0}^\infty g_s j_{t-s}}{j_t}, \tag{16}$$

the equation (16) has on the right-hand side of equation in the numerator the total numbers of effective contacts made by potential primary cases in year t with equal potential for resulting in a secondary transmission, and in the denominator the number of secondary cases in year t.

Then Nishiura says that, the number of effective contacts have equal potential to result in a secondary transmission, in other words, the probability of generation of secondary cases is linked to effective contacts made by a single primary case in year t, which is a processes given by $\frac{1}{R_0}$ in the equation (16), a binomial sampling process, then R_0 can be estimated by the likelihood function,

$$L(R_0) = \prod_{t=1}^{T} \left(\frac{j_t}{\sum_{s=0}^{\infty} g_s j_{t-s}} \right) \left(\frac{1}{R_0} \right)^{\sum_{s=0}^{\infty} g_s j_{t-s}} \left(1 - \frac{1}{R_0} \right)^{j_t - \sum_{s=0}^{\infty} g_s j_{t-s}}, \tag{17}$$

where T is the time point of observation an early epidemic growth stage of the disease, and the maximum likelihood of R_0 is obtained by minimizing the negative logarithm of (17).

3. Results

In this section we present the calculation of the value of R_0, for city of São Paulo in 2001. The total number of dengue cases in the state of São Paulo in 2001 was 48,698, in which 348 was recorded in the city of São Paulo[17], the value of R_0 for dengue was estimated in 6.59 [22]. We used the incidence data for the first 10 weeks of the year[17].

3.1. *R_0 studied by Massad et al. (2010)*

For estimating the value of R_0 by Macdonald's Method[3], we need information about some parameters, like death rate of humans, recovery rate of dengue, etc, however some parameters are hard to estimate, like biting rate (a), probability that a host to get infected (b), probability that a vector to get infected(c).

In order to determine which values best fits this parameters, we fitted the curve of model given by system (1) to real data[17], thereunto we used the maximal likelihood function [1]:

$$L(m(x), \sigma^2 | d) = \Pi_i \frac{1}{\sqrt{2\pi\sigma^2}} exp \frac{-(d_i - m_i)^2}{2\sigma^2} \tag{18}$$

where d is a vector of real data, and m is a vector of model. The parameters fitted a, b, c ranging among $0 - 1$, the value of σ^2 was calculated from real data, and the parameter space was varied until (18) was maximized.

The system (1) was implemented using the 4th Order Runge-Kutta method, and C language. The fixed parameters used in the simulation are $\gamma = 0.14$ days^{-1}(source[1]), $\tau = 7$ days (source[1]), $\mu_V = 0.025$ days^{-1}(source[1]), $\mu_H = 3.4 \times 10^{-5}$ days^{-1} (source[1]), and $m = 2$(source[23]).

Figure **1**, shows the curve obtained for system (1) using the maximal likelihood function (18), where the values best fits the parameters are: $a = 0.74$ days^{-1}, $b = 0.5$ and $c = 0.04$.

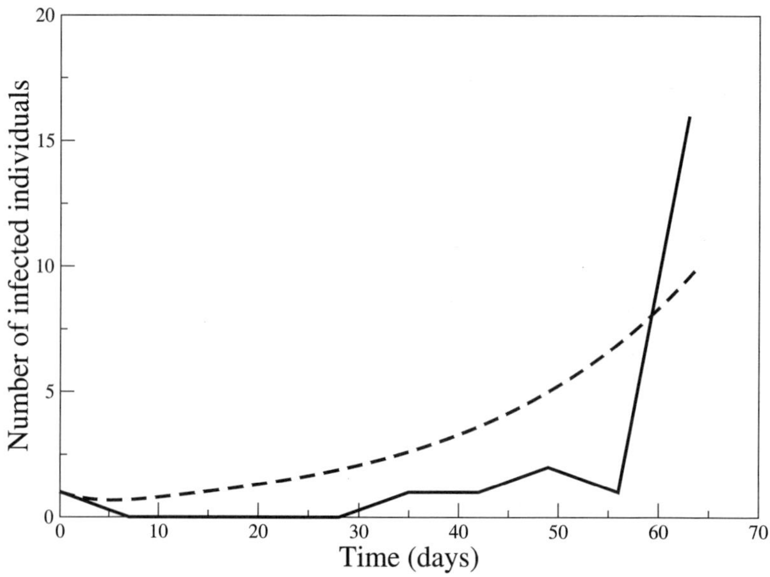

Figure 1. In the continuous line the real data, and in dashed line the fitted data.

3.2. R_0 *given by Nishiura's Method*

We calculated the maximum likelihood for equation (17), we had the following expression:

$$R_0 = \frac{\sum_{t=1}^{T} j_t}{\sum_{t=1}^{T} \sum_{s=0}^{\infty} g_s j_{t-s}}, \tag{19}$$

We replaced the sum $\sum_{s=0}^{\infty}$ by $\sum_{s=0}^{min(k,t)}$, where k is the maximal length of g_s [21],

$$R_0 = \frac{\sum_{t=1}^{T} j_t}{\sum_{t=1}^{T} \sum_{s=0}^{min(k,t)} g_s j_{t-s}}, \tag{20}$$

We assumed that g_s is given by a cumulative gamma distribution,

$$g_s = \frac{\lambda^\alpha}{\Gamma(\alpha)} \int_s^{s+n} x^{\alpha-1} e^{-\lambda x} dx. \tag{21}$$

The value of k is finite and must be selected large enough, to ensure that g_s follow a gamma distribution, we normalize g_s to ensure that its sum to one and represents a density. The limits of integration $(s, s+n)$ of (21) varies according with k, and it depends on the available data[21].

The generation interval for dengue ranges from 14 to 18 days [24,25], we took the generation interval of dengue virus in 14 days comprising intrinsic (within human) and extrinsic (within mosquito) periods with 7 days duration each[24],and we assumed the standard deviation of 5 days. The gamma distribution of generation interval was calculated using the software Mathematica 9.0.

We assumed that $k = 40$, and we divided the gamma distribution in 10 subintervals, because our incidence data are the first 10 weeks of the year, thus we ensured that g_s follows a gamma distribution.

The results are present in the Table 2.

Table 2. Estimation of R_0 by Nishiura's Method and Macdonald's Method.

R_0[22]	Nishiura's Method[2]	Macdonald's Method[3].
6.59	6.33	5.25

4. Discussion

In this article we presented two methods for estimating R_0 from the initial growth phase of the outbreak in diseases caused by vectors, without assuming exponential growth of cases. The method studied by Massad et. al.(2010)[1], a method applied to diseases caused by vectors, and the method proposed by Nishiura(2010)[2], a method applied to diseases of human-to-human transmission. We observed the differences between two methods, and we compared with the values estimated by Massad et. al.(2003)[22].

To estimate the value of R_0 by Macdonald's Method[3], besides incidence data, we also need some rates about the disease, as we mentioned earlier. Nevertheless, fitting the parameters a, b and c is required a hard computational effort.

To estimate the value of R_0 by Nishiura's Method[2], besides incidence data, we only need information about the average and standard deviation of the generation interval. However, we have few information about generation interval of dengue, in our study we performed several tests, and we estimate the generation interval with average of 14 days[24] and standard deviation of 5 days.

When we observe the results presented in Table 2, we may conclude that the method proposed by Nishiura (2010)[2] is better than method studied by Massad et. al. (2010)[1], finally the value of R_0 estimated by Nishiura's Method[2] is closer to value of R_0 estimated by Massad et. al.(2003) [22] than the value of R_0 etimated by Macdonal's Method[3], but it does not means that the method studied by Massad et. al.(2010)[1] is bad.

The method proposed by Nishiura(2010)[2] does not require a large computational effort, it only requires tools for calculating the generation interval (the problem is, making this calculation is not always an easy task), in the case of the method studied by Massad et. al. (2010)[1], are necessary few information but it demands a hard computational effort. Therefore we can conclude in this first study that: what it may define which is the best method to estimate the value of R_0, are the information we have available for making this calculation.

Acknowledgments

R.P. Sanches is supported by CAPES(Brazil).

References

1. E. Massad, F.A.B. Coutinho, M.N. Buratini, M. Amaku, *Tropical Medicine and Internacional Health.* **15(1)**,120 (2010).
2. N. Nishiura, *J. Environ. Res. Public Health.* **7** 291 (2010)
3. G. Macdonald, *Tropical Diseases Bulletin.* **49** 813 (1952)
4. F.A.B. Coutinho, M.N. Buratini, L.F. Lopez, E. Massad, *Bulletin of Mathematical Biology.* **68** 2263 (2006).
5. L.F. Lopez, F.A.B. Coutinho,M.N. Buratini,E. Massad, *C.R. Biologies* **325** 1073 (2002).
6. O. Diekmann,J.A.P. Heesterbeek, *J. Math. Biol.* **28** 365 (1990).
7. P. van den Driessche, J. Watmough, *Mathematical Biosciences* **180** 29 (2002).

8. R. Smith, *American Institute of Mathematical Sciences.* **2** (2008).
9. R.M. Anderson, R.M. May,*Oxford University Press* (1991).
10. J.M. Heffernan, R.J. Smith, L.M. Wahl,*Journal of the Royal Society Interface.* **2** 281 (2005).
11. M.G. Roberts, J.A.Heesterbeek,*Proc. R. Soc. Lond.* **270** 1359 (2003).
12. D. J. Gubler, G. G. Clark, *Emerging Infectious Diseases.* **1(2)** (1995).
13. A. Guzman,R. E. Istriz,*International Journal of Antimicrobial Agents.* **36S** S40 (2010).
14. J.R. Coura, *Guanabara-Koogan* (2005).
15. Ministério da Saúde.*MS.* **7** (2008).
16. M. G. Teixeira,M. L. Barreto,*BMJ.* **339(b4338)** (2009).
17. CVE. Centro de Vigilângia Epidemiológica *http : //www.cve.saude.sp.gov.br/htm/cve_dengue.html.*
18. P.J. White, H. Ward,G.P. Garnett,*AIDS 2006* **20(14)** 1898 (2006).
19. G. Chowell,H. Nishiura, *Physics of Life Reviews.* **5** 50 (2008).
20. O. Diekmann, J.A.P. Heesterbeek *John Wiley & Son* (2000).
21. L.F. White, M. Pagano, *Statistics in Medicine.* **27**,2999 (2008).
22. E. Massad, M. N. Burattini, F. A. B. Coutinho, L. F. Lopez,*Rev. Saúde Pública* **37(4)**,477 (2003).
23. C. Favier, N. Degallier, M. G. Rosa-Freitas,J. P. Boulanger, J. R. Costa Lima, J. F. Luitgards-Moura, C. E. Menke's, B. Mondet, C. Oliveira, E. T. S. Weimann, P. Tsouris,*Tropical Medicine and International Health* **II(4)** 343 (2006).
24. S. N. Bennett, E.C. Holmes,M. Chirivella,D.M. Rodriguez,M. Beltran, V. Vorndam, D. J. Gubler, W. O. McMillan, *Mol. Biol. Evol.* **20(10)**,1650 (2003).
25. J. Aldstadt, I. Yoon,D. Tannitisupawong, R. G. Jarman, S. J. Thomas,R. V. Gibbons, A. Uppapong, S. Iamsirithaworn, A. L. Rothman, T. W. Scott, T. Endy, *Tropical Medicine and International Health* **17(9)**,1076 (2012).

PARAMETER ESTIMATION OF A TUBERCULOSIS MODEL IN A PATCHY ENVIRONMENT: CASE OF CAMEROON*

D. P. MOUALEU

Computational Systems Biology Group, Zuse Institute Berlin (ZIB),
Takustr. 7, 14195 Berlin, Germany
moualeu@zib.de

S. BOWONG†

Department of Mathematics and Computer Science, University of Douala,
PO Box 24157 Douala, Cameroon
UMI 209 IRD/UPMC UMMISCO, Bondy, France and
Project-Team GRIMCAPE, LIRIMA, University of Yaounde I, Cameroon,
E-mail: sbowong@gmail.com

J. KURTHS

Postdam Institute for Climate Impact Research (PIK),
Telegraphenberg A 31, 14412 Potsdam, Germany

Tuberculosis (TB) is a common lethal infectious disease usually caused by Mycobacterium tuberculosis. According to the WHO, TB to date, claims the second largest number of victims due to a single infectious agent right after HIV/AIDS. Although a widespread implementation of control measures focus on case finding and short-course chemotherapy, the global burden of TB has increased over the past two decades. A deterministic TB model with a demographic structure for the population living in a multi-patch environment (cities, towns, regions or countries, etc.) is formulated. The basic reproduction number $\mathcal{R}_0$ is computed. A sensitivity analysis of the model parameters is performed to estimate the most influential parameters in TB dynamics. Using the iterative Gauss-Newton method to solve the inverse problem of parameter identification, estimability of parameters have been studied, and estimable unknown parameters have been computed using real data of TB in Cameroon, subdivided into four regions. Numerical simulations showed the model to reproduce the TB dynamics in Cameroon and predict a short term increase in the number of TB active cases over next years.

*This work is supported by imu einstein foundation

†Work partially supported by ictp in Trieste-Italy under the Associate Federation Scheme.

1. Introduction

Tuberculosis (abbreviated TB for *Tubercle Bacillus*) is the world's seventh-leading cause of death. It is the second largest cause of death from an infectious agent after HIV/AIDS in developing countries [1]. In the modern era, TB is recognized as a disease that preys upon social disadvantage [2]. It remains a worldwide emergency mostly affecting poor countries and to this old and persistent threat, a new emergency is adding further challenge, i.e., the multidrug-resistant TB [3]. Despite the development of a number of effective treatments over the past half century, TB remains one of the most destructive bacteria infections in humans. A third of the world's population carries the bacterium. More than 36 million patients have been successfully treated via the World Health Organization (WHO) strategy for tuberculosis control since 1995. Despite predictions of a decline in global incidence, the number of new cases continues to grow, approaching 10 million in 2010 [4].

Unfortunately, spatial plays an important role in many infectious disease processes. For instance, spatial heterogeneity is a strong determinant of host-parasite relationships [5]. In modelling spatial or geographic effects on the spread of a disease, a distinction is usually made between diffusion and dispersal models. Diffusion models traditionally use partial differential equations (PDE) and, spread is immediately to adjacent zones, hence the phenomenon of travelling waves can appear. However, there are some important situations that cannot be modelled by PDEs. This is the case when the space considered is discrete. For such situations, dispersal models are more appropriate where human can travel over long distance in a short period of time. When considering such dispersal models, there is an approach based on the metapopulation concept. The population is subdivided into a number of discrete patches which are supposed to be well mixed. Then in each patch, the population is subdivided into compartments corresponding to different epidemic status [6,7,8]. The basic and important research subjects for these systems are the existence of a threshold value which distinguishes whether an infectious disease will die out, the local and global stability of the disease-free equilibrium and endemic equilibria, the existence of periodic solutions, the persistence and extinction of the disease, etc. Although these approaches, and their more complex variants, have been instrumental in understanding several features of infectious diseases in metapopulation networks over the past three decades [5], a systematic construction of an epidemiological model that best matches real data is still an unsolved problem. However, relating such models to observed data is not a straightforward matter due to the complexity of the model and the noise in the data. A brief survey on previous studies provides the context of this paper. Huang and Wu [9] have proposed a Bayesian approach to estimate the parameters in HIV dynamics models for a population of patients. Their method is highly dependent on informative priors for most of the parameters and is computationally intensive, which may limit its use. Jeffrey and Xia [10] investigated the identifiability of HIV dynamics models, but the solution of the identifiability problem was not provided for the case where only measurements of viral load are available. Wu et al. [11] used the concept of an identification function and identification equation to study the parameter identifiability of a three-dimensional HIV/AIDS dynamics, but two parameters are not identifiable (i.e., are indistinguishable). Bowong and Kurths [12] used the concept of synchronization to estimate all parameters of a simple TB model using real data of the region of Center in Cameroon. Bowong and Kurths [12] have estimated the unknown parameter in a TB model using real data of Cameroon and showed that there is a seasonal pattern in new TB cases in the mainland of Cameron. The drawback of these studies is that they do not take into account the spatial spread of the disease. Unfortunately, nothing has been done in term of estimating parameters of TB in a metapopulation network.

In this paper, the problem of estimating unknown parameters of a TB metapopulation model from real data of Cameroon is presented. We first consider the spread of TB on complex networks of populations connected by migratory flows. We distinguish n patches. For overall patches, we divided the infective class into two subgroups with different properties: i) diagnosed infectious and ii) undiagnosed infectious. Based on the fact that diagnosed infectious are educated from the disease and its infectivity, and that undiagnosed infectious in many villages or regions in Africa do not know their TB-status due to the absence of hospital in rural areas, education, mentalities, ignorance, etc., we assumed that the transmission of the infection from diagnosed infectious to susceptible individuals is frequency-dependent (standard incidence), while the transmission of the infection from undiagnosed infectious to susceptible individuals is density-dependent (simple mass action), i.e., non-limited. The model has been analyzed. We compute the disease-free equilibrium (DFE) and the basic reproduction number $\mathcal{R}_0$. We show that the disease-free equilibrium is globally asymptotically stable when $\mathcal{R}_0 < 1$. Using the reported data (2000-2010) of WHO [13] and the National Committee to Fight against Tuberculosis of Cameroon [14] where the average TB notification rate is close to 93 per 100,000 habitants per year, we suggest a numerical method based on an iterative Gauss-Newton Method implemented in the software POEM to estimate unknown parameters using real data of TB in Cameroon. The Cameroonian population has been divided into four geographical zones or patches according to main roads, big cities and, social distance, cultural and sociology facts of the peoples of these regions. Simulation results have been presented to predict the evolution of the disease in each considered regions of Cameroon in the next future year.

2. The model

2.1. *Human metapopulation model*

The transfer rate from patch i to patch j, for $i \neq j$, is denoted by $m_{ji} \geq 0$. We assume that the infection and death due to the disease do not occur during the travel. The total human population in patch i is denoted by N_i. Hence, for $i = 1, \ldots, n$, the dynamics is given by

$$\dot{N}_i = \psi_i - \mu_i N_i + \sum_{\substack{j=1 \\ j \neq i}}^{n} m_{ij} N_j - N_i \sum_{\substack{j=1 \\ j \neq i}}^{n} m_{ji}, \tag{1}$$

where μ_i denotes the natural mortality of individuals in patch i and ψ_i the recruitment inside the population of patch i. Model system (1) can be written in the following compact form:

$$\dot{N} = \psi - \text{diag}(\mu)N + \mathcal{M}N, \tag{2}$$

where $N = (N_1, \ldots, N_n)^T$, $\psi = (\psi_1, \ldots, \psi_n)^T$, $\mu = (\mu_1, \ldots, \mu_n)^T$, the superscript T denoting the transpose, the matrix $\mathcal{M}$ is defined by $\mathcal{M}(i,j) = m_{ij}$ for $i \neq j$ and $\mathcal{M}(i,i) = -\sum_{j \neq i} m_{ji}$ and $\text{diag}(\mu)$ the diagonal matrix with μ_i as its (i,i) entry.

2.2. *Metapopulation TB model*

The sub-population in each patch is subdivided into five mutually exclusive classes of susceptible, exposed, infective and recovered individuals, and hence are known as SEIR (Susceptible-Exposed-Infective-Recovered) models [15,16]. At any given time, an individual is in one of the following states: susceptible, latently infected (exposed to TB but not infectious), diagnosed infectious (has an active TB confirmed after a sputum examination at the hospital), undiagnosed infectious (has an active TB not confirmed by a sputum examination at the hospital), and recovered individuals (cured after a therapy of treatment in the hospital). For each patch i, these states are denoted by S_i, E_i, I_i, J_i and R_i, respectively. Since diagnosed infectious have been detected in the health centers, we assume that the transmission of the infection from diagnosed infectious to susceptible individuals is assumed to be frequency-dependent (i.e., limited), while the TB transmission from undiagnosed infectious to susceptible individuals is assumed to be density-dependent (i.e., non-limited). Indeed, diagnosed infectious are educated from TB so that they can avoid contacts with susceptible individuals. This is why we assume that the TB transmission of the infection from diagnosed infectious to susceptible individuals is limited. Moreover, since some undiagnosed infectious do not know their TB-status, people around them are not aware of their TB-status, and thus interact with them naturally. So, we assume that the transmission of the infection of undetected infectious to susceptible individuals is non-limited.

All recruitment is into the susceptible classes and occurs at rate ψ_i. The constant rate for non-disease related death is μ_i. Diagnosed and undiagnosed infectious have additional death rates due to disease with rates d_i and δ_i, respectively. Transmission of TB bacteria occurs after adequate contacts between susceptible individuals, diagnosed and undiagnosed infectious. Then, a susceptible individual of patch i acquires TB infection from individuals with active TB at rate λ_i, given by

$$\lambda_i = \beta_i^I \frac{I_i}{N_i} + \beta_i^J J_i, \tag{3}$$

where β_i^I and β_i^J are effective contact rates of diagnosed and undiagnosed infectious that are sufficient to transmit the infection to a susceptible individual in patch i. Proportions p_i^I and p_i^J of new infected among susceptible individuals are assumed to undergo a fast progression of the disease and will transfer directly to the classes of diagnosed infectious I_i and undiagnosed infectious J_i, respectively, while the remainder $(1 - p_i^I - p_i^J)$ develops a latent TB and enter the latent class E_i. Latently infected individuals are assumed to acquire some immunity as a result of the infection, which reduces the risk of subsequent infection but does not fully prevent it. Once latently infected, an individual can follow a chemoprophylaxis which reduces its risk of reactivation. We denote by α_i the chemoprophylaxis rate of latently-infected individuals in patch i. Due to endogenous reactivation, a rate $1 - \alpha_i$ of latently infected individuals who did not received effective chemoprophylaxis becomes infectious at rate k_i. Among latently infected individuals who become infectious, a proportion h_i of them is diagnosed and treated, while the remaining $1 - h_i$ is not diagnosed and enter the class of undiagnosed infectious J_i. The model description is summarized in the flow diagram shown in Fig. 1.

This yields the following differential equations for $i = 1, ..., n$:

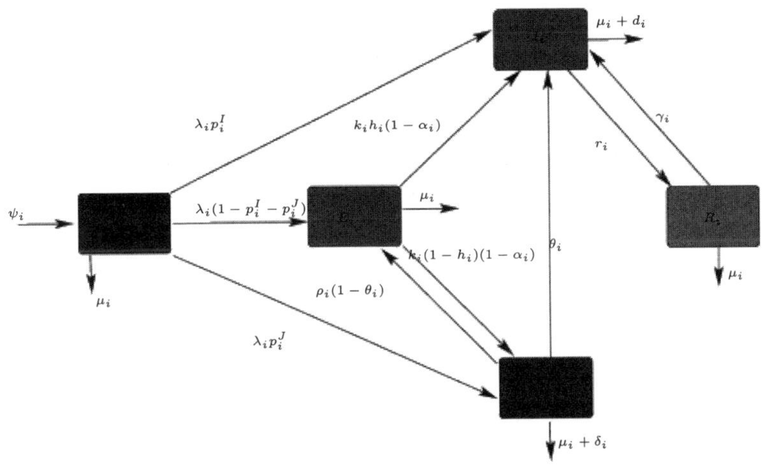

Figure 1. Flow diagram for the dynamics transmission of tuberculosis where $\lambda_i = \beta_i^I \dfrac{I_i}{N_i} + \beta_i^J J_i$ in patch i.

$$\begin{cases} \dot{S}_i = \psi_i - \lambda_i S_i - \mu_i S_i + \sum\limits_{\substack{j=1 \\ j \neq i}}^{n} m_{ij} S_j - S_i \sum\limits_{\substack{j=1 \\ j \neq i}}^{n} m_{ji}, \\ \dot{E}_i = (1 - p_i^I - p_i^J) \lambda_i S_i + \rho_i J_i - A_{E_i} E_i + \sum\limits_{\substack{j=1 \\ j \neq i}}^{n} m_{ij} E_j - E_i \sum\limits_{\substack{j=1 \\ j \neq i}}^{n} m_{ji}, \\ \dot{I}_i = p_i^I \lambda_i S_i + \theta_i J_i + \gamma_i R_i + h_i (1 - \alpha_i) k_i E_i - A_{I_i} I_i \\ \qquad + \eta \left(\sum\limits_{\substack{j=1 \\ j \neq i}}^{n} m_{ij} I_j - I_i \sum\limits_{\substack{j=1 \\ j \neq i}}^{n} m_{ji} \right), \\ \dot{J}_i = p_i^J \lambda_i S_i + (1 - h_i)(1 - \alpha_i) k_i E_i \\ \qquad - A_{J_i} J_i + \eta \left(\sum\limits_{\substack{j=1 \\ j \neq i}}^{n} m_{ij} J_j - J_i \sum\limits_{\substack{j=1 \\ j \neq i}}^{n} m_{ji} \right), \\ \dot{R}_i = r_i I_i - A_{R_i} R_i + \sum\limits_{\substack{j=1 \\ j \neq i}}^{n} m_{ij} R_j - R_i \sum\limits_{\substack{j=1 \\ j \neq i}}^{n} m_{ji}, \end{cases} \tag{4}$$

where m_{ij} and m_{ji} are the migration rates from patch j to patch i and from patch i to patch j (with $i \neq j$), respectively, η is a modification parameter which captures the reduced movements of diagnosed and undiagnosed infectious because of the disease,

$$\begin{aligned} A_{E_i} &= \mu_i + k_i(1-\alpha_i), \qquad A_{I_i} = \mu_i + d_i + r_i, \\ A_{J_i} &= \mu_i + \delta_i + \theta_i + \rho_i(1-\theta_i) \qquad \text{and} \qquad A_{R_i} = \gamma_i + \mu_i. \end{aligned}$$

The total human population of patch i is denoted by N_i. Hence for $i = 1, \ldots, n$, the dynamics of the total population is given by

$$\begin{aligned} \dot{N}_i = \psi_i - \mu_i N_i &+ \sum_{\substack{j=1 \\ j\neq i}}^{n} m_{ij}(S_j + E_j + R_j + \eta(I_j + J_j)) \\ &- (S_i + E_i + R_i + \eta(I_i + J_i)) \sum_{\substack{j=1 \\ j\neq i}}^{n} m_{ji} - (\delta_i J_i + d_i I_i). \end{aligned} \tag{5}$$

The model variables is recapitulated in Table 1.

Table 1. State variables of model system (4).

Variables	Symbol
Susceptible population	S_i
Latently infected population	E_i
Diagnosed infectious population	I_i
Undiagnosed infectious population	J_i
Recovered population	R_i
Total human population in patch i	N_i
Global human population	H

The model parameters are defined in Table 2. Setting $H(t) = \sum_{i=1}^{n} N_i(t)$, the total population dynamics is

$$\dot{H} = \Psi - \sum_{i=1}^{n} \mu_i N_i - \sum_{i=1}^{n} (\delta_i J_i + d_i I_i), \tag{6}$$

where $\Psi = \sum_{i=1}^{n} \psi_i$. Setting $S = (S_1, \ldots, S_n)^T$, $E = (E_1, \ldots, E_n)^T$, $I = (I_1, \ldots, I_n)^T$, $J = (J_1, \ldots, J_n)^T$ and $R = (R_1, \ldots, R_n)^T$, the model system

Table 2. Parameters of model system (4).

Parameters	Symbol
Recruitment rate of susceptible	ψ_i
Contact rate of diagnosed and undiagnosed infectious	β_i^I, β_i^J
Fast route to diagnosed infectious class	p_i^I
Fast route to undiagnosed infectious class	p_i^J
Slow route to active TB	k_i
Natural mortality	μ_i
TB mortality of diagnosed infectious	d_i
TB mortality of undiagnosed infectious	δ_i
Chemoprophylaxis of latently infected individuals	α_i
Diagnosis rate of active TB	h_i
Recovery rate of diagnosed infectious	r_i
Natural recovery rate of undiagnosed infectious	ρ_i
Relapse of recovered individuals	γ_i
Diagnosis rate of undiagnosed infectious	θ_i
Migration rate from patch i to patch j	m_{ji}
Migration rate from patch j to patch i	m_{ij}
Reduction factor for migration	η

(4) becomes:

$$\begin{cases} \dot{S} = \psi - \mathrm{diag}(\lambda)S - \mathrm{diag}(\mu)S + \mathcal{M}S, \\ \dot{E} = \mathrm{diag}((\mathbf{1} - p^I - p^J)\cdot\lambda)S + \mathrm{diag}(\rho\cdot(\mathbf{1}-\theta))J - \mathrm{diag}(A_E)E + \mathcal{M}E, \\ \dot{I} = \mathrm{diag}(p^I\cdot\lambda)S + \mathrm{diag}(\theta)J + \mathrm{diag}(\gamma)R + \mathrm{diag}(h\cdot(\mathbf{1}-\alpha)\cdot k)E \\ \quad - \mathrm{diag}(A_I)I + \eta\mathcal{M}I, \\ \dot{J} = \mathrm{diag}((p^J\cdot\lambda)S + \mathrm{diag}(\mathbf{1}-h)\cdot(\mathbf{1}-\alpha)\cdot k)E - \mathrm{diag}(A_J)J + \eta\mathcal{M}J, \\ \dot{R} = \mathrm{diag}(r)I - \mathrm{diag}(A_R)R + \mathcal{M}R, \end{cases} \tag{7}$$

where

$\psi = (\psi_1, \ldots, \psi_n)^T$, $\lambda = (\lambda_1, \ldots, \lambda_n)^T$, $\mu = (\mu_1, \ldots, \mu_n)^T$,
$p^I = (p_1^I, \ldots, p_n^I)^T$, $p^J = (p_1^J, \ldots, p_n^J)^T$, $\rho = (\rho_1, \ldots, \rho_n)^T$,
$\theta = (\theta_1, \ldots, \theta_n)^T$, $\alpha = (\alpha_1, \ldots, \alpha_n)^T$, $A_E = (A_{E_1}, \ldots, A_{E_n})^T$,
$A_I = (A_{I_1}, \ldots, A_{I_n})^T$, $A_J = (A_{J_1}, \ldots, A_{J_n})^T$, $A_R = (A_{R_1}, \ldots, A_{R_n})^T$,
$k = (k_1, \ldots, k_n)^T$, $h = (h_1, \ldots, h_n)^T$, $r = (r_1, \ldots, r_n)^T$,
and $\mathbf{1} = (1, \ldots, 1)^T$.

In model system (7), the notation "$a \cdot b$", where a and b are vectors with the same dimension, denote the component-wise vector multiplication, $\mathcal{M}$ is defined as in Eq. (2) and $N = S + E + I + J + R$ is the vector-size of the total population.Adding all equations in model system (7) gives

$$\dot{N} = \psi - \text{diag}(\mu)N + \mathcal{M}[S+E+R+\eta(I+J)] - [\text{diag}(\delta)J + \text{diag}(d)I], \quad (8)$$

where $\mathcal{M}$ is defined as Eq. (2)

Model system (7) can be written in the following compact form:

$$\begin{cases} \dot{x} = \psi - \text{diag}(\lambda)x + (\mathcal{M} - \text{diag}(\mu))x, \\ \dot{y} = B\text{diag}(\lambda)x + V_y\, y, \end{cases} \quad (9)$$

where $x = S \in \mathbb{R}^n_+$ stands for the susceptible individuals and $y = (E, I, J, R)^T \in \mathbb{R}^{4n}_+$ is the vector representing the state of infected individuals (latently infected, diagnosed and undiagnosed infectious and recovered individuals). In Eq. (9),

$$B = (\text{diag}(\mathbf{1} - p^I - p^J), \text{diag}(p^I), \text{diag}(p^J), 0)^T, \quad (10)$$

is a $(4n \cdot n)$ block matrix, and V_y is a constant matrix, given by

$$V_y = \begin{bmatrix} \mathcal{M} - \text{diag}(A_E) & 0 & \text{diag}(\rho \cdot (\mathbf{1} - \theta)) & 0 \\ \text{diag}(k \cdot h \cdot (\mathbf{1} - \alpha)) & \eta\mathcal{M} - \text{diag}(A_I) & \text{diag}(\theta) & \text{diag}(\gamma) \\ \text{diag}(k \cdot (\mathbf{1} - h) \cdot (\mathbf{1} - \alpha)) & 0 & \eta\mathcal{M} - \text{diag}(A_J) & 0 \\ 0 & \text{diag}(r) & 0 & \mathcal{M} - \text{diag}(A_R) \end{bmatrix},$$

with A_E, A_I, A_J and A_R defined as in Eq. (7).

We point out that V_y and $[\mathcal{M} - \text{diag}(\mu)]$ are Metzler matrices since their off-diagonal entries are non-negative [17,18]. Indeed, they are stable and invertible matrices. We have the following result on matrix V_y. Matrices V_y, $[\text{diag}(\mu) - \mathcal{M}]$ and $-[\text{diag}(\mu) - \mathcal{M}]^{-1}$ are stable matrices, such that all its eigenvalues have negative real parts and diagonally column dominant [17].

3. Basic reproduction number

The basic reproduction number $\mathcal{R}_0$ is computed using the next generation approach [19]. The infected classes are the classes of the components of y. From the second equation of (9), the jacobian at the DFE P_0 is

$$J = F - V, \quad (11)$$

where

$$F = [F_1 + \text{diag}(S^0)F_2]B \qquad \text{and} \qquad V = -V_y,$$

with $F_1 = [0, \text{diag}(\beta^I), 0, 0]$ and $F_2 = [0, 0, \text{diag}(\beta^J), 0]$. Then, the next generation matrix is $F\,V^{-1}$:

$$F\,V^{-1} = \left[F_1 + \text{diag}(S^0)\; F_2\right](-V_y^{-1})B.$$

Following [19] and using the transformation the same transformation as in [20,21], the basic reproduction number is given by

$$\mathcal{R}_0 = \rho(FV^{-1}) = \rho\left(\left[F_1 + \text{diag}(S^0)\; F_2\right](-V_y^{-1})B\right), \tag{12}$$

where ρ is the spectral radius of the matrix FV^{-1}.

$\mathcal{R}_0$ is the average number of secondary cases produced by a single infective individual which is introduced into an entirely susceptible population. The following result comes from the [19].

Lemma 3.1. : *The disease-free equilibrium is locally asymptotically stable whenever $\mathcal{R}_0 < 1$, and instable if $\mathcal{R}_0 > 1$.*

4. Parameter sensitivity and estimation

A sensitivity analysis generates essential information for parameters estimation, optimization, model simplification and experimental design. Through a sensitivity analysis, a systematic study of the effects of parameter values on the predictions of a mathematical model is a valuable tool for model evaluation and validation as well as for quantifying the effect of parametric uncertainty and variability. Model system (4) can be written as the following compact form:

$$\begin{cases} \dfrac{d}{dt}y(t,\mathbf{p}) = f(t,y,\mathbf{p}), & t \geq 0, \\ y(0,\mathbf{p}) \quad = y_0, & \end{cases} \tag{13}$$

where $y(t,\mathbf{p}) = (S(t,\mathbf{p}), E(t,\mathbf{p}), I(t,\mathbf{p}), J(t,\mathbf{p}), R(t,\mathbf{p})) \in \mathbb{R}^{20}$ and $\mathbf{p} = (\psi, \beta^I, \beta^J, \cdots, \mu) \in \mathbb{R}^q$, with $q = 76$.

Let us assume that there are given m experimental data points $\tau_1, \cdots, \tau_m$, and the corresponding data values in different components z_k, $k = 1, \cdots, m$ associated with corresponding tolerances δz_k. The parameter identification problem consists of solving the least-square minimization problem:

$$g(\mathbf{p}) = \frac{1}{m}\sum_{j=1}^{m} \parallel D_j^{-1} \cdot (y(\tau_j,\mathbf{p}) - z_j) \parallel_2^2 \longrightarrow \min_{\mathbf{p}}, \tag{14}$$

with a diagonal weighting

$$D_j := diag((\delta z_j)_1, \ldots, (\delta z_j)_n) \in \mathcal{M}_n(\mathbb{R}), \qquad j = 1, \cdots, m. \tag{15}$$

In Eq. (15), $\mathcal{M}_n(\mathbb{R})$ is the set of square matrices of order n with real coefficients. The minimization problem (14) can be written in an equivalent least squares minimization problem defined as

$$g(\mathbf{p}) := F(\mathbf{p})^T \cdot F(\mathbf{p}) \to \min_{\mathbf{p}}, \tag{16}$$

where $F(\mathbf{p}) = (F_1(\mathbf{p}), \ldots, F_m(\mathbf{p}))$ is a vector of length m with entries defined by

$$F(\mathbf{p}) = \begin{bmatrix} D_1^{-1} \cdot (y(\tau_1, \mathbf{p}) - z_1), \\ \vdots \\ D_m^{-1} \cdot (y(\tau_m, \mathbf{p}) - z_m) \end{bmatrix}. \tag{17}$$

$F : \mathbb{R}^q \longrightarrow \mathbb{R}^N$ for $N = nm$ is a non-linear mapping and structured as a stacked vector. If not all components of a measurement z_j are given, the number q is accordingly made smaller $q < nm$. Thus, we want to minimize the relative deviation of model and data at the measurement time points τ_k. The above problem, which is highly nonlinear in $\mathbf{p}$, can be solved by the affine covariant Gauss-Newton iteration [22] where each iteration step i requires the solution of a linear least-squares problem:

$$\begin{aligned} &\|J(\mathbf{p}^i) \cdot \Delta\mathbf{p}^i + F(\mathbf{p}^i)\|_2 \to \min_{\mathbf{p}^i}, \\ &\mathbf{p}^{i+1} = \mathbf{p}^i + \Delta\mathbf{p}^i, \end{aligned} \tag{18}$$

where $J(\mathbf{p}^i) = F'(\mathbf{p}^i)$ represents the sensitivity of the solution y with respect to the parameters $\mathbf{p}$ at the time points of measurements.

4.1. *Sensitivity analysis*

Sensitivity here describes the influence of changes in the parameters vector $\mathbf{p}$ on the solution y. A parameter is called sensitive if small changes in its value lead to large changes of the solutions. The parameter vector $\mathbf{p}$ may include initial conditions y_0, although y_0 does not enter explicitly in f.

The sensitivities

$$s_{ij}(t) = \frac{\partial y_i(t)}{\partial \mathbf{p}_j}, \qquad i = 1, \ldots, 20, \qquad j = 1, \ldots, 76. \tag{19}$$

are computed for all data time-points and stacked into one large sensitivity matrix. In Eq. (19), $J(\mathbf{p}) = (s_{ij})$ represents the sensitivity of the solution

y with respect to the parameters $\mathbf{p}$ at the time points of measurements. An analysis of the matrix $J(\mathbf{p})$ gives some hints whether the current combination of model and data will allow the reliable identification of a given parameter. Parameters with very small sensitivity have nearly no influence on the solution and can therefore not be estimated. In this case, the entries of the corresponding column in $J(\mathbf{p})$ (and thus, the weighted l^2 column norm) are almost zero. Furthermore, some parameters might be linearly dependent, which leads to nearly identical columns in $J(\mathbf{p})$. In both cases, the matrix $J(\mathbf{p})$ will be singular or, from a numerical point of view, nearly singular.

Figure 2 shows the column norm $(\|DF(:,j)\|)$ of this sensitivity matrix with respect to the parameter values. This figure illustrates that the parameter values with the largest column norm are γ_i, $\mathbf{p}^i_J$ and ρ_i. These parameters have the largest influence on the trajectory at the measurement time-points.

4.2. *Parameter estimation*

Epidemiology is fundamentally a data-driven discipline, and a key element in this research field is to establish links of mathematical models to data. For the model system (4), some important parameters are well-known for their specificity to the Cameroonian population. The values of these parameters will be fixed. The main difficulty with the model system (4) is to identify the unknown parameters and estimate keys parameters for the disease dynamics in each patch. The physical map of Cameroon with all roads and cities is depicted in Fig. 3. The present study focuses on four regions of Cameroon according to roads, big cities and, social distance, cultural and sociology facts of the people of these regions, even if we have presented the migration model in a general setting, in order to have a generic modelling. All regions are more or less connected, and in principle, people can move from one region to other (see Fig. 3).

Figure 3 presents the migration model throughout Cameroon. In this figure, the patches 1, 2, 3 and 4 include Yaounde, Douala, Bertoua and Garoua which are the main big cities in Center, West, East and North of Cameroon (see Fig. 3). The linear least squares problem (18) is solved by QR factorization with column pivoting [22]. By a suitable permutation of the matrix columns of $J(\mathbf{p})$, the diagonal elements of the upper triangular matrix R can be ordered in the form $r_{11} \geq r_{22} \geq \cdots \geq r_{qq}$. The sub-

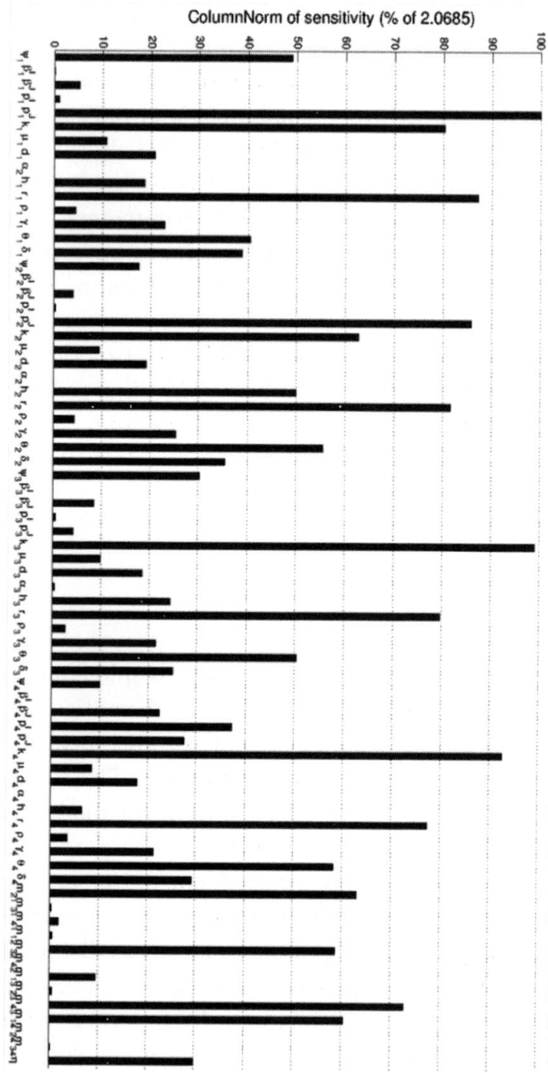

Figure 2. Sensitivity norms representing all parameters of the model.

condition of parameter $\mathbf{p}_j$ is defined by

$$sc_j = \frac{r_{11}}{r_{jj}}.$$

Thus, the permutation of matrix columns corresponds to a new ordering of parameters according to the increasing sub-condition. The sub-condition

Figure 3. Detailed physical map of Cameroon with all roads, cities and airports.

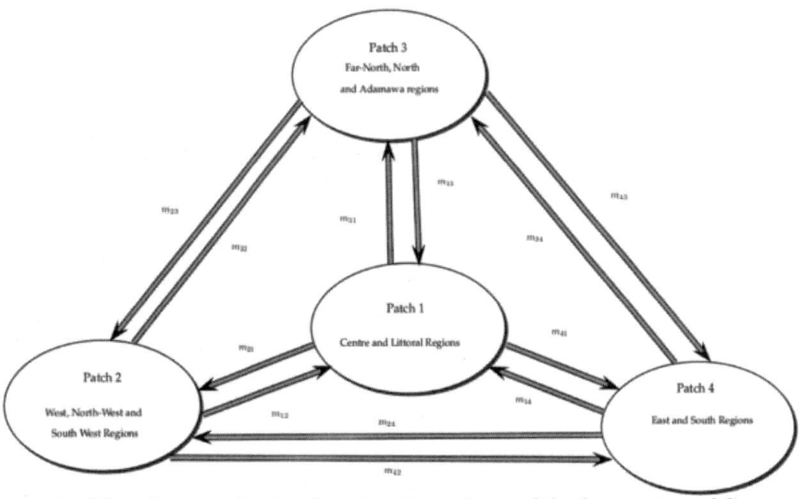

Figure 4. Flow diagram showing the migration scheme of the four regions of Cameroon.

indicates whether a parameter can be estimated from the given data or not. Only parameters for which

$$sc_j \leq 1/\varepsilon,$$

where ε is the relative precision of the Jacobian $J(\mathbf{p})$ can be estimated [23].

In the global case, one has

$$\mathbf{p}^{i+1} = \mathbf{p}^{i} + \lambda_i \Delta \mathbf{p}^{i}; \quad 0 < \lambda_i \leq 1.$$

The step length $0 < \lambda_i < 1$ is computed successively in each iteration by a trust-region method [22]. The above described method for solving a nonlinear least squares problem is implemented in the software code NLSCON [22] and part of the software package BioPARKIN [24]. Here, a Matlab-based version of this software package, named POEM 2.0, which is especially adapted to the parameter identification of ordinary differential equations, has been used throughout the study. Figure 5 shows the sub-condition of the sensitivity matrix with respect to the parameter values. The sensitivity analysis reveals that 46 of 76 parameters can be estimated. Sensitivity analysis of parameters is not only critical to model verification and validation in the process of model development and refinement, but it also provides on insight into the robustness of the model results when making decisions [25]. Indeed, Fig. 5 shows that γ_i, p^i_J and ρ_i are among the parameters with lowest sub-condition, i.e. they can be identified from the given measurements. Fortunately, for some of the non-identifiable parameters, values were already specified for Cameroon or can be derived from other sources.

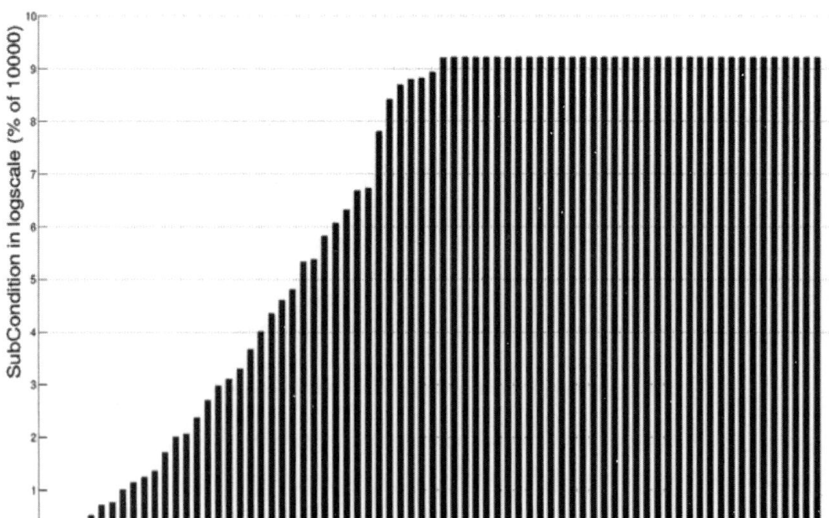

Figure 5. Sub-condition of estimated parameters

Table 3. Initial values of model system (4).

Symbol	Initial value	Source	Symbol	Initial value	Source
S_0^1	$3.0998 \cdot 10^6$	Estimated	E_0^1	$2.0601 \cdot 10^6$	Estimated
I_0^1	1843	WHO	J_0^1	3,685	Estimated
R_0^1	921	Estimated	N_0^1	5166279	[26]
S_0^2	$2.7116 \cdot 10^6$	Estimated	E_0^2	$1.8048 \cdot 10^6$	Estimated
I_0^2	688	WHO	J_0^2	1,375	Estimated
L_0^2	921	Estimated	N_0^2	4519380	[26]
S_0^3	$3.2326 \cdot 10^6$	Estimated	E_0^3	$2.1535 \cdot 10^6$	Estimated
I_0^3	224	WHO	J_0^3	448	Estimated
R_0^3	921	Estimated	N_0^3	5387716	[26]
S_0^4	$7.218 \cdot 10^5$	Estimated	E_0^4	$4.8000 \cdot 10^5$	Estimated
I_0^4	94	WHO	J_0^4	188	Estimated
R_0^4	921	Estimated	N_0^4	1202998	[26]

Using the method described above, the unknown parameters of the model system (4) are estimated using the software POEM. Table 4 gives the parameter values of the model system (4). As denoted in the last column, some values of parameters have been taken from the literature [26,27,28] and the rest was estimated using POEM.

4.3. *Model Validation*

In order, to illustrate the theoretical results of the foregoing analysis, numerical simulations of the model system (4) are carried out using the Matlab version of the software POEM. A comparison of the results of the numerical simulation with the measurement data allows to test the suitability of the model system (4) to effectively enable the assessment of targeted public health education strategies against TB spread in a population. The model is fitted using data from Cameroon as follows. Parameter values are given in Table 4. We take the year 1994 as the starting time of the numerical simulation. According to the National Institute of Statistics [26], the total of the Cameroonian population in 1994 in the four regions considered were $N_0^1 = 5,166,279$, $N_0^2 = 4,519,380$, $N_0^3 = 5,387,716$ and $N_0^4 = 1,202,998$. The initial values of other variables used for numerical simulations are given in Table 3. Using data of Cameroon, the model system (4) gives a very good fit of the Cameroonian data for the period 2002-2010, as depicted in Figs. 6-10. For each patch, we present the time evolution of the state

Table 4. Numerical values for the parameters of model system (4) in each patch i.

Symbol	Value/yr	Source
ψ_i	$3.21 \cdot 10^5, 1.06 \cdot 10^5, 2.55 \cdot 10^5, 1.94 \cdot 10^4$	Estimated
β_i^I	$9.99 \cdot 10^{-1}, 9.87 \cdot 10^{-1}, 1.34, 9.98 \cdot 10^{-1}$	Estimated
p_i^I	$6.70 \cdot 10^{-5}, 1.14 \cdot 10^{-3}, 1.06 \cdot 10^{-3}, 1.97 \cdot 10^{-3}$	Estimated
p_i^J	$6.70 \cdot 10^{-5}, 1.14 \cdot 10^{-3}, 1.20 \cdot 10^{-2}, 4.88 \cdot 10^{-3}$	Estimated
β_i^J	$4.55 \cdot 10^{-5}, 4.24 \cdot 10^{-5}, 4.85 \cdot 10^{-5}, 5.57 \cdot 10^{-4}$	Estimated
k_i	$1.32 \cdot 10^{-3}, 9.81 \cdot 10^{-4}, 9.41 \cdot 10^{-4}, 2.39 \cdot 10^{-3}$	Estimated
μ_i	$1/53.4$	[26]
d_i	0.139	[29]
α_i	$1.02 \cdot 10^{-3}, 1.47 \cdot 10^{-3}, 5.37 \cdot 10^{-3}, 1.00 \cdot 10^{-3}$	Estimated
h_i	$9.98 \cdot 10^{-1}, 1.50 \cdot 10^{-2}, 4.56 \cdot 10^{-1}, 3.04 \cdot 10^{-2}$	Estimated
r_i	0.7553	[14]
ρ_i	$1.34 \cdot 10^{-1}, 1.69 \cdot 10^{-1}, 2.24 \cdot 10^{-1}, 1.78605 \cdot 10^{-1}$	Estimated
γ_i	$0.0986, 0.0986, 0.0986, 0.0986$	Estimated
θ_i	$2.16 \cdot 10^{-1}, 3.67 \cdot 10^{-1}, 8.29 \cdot 10^{-2}, 8.59 \cdot 10^{-3}$	Estimated
η	1	Estimated
δ_i	$0.413, 0.413, 0.413, 0.413$	[30]
m_{i1}	0, $1.91 \cdot 10^{-3}$,0.009, 0.0099	Estimated
m_{i2}	$8.42 \cdot 10^{-2}$,0, $9.99 \cdot 10^{-2}$, 0	Estimated
m_{i3}	0.0099, $6.26 \cdot 10^{-4}$, 0, $8.95 \cdot 10^{-3}$	Estimated
m_{i4}	0.0099, 0, $3.44 \cdot 10^{-1}$, 0	Estimated

trajectories of susceptible individuals, latently-infected individuals, diagnosed infectious, undiagnosed infectious, recovered individuals and the total population. With the estimated transmission parameters, the deterministic model appears to capture all the qualitative properties of the observed pattern. Hence, model system (4) can be used to gain insight into tuberculosis transmission dynamics at least for a limited period. Figure 6 (a)-(f) shows the time evolution of all populations in patch 1.

The time evolution of the trajectories of classes of population in patch 2 are depicted in Figs. 7 (a)-(f). Compare to numerical results in patch 1, one can see that the number of undiagnosed infectious is larger than the number of undiagnosed infectious in patch 1. This is because the patch 2 is the West part of Cameroon which is the most populated area of Cameroon and people of this region has a strong interaction due to their tradition and culture. Figure 8 gives the trends of different classes population in patch 3.

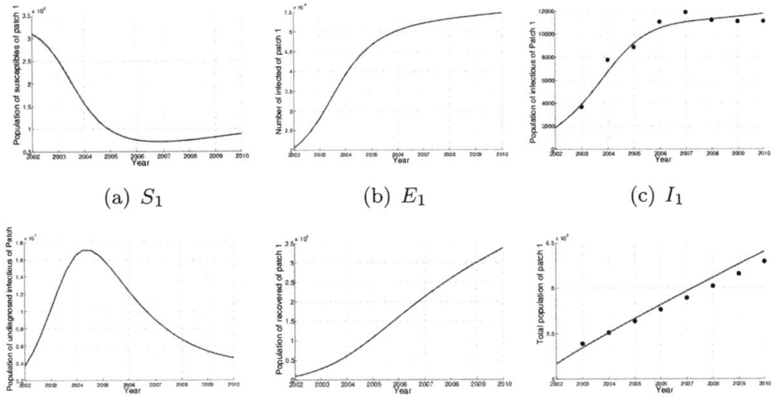

(a) S_1 (b) E_1 (c) I_1

(d) J_1 (e) R_1 (f) N_1

Figure 6. Time series of populations in patch 1. The dot plots represent the yearly trend and variability in yearly case reports over the period 2002-2010.

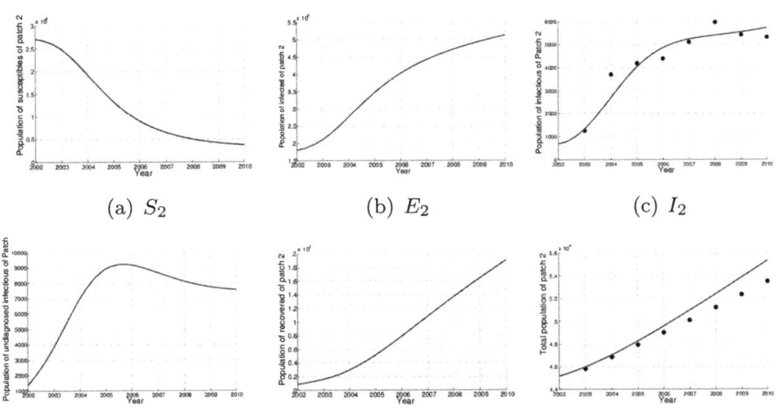

(a) S_2 (b) E_2 (c) I_2

(d) J_2 (e) R_2 (f) N_2

Figure 7. Time evolution of populations in patch 2. The dot plots represent the yearly trend and variability in yearly case reports over the period 2002-2010. Parameter values and initial values are defined, respectively in Tables 4 and 3.

As in the previous cases, there are no obvious discrepancies between data and prediction. Figure 8(d) clearly shows that the number of undiagnosed infectious is great than the number of undiagnosed infectious in patch 2. This can be explained by the fact that the patch 3 includes the East region of Cameroon which is the lest educated. The results of numerical simulation

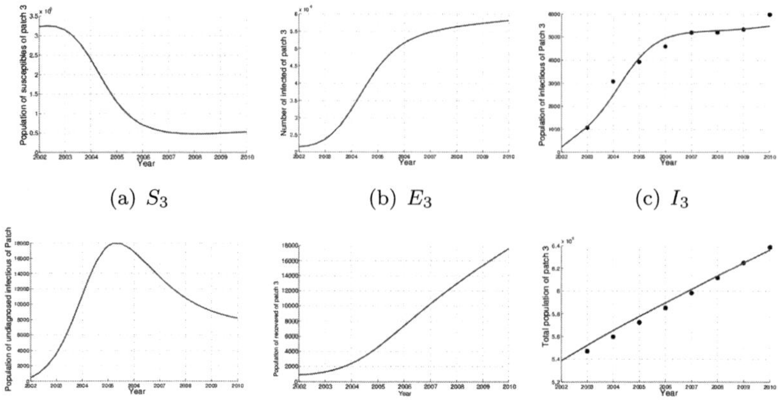

(a) S_3 (b) E_3 (c) I_3

(d) J_3 (e) R_3 (f) N_3

Figure 8. Time series of the populations in patch 3. The dot plots represent the yearly trend and variability in yearly case reports over the period 2002-2010. Parameter values and initial values are defined, respectively in Tables 4 and 3.

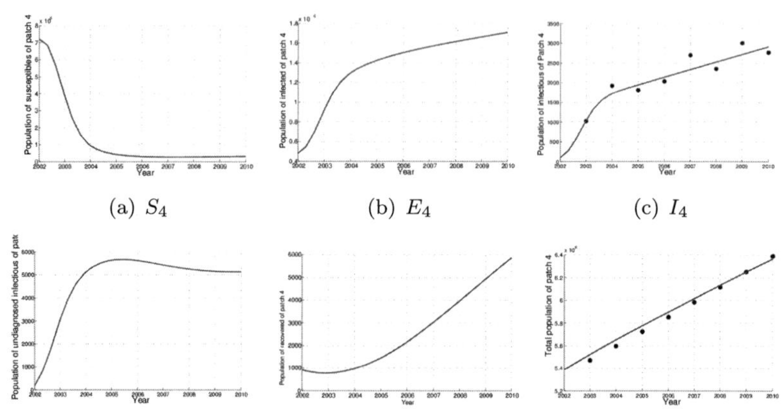

(a) S_4 (b) E_4 (c) I_4

(d) J_4 (e) R_4 (f) N_4

Figure 9. Time evolution of populations in patch 4. The dot plots represent the yearly trend and variability in yearly case reports over the period 2002-2010. Parameter values and initial values are defined, respectively in Tables 4 and 3.

in patch 4 are reported in Fig. 9. It appears that there are less undiagnosed infectious people than in patches 1, 2 and 3.

Figure 10 presents the comparison between the data and the estimated trajectories of the diagnosed infectious population and the total human

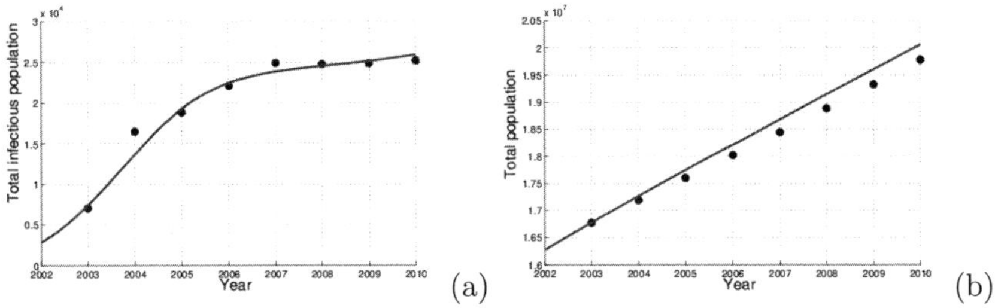

Figure 10. Comparison data and the model prediction in all patches. (a) reported number of diagnosed infectious and its simulation curve and (b) reported number of the total population and its fitted curve. The dot plots represent the yearly trend and variability in yearly case reports over the period 2002-2010. Parameter values and initial values are defined, respectively in Tables 4 and 3.

population in Cameroon. It illustrates that the model system (4) reproduces the TB dynamics in the whole country. Parameter values and initial values are defined in Tables 4 and 3, respectively.

5. Conclusion

In this paper, a deterministic human metapopulation model for the transmission dynamics of TB is presented and analyzed. We have shown that the model is biologically well-posed. A mathematical analysis of the model is presented. We have computed the basic reproduction number $\mathcal{R}_0$ and numerical data gave the value $\mathcal{R}_0 = 7.32$. An sensitivity analysis of the model's parameters have been performed. This study can be used to measure successfully intervention strategies to fight TB by decreasing the number of undiagnosed infectious and implement efforts to that aim. The unknown parameters of the proposed metapopulation model have been estimated using an iterative state of the art Gauss-Newton Method. We divided the entire territory of Cameroon into four regions according to roads, big cities, social distance, cultural and sociology facts of the peoples of these regions. We also validated the proposed metapopulation model by using real data of TB in Cameroon. An important result of this analysis is that the number of infectious will increase in next years in Cameroon. This indicates that TB will remain endemic in Cameroon in the next years. An educational campaigns that encourage individuals to go for TB screening could be a solution to eradicate TB in Cameroon, because it may obviously contribute to decrease the undiagnosed population. An optimal control problem can therefore been addressed with this model. Although, our analysis has been applied on a TB model, the basic idea and the proposed methodology can be applied to other models of infectious diseases as well.

6. Acknowledgement

Authors would like to thanks anonymous referees for their constructive comments which allow to improve the quality of this work.

References

1. C. Dye, S. Schele, P. Dolin, V. Pathania, and M. Raviglione. For the who global surveillance and monitoring project. global burden of tuberculosis estimated incidence and prevalence and mortality by country. *JAMA*, 282:677–686, 1999.
2. S. R. Benatar, A. S. Daar, and P. A. Singer. Global health challenges: The need for an expanded discourse on bioethics. *Public Library of Science Medicine PLOS*, 2(7):100–102, 2005.
3. S J Kim, M A Espinal, C Abe, G H Bai, F Boulahbal, L Fattorin, C Gilpin, S Hoffner, K M Kam, N Martin-Casabona, L Rigouts, and V Vincent. Is second-line anti-tuberculosis drug susceptibility testing reliable? *Int. J. Tuberc. Lung Dis.*, 8(9):1157–8, September 2004.
4. C. Dye and B. G Williams. The population dynamics and control of tuberculosis. *Science*, 328(5980):856–61, May 2010.
5. P. Auger, E. Kouokam, G. Sallet, M. Tchuente, and B. Tsanou. The Ross-Macdonald model in a patchy environment. *Math Biosci*, 216(2):123–31, December 2008.
6. H. W. Hethcote and H. R. Thieme. Stability of the endemic equilibrium in epidemic models with subpopulations. *Math. Biosci.*, 75:205–227, 1985.
7. A. Lajmanovich and J. A. Yorke. A deterministic model for gonorrhea in a nonhomogeneous population. *Math. Biosci.*, 28:221–236, 1976.
8. H. W. Hethcote. Mathematical models for the spread of infectious diseases. In D. Ludwig and K. L. Cooke, editors, *Epidemiology*, pages 121–131. SIAM, Philadephia, 1975.
9. Y. Huang and H. Wu. A bayesian approach for estimating antiviral efficacy in hiv dynamic models,. *Journal of Applied Statistics*, 33:155–174, 2006.
10. A.M. Jeffrey and X. Xia. *Deterministics and Stochastic Models of AIDS Epidemics and HIV Infections with Intervention*, chapter Identifiability of HIV/AIDS model, page p255. World Scientific, Singapore, 2005.
11. H. Wu, H. Zhu, H. Miao, and A. S. Perelson. Parameter identifiability and estimation of hiv/aids dynamic models. *Bulletin of Mathematical Biology*, 70(3):785–799, 2008.
12. S. Bowong and J. Kurths. Parameter estimation based synchronization for an epidemic model with application to tuberculosis in cameroon. *Physics Letters A*, 374:4496–4505, 2010.
13. WHO. Fact sheet number 104. Technical report, World Health Organization, Geneva, Switzerland, 2012.

14. NCFT. *Guide de personnel de la santé*. National Comity of Fight Againts Tuberculosis, Ministère de la Santé Publique du Cameroun, 2001.
15. R.M. Anderson and R.M. May. *Infectious Diseases of Humans: Dynamics and Control*. Oxford University Press, 1991.
16. V. Capasso. *Mathematical structures of epidemic systems*. Lecture notes in biomathematics, Berlin Springer, 1993.
17. A. Berman and R. J. Plemmons. *Nonnegative matrices in the mathematical sciences*. SIAM, 1994.
18. J.A. Jacquez and C. P. Simon. Qualitative theory of compartmental systems. *SIAM Rev.*, 35(4):43–79, 1993.
19. P. van den Driessche and J. Watmough. Reproduction numbers and subthreshold endemic equilibria for compartmental models of disease transmission. *Math. Biosc.*, 180:29–28, 2002.
20. P. Adda, J. L. Dimi, A. Iggidr, J. C. Kamgang, G. Sallet, and J. J. Tewa. General models of host-parasite systems, Global analysis. *Discrete Contin. Dyn. Syst. Ser. B*, 8:1–17, 2007.
21. A. Iggidr, J. Mbang, G. Sallet, and J. J. Tewa. Multi-compartment models. *Discrete Contin. Dyn. Syst. Ser. B*, 1:506–519, 2007.
22. P. Deuflhard. *Newton Methods for Nonlinear Problems: Affne Invariance and Adaptive Algorithms*, volume 35. Springer Series in Computational Mathematics. Springer Verlag, Berlin, 2004.
23. P. Deuflhard and W. Sautter. On rank-deficient pseudoinverses. *Lin. Alg. Appl.*, 29:91–111, 1980.
24. T. Dierkes, M. Wade, U. Nowak, and S. Röblitz. Bioparkin- biology-related parameter identification in large kinetic networks. arXiv:1303.4928, December 2011.
25. A. Saltelli, K. Chan, and M. Scott. *Sensitivity Analysis, Probability and StatisticsSeries*. Wiley New York., 2000.
26. NIS. Rapport sur la présentation des résultats définitifs. Technical report, National Institute of Statistics, Bureau Central des Recensements et des Etudes de Population, 2010.
27. S. M. Blower, P. Small, and P. Hopewell. Control strategies for tuberculosis epidemics: new method for old problem. *Science*, 273:497–500, 1996.
28. N. Bacaër, R. Ouifki, C. Pretorius, R. Wood, and B. Williams. Modeling the joint epidemics of TB and HIV in a South African township. *J Math Biol*, 57(4):557–93, October 2008.
29. S. M. Blower, P. M. Small, and P. C. Hopwell. Control strategies for tuberculosis epidemics: New models for old problems. *Science*, 273:497–520, 1996.
30. S. M. Blower, A. R. McLean, T. C. Porco, P. M. Small, P. C. Hopwell, M. A. Sanchez, and A. R. Ross. The intrinsic transmission dynamics of tuberculosis epidemics, nature medicine. *Histoire de l'Acad. Roy. Sci. (Paris) avec Mém. des Math. et Phys. and Mém.*, 1(2):815–821, 1995.

AN AGENT-BASED MODELLING FRAMEWORK FOR TUBERCULOSIS INFECTION WITH DRUG-RESISTANCE

AQUINO L. ESPÍNDOLA

Departamento de Física,
Instituto de Ciências Exatas - ICEx,
Universidade Federal Fluminense, Volta Redonda, 27.213-145, Brazil

ALEXANDRE SOUTO MARTINEZ

Faculdade de Filosofia, Ciências e Letras de Ribeirão Preto,
Universidade de São Paulo,
Avenida dos Bandeirantes, 3900, 14.040-901, São Paulo, Brazil

SEYED M. MOGHADAS

Agent-Based Modelling Laboratory
York University, 4700 Keele St., Toronto, M3J 1P3, Canada
E-mail: moghadas@yorku.ca

Tuberculosis is an old disease which co-exists with humans. It claims for over 10 million new infections and deaths annually. According to the World Health Organization TB facts, about one-third of the world's population has latent TB, and drug resistant strains have been found in all countries surveyed.[1,2] With rapid spread of multi-drug resistance, TB has become a global public health challenge. Understanding the dynamics of TB infection and the interplay between confounding factors involved in the emergence and spread of drug-resistance is crucial for addressing this growing challenge. While mathematical models have contributed to enhance our understanding of the TB infection mechanisms and control,[3–6] the perplexity of this disease calls for application of more advanced and computationally sophisticated methods.[7] Here, we propose an agent-based modelling computational system as a general framework to study TB infection and interrelated processes. This framework includes two strains of TB, namely drug-sensitive and drug-resistant, and encapsulates principal mechanisms of latent, endogenous, exogenous, and relapse infection. This paper is concerned with the development of this framework both epidemiologically and computationally, with methodologies for model implementation. Using this model, we provide a new pathway towards evaluating treatment strategies for curtailing TB drug-resistant infection.

1. Introduction

Advances in information and communications technologies, along with computational infrastructure, during the past two decades have led to major scientific discoveries, and provided a new conduit for knowledge exchange and application. These advances have had profound impacts on many aspects of real life, and led to dramatic changes in approaches to addressing societal issues. For example, applied mathematical sciences, in general, and dynamic modelling and simulations in particular, are being systematically employed to analyze the complex network of human interactions and mobility patterns within and between societies.[8] Understanding such connectivity and individual movements has had readily identifiable impacts on public health at the global scale, in particular for the development of intervention strategies against persistent infections and emerging diseases.[9,10]

Among modelling strategies that have been developed to-date for describing complex social dynamics of human populations, agent-based modelling (ABM) is the only advanced methodology capable of grasping the intricacies pertinent to the interplay between biological and demographic parameters with the inclusion of social behaviours inherent to human societies.[7] ABMs can be developed to be computer representations of human populations in which independent individuals (autonomous agents) perceive, make decisions, interact among themselves, and are bonded by social rules. Using ABM computational systems, it is possible to systematically test different hypotheses related to attributes of the agents, and investigate how macro phenomena emerge from individual level behaviour among a heterogeneous set of interacting agents. ABM frameworks, which represent a bottom-up approach in a modelling context, generally start from well-established social networks with reasonable assumptions. The construction of an ABM, for example, for simulating a disease outbreak, relies on modelling of three main components that constitute the core of simulations, namely: agents, environment, and disease.[7] Within this modelling framework, interactions, relationships, and interventions can be investigated. Each component has its own databases that make it specific to the population and epidemic situation. Agents form a population that resembles a certain region are modelled based on demographic characteristics described by census data. The environment offers the geographical location where agents are placed and interact. And the disease is characterized by its properties, such as epidemiological and clinical manifestations, and associated parameters (e.g., transmissibility, infectious period, attack rates, and evolutionary responses of the pathogens).

Understanding the interplay between demographic, health, infection and control parameters requires the development of a modelling framework that can identify individuals with their assigned information, and describe disease spread in the population *in-silico* (i.e., via computer simulations). ABMs which specifically encapsulate individuals characteristics and behavioural responses provide such a framework capable of reproducing observed scenarios in epidemics and evaluating plausible intervention policies against future emerging diseases.[11–14]

In this paper, we describe the construction of an ABM simulation framework for tuberculosis (TB), which remains a major public health problem globally. TB has co-existed with humans for many centuries, and currently claims for over 10 million new infections and deaths annually.[1,2] It is an airborne disease caused by *mycobacterium tuberculosis*, which can be transmitted through several modes, including cough, sneeze or speak in close contacts with an actively infected person. In construction of the ABM framework, we describe several epidemiological and clinical aspects of TB, based on which the population is divided into several compartments. These compartments and associated properties are described in the next sections. Following the description of the disease compartments, we develop the ABM computational framework for simulating TB dynamics *in-silico*. We further discuss the computational implementation of this framework, and set the stage for future studies in this direction.

2. Disease compartments

The basic structure for TB dynamics includes three main compartments,[15] referred to as disease states in the ABM model: susceptible (pathogen-free); latent (pathogen-harboured but not infectious); and infectious (active TB and infectious). Latent and active TB states may be further divided into several sub-states depending on the nature of disease and involved epidemiological and clinical processes. For example, active TB may be identified and offered treatment to reduce infectiousness, which may eventually bring the amount of pathogen sufficiently low to levels regarded as latent with no active disease. One possible scenario for treatment failure, which introduces a new disease state in the model structure, is the development of drug resistance due to a strong selective pressure induced by treatment. It is also possible that a treated active TB becomes reactivated during latency, a mechanism referred to as relapse. Activation (during latency of primary exposure), or reactivation (during latency after effective treatment of active TB) could be influenced by re-exposure to another active TB, a mechanism known as exogenous re-infection, which could accelerate the process of developing active TB. These disease-related processes will require independent model states to accurately portray the long-term dynamics of infection and the effect of intervention measures.

We have also considered treatment of drug resistant infections with a second drug. Failure during treatment with the second drug may represent the scenario of multi-drug resistance, which has now become a major global public health issue. Disease-related states considered in our main framework are listed in Table 1, and we detail each state in the following sections.

Table 1. Disease states with their epidemiological description used for model compartments. Note: S and R represents states related to drug sensitive and drug resistant pathogens, respectively; D_1 and D_2 correspond to the treatment with the first and second drugs, respectively.

State	Description
X	susceptible
L_S	latent infected with sensitive pathogens only
L_R	latent infected with resistant pathogens only
L_{SR}	latent co-infected with both sensitive and resistant pathogens
T_S	active TB infectious with sensitive pathogens only
T_R	active TB infectious with resistant pathogens only
T_{SR}^S	co-infected TB active with sensitive pathogens only
T_{SR}^R	co-infected TB active with resistant pathogens only
$L_S^{D_1}$	effectively treated sensitive active TB
$L_R^{D_2}$	effectively treated resistant active TB
$L_{RS}^{D_1}$	effectively treated co-infected active TB with sensitive pathogens only
$L_{RS}^{D_2}$	effectively treated co-infected active TB with resistant pathogens only

2.1. *Latent TB*

Latency is the state where individuals carry TB pathogens without showing illness symptoms or being infectious. Majority of individuals infected with TB will pass through latency before developing active TB, with often a prolonged period and some individuals never develop active TB during life-time despite being latent.[16,17] In a modelling context, an individual will move to the L state (latent) following exposure to active TB. The development of active TB will decrease the population of individuals in the latent state. This can occur through endogenous activation, exogenous re-infection, or relapse (activation after effective treatment of active TB).[15] The population of this state will increase through exposure of susceptible individuals to active TB, or effective treatment of active TB. Exposure to drug sensitive or drug resistant pathogens will have different latent states and associated movements (Figure 1).

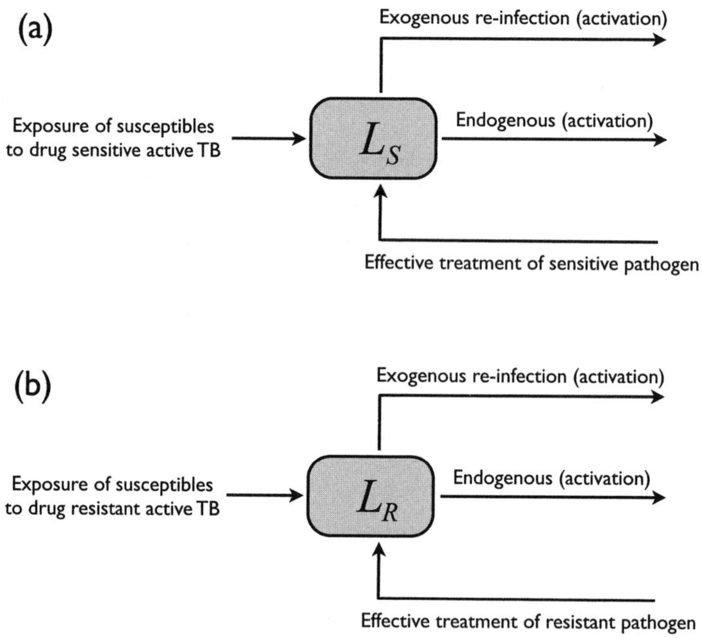

Figure 1. Influx and outflow of the latent states with drug sensitive and drug resistant pathogens.

During latency followed by primary exposure, it is possible to have a secondary exposure to a different pathogen strain, causing co-infected latency harbouring both drug sensitive and drug resistant pathogens (L_{SR}). This state will be increased through effective treatment of active TB, and will be depleted by activation or reactivation (i.e., endogenous, exogenous, and relapse).

2.2. *Active TB*

Three distinct mechanisms can lead to active TB, namely: endogenous activation, exogenous re-infection, and relapse.[15] Once active TB has developed, disease transmission can occur. During this state, individuals may be identified for treatment, and effectiveness of treatment depends on the sensitivity of the pathogen to the type of drug being used. We assumed that the first drug is effective against sensitive pathogens, and the second drug is effective against resistant pathogens. We also assumed that active TB occurs with one type of the pathogen only (i.e., sensitive or resistant) regardless of co-infected latency with both types of pathogen. Effective treatment (following a course of drug therapy) will lead to a decrease in the active TB population by moving the treated individuals back to the latency. The development of drug resistance during the treatment of the sensitive pathogen with the first drug will increase the population of active TB with resistant pathogen, for which the second drug is used for treatment. The processes involved in the active TB states are presented in Figure 2.

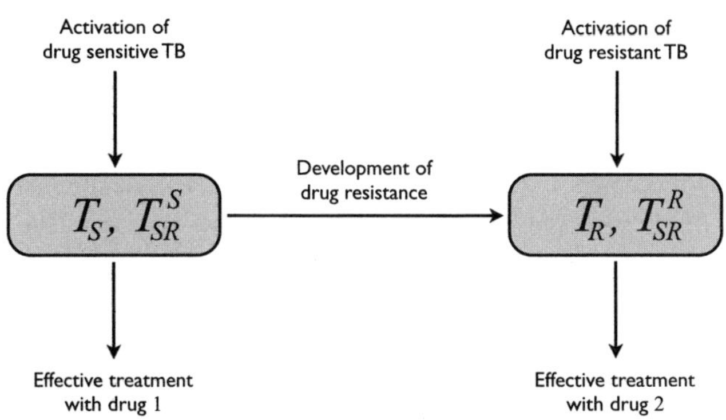

Figure 2. Influx and outflow of the active TB states with drug sensitive and drug resistant pathogens.

2.3. *Exogenous re-infection and relapse*

Exogenous re-infection refers to re-exposure of an individual in the latent state to the same type of pathogen, which may accelerate the rate of developing active TB.[17] There is much evidence to support the occurrence of exogenous re-infection, and clinical studies highlight the importance of this mechanism in determining the transmission dynamics of TB and its resurgence in developed countries.[18–22] A report addressing this issue in South Africa as a country with a high incidence of TB indicated that 75% of recurrent cases had developed active TB as a consequence of exogenous re-infection.[19] However, the relative contribution of re-infection to the overall burden of TB resurgence is still unclear and can have important public health implications for the design of appropriate control strategies. Another important pathway towards developing active TB is relapse, which refers to reactivation of TB disease during latency following an effective course of treatment.

2.4. *Treatment and development of resistance*

Drug resistance is a major concern in treatment of many infectious diseases, in particular TB for which multi-drug resistance has begun to spread worldwide.[23] For pathogens that evolve rapidly, such as *mycobacterium tuberculosis*, replication and mutation processes are generally error-prone, resulting in the generation of pathogens that are resistant to one or more types of drugs .[24,25] In a drug-free environment, sensitive strains of the pathogen prevail and resistant strains are soon out-competed due to their impaired fitness compared to sensitive strains. However, in the presence of drugs, replication of sensitive strains are largely blocked, providing an opportunity for resistant strains to replicate and restore their fitness to levels required for successful transmission. In our model, we assumed that the development of resistance can occur during treatment with the first drug, and included a second drug for treatment of the resistant infection. In addition to resistance emergence during treatment, we considered its direct transmission which may result in co-infection of individuals previously infected with the sensitive pathogen strains.

3. Model integration and implementation

To capture the dynamics of TB infection, we considered an *in-silico* environment in the form of a scaled lattice, in which each squared cell represents a location of a model agent (Figure 3). In this Moore-type representation, each agent can have up to 8 neighbours, with an identified disease state at any time during the simulation scenarios. Each agent is randomly placed in one of the cells with coordinates (i, j) of the linearly-sized M squared lattice. This lattice can therefore host a total population of size $N = M \times M$. We assumed that the total population is constant during the entire simulation.

At $t = 0$ (start of simulations), only two model states are present: susceptible (X), and active TB with sensitive strains (T_S). We assumed that active TB with resistant strains may develop during treatment of active TB with sensitive strains of the pathogen. Other disease states discussed above will naturally develop once the model is computationally implemented.

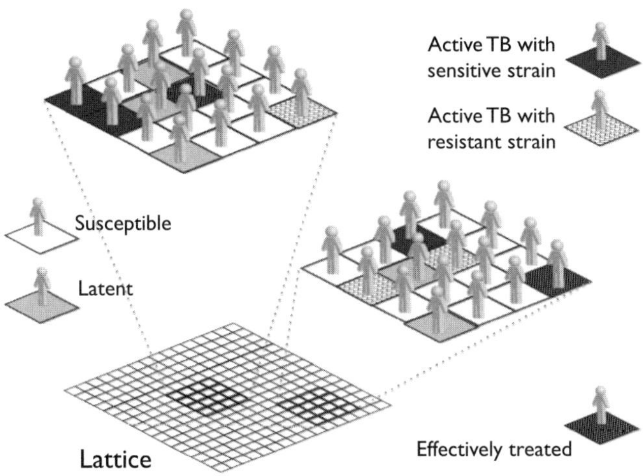

Figure 3. Schematic diagram of the lattice environment in the ABM model with neighbouring individuals in different states.

In the model developed here, agents interact with only neighbours of the adjacent sites in the lattice. This may be considered as a limitation of this framework for capturing the movements of agents; however, a random re-distributions of agents into the lattice with appropriate timing could overcome the limitation associated with a fixed location of each agent. We considered periodic boundary conditions to reduce border effects in the lattice sites.

The ABM simulation model is implemented in time-steps of one day, during which transitions between states of the model can take place based on the changes that occur in the status of agents as a result of their interactions, interventions, or other disease-related mechanisms. The lattice sites are updated synchronously, i.e., state transitions of all sites are modified and stored at the same time. Some of these state transitions depend on the underlying stochastic processes, with outcomes that are associated with some probabilities described below.

In the start of simulations, the likely scenarios include infection of susceptible agents through contacts with their adjacent neighbours in any of the active TB states of T_S, T_R, T_{SR}^S and T_{SR}^R. The transition of a susceptible agent to the latent state may occur if transmission takes place, which depends on the treatment and the probability of transmission associated with different active TB states. The probability of event $X \to L_S$ due to exposure of a susceptible to an active TB with the sensitive strain is calculated by $P_S = 1 - (1 - \beta_S)^{K_S}$, where β_S is the baseline infectivity of the sensitive strain and K_S is the number of neighbours having active TB with the sensitive strain. Similarly, the probability of event $X \to L_R$ is given by $P_R = 1 - (1 - \beta_R)^{K_R}$, where β_R is the baseline infectivity of the resistant strain and K_R is the number of neighbours having active TB with the resistant strain. With these probabilities, disease transmission (i.e., the event) will occur as a result of rejection sampling based trials.[7]

Given the structure of the lattice and disease states, the model was computationally implemented *in-silico* to monitor the spread of disease in the model population for 25 years following the start of treatment. Figures 4 and 5 illustrate the evolution of the lattice in every five-year cycle using parameters values and ranges extracted from the published literature,[3,4,15] with a population size 100×100. As evident from these simulations, the presence of relapse in the model corresponds to a significantly higher incidence of active TB compared with the scenario in which the mechanism of relapse is absent (Table 4). These simulations are not meant to provide any specific quantitative projections, but rather to show the implementation of the model and its potential to simulate outcomes for different scenarios.

4. Discussion and Future Direction

In this paper, we developed a general framework that can be used for the study of complex TB dynamics by considering two strains of the pathogen, namely drug sensitive and drug resistant. In addition to the possibility of treatment with two types of antimicrobial agents that are effective against sensitive and resistant infections, our model encapsulates mechanisms of exogenous re-infection and relapse. Moreover, this ABM framework captures the stochasticity inherent to the mechanisms of infection by including critical population and disease-specific parameters and variables. This framework has the potential to be customized for specific population demographics and disease prevalence to address the effect of treatment, exogenous re-infection, relapse, and drug resistance on the control of TB.

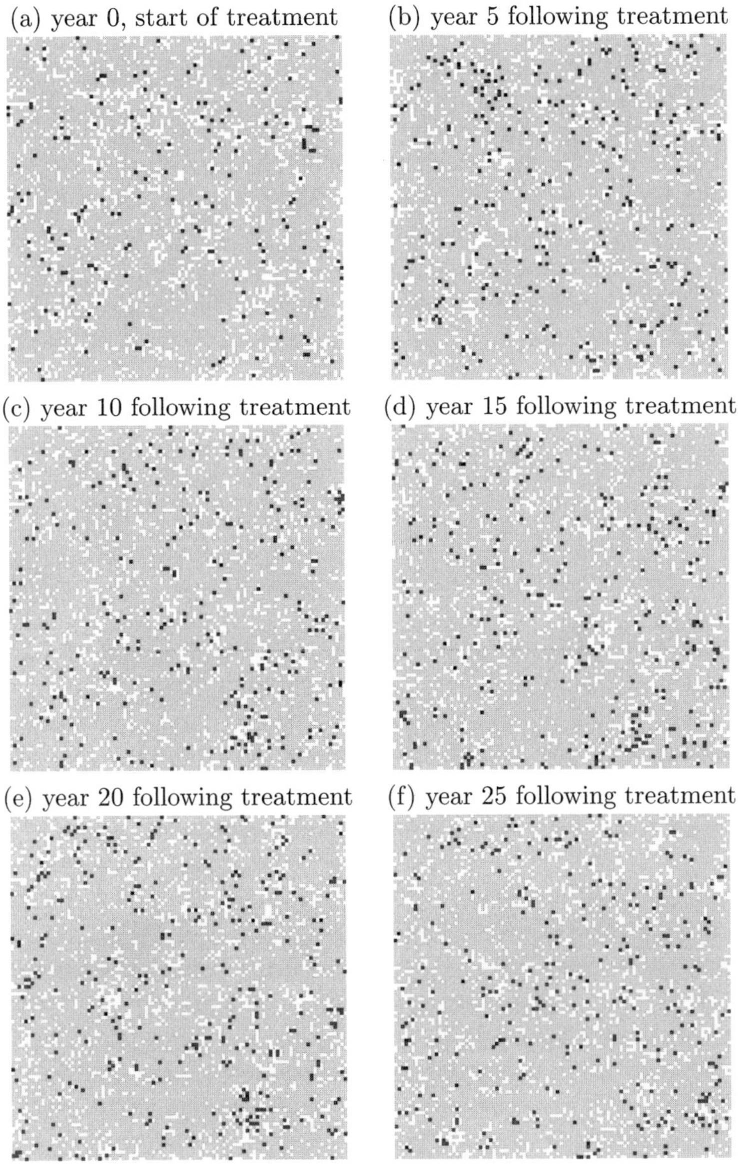

Figure 4. Simulations illustrating the evolution of lattice for population *in-silico* for 25 years following the start of treatment in the presence of drug resistance, exogenous re-infection, and relapse. Simulations were run with parameter values extracted from literature, where an active TB generates on average 5 new infected cases per year. Colour scheme corresponds to susceptible (grey), latent, active TB with the sensitive strain, and active TB with the resistant strain (dark), and effectively treated (white).

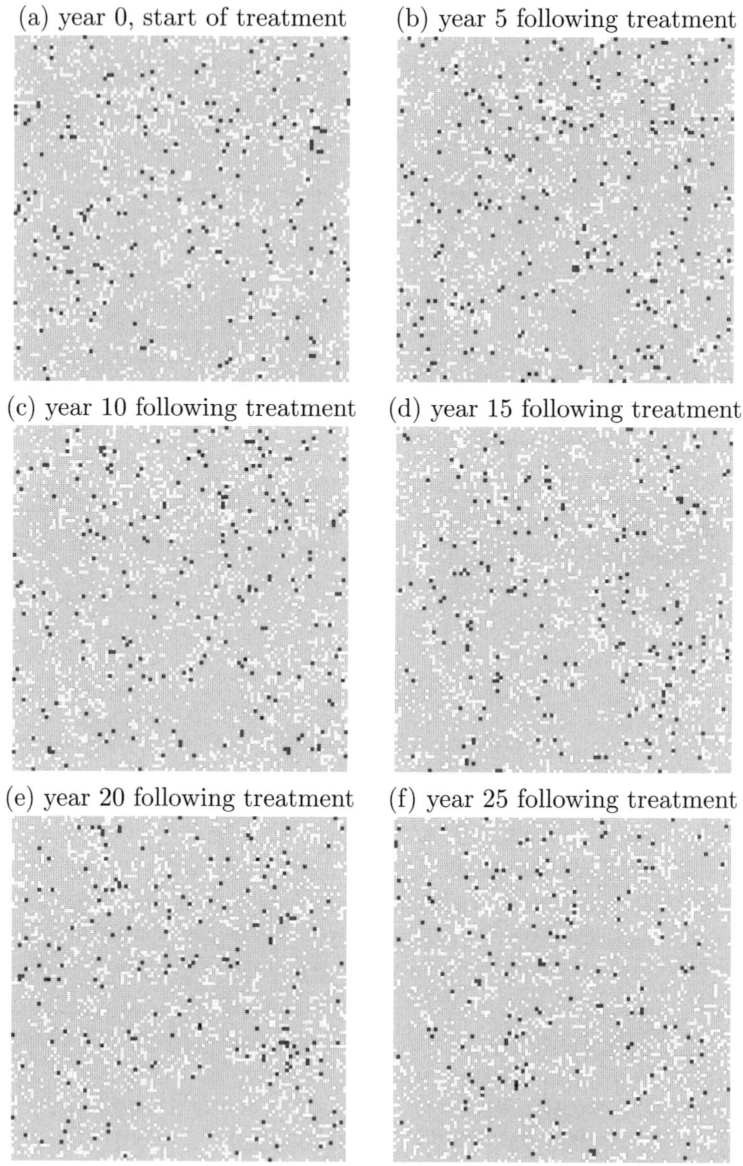

Figure 5. Simulations illustrating the evolution of lattice for population *in-silico* for 25 years following the start of treatment in the presence of drug resistance and exogenous re-infection, without relapse. Simulations were run with parameter values extracted from literature, where an active TB generates on average 5 new infected cases per year. Color scheme corresponds to susceptible (grey), latent, active TB with the sensitive strain, and active TB with the resistant strain (dark), and effectively treated (white).

Table 2. Fractions of the sub-populations for 25 years following the start of treatment at time $t = 0$ (corresponding to year 0). R: relapse; No-R: no relapse.

	% of the population in different states											
	year 0		year 5		year 10		year 15		year 20		year 25	
States	R	No-R	R	No-R	R	No-R	R	No-R	R	No-R	R	No-R
Susceptible	78.50	78.50	79.26	78.45	79.73	78.63	78.93	78.62	78.61	79.00	78.93	79.55
Latent	19.67	19.67	17.85	16.69	17.41	15.16	18.00	14.07	17.88	13.00	17.98	12.00
Active TB (sensitive)	1.83	1.83	2.62	2.21	2.65	2.05	2.75	1.78	3.15	1.85	2.70	1.61
Active TB (resistant)	0.00	0.00	0.14	0.02	0.12	0.06	0.14	0.02	0.22	0.03	0.29	0.01
Effectively treated	0.00	0.00	0.13	2.63	0.09	4.10	0.18	5.51	0.14	5.51	0.10	6.83

An advantage of this general framework compared with existing dynamic models rests on its capability to explicitly evaluate the impact of various factors and mechanisms of infection spread on TB dynamics. One can, for instance, disregard the effect of relapse and test various scenarios where all other characteristics are present. In this case, comparison with scenarios where relapse is present could be used to quantify the effect of this mechanism on disease prevalence and measure the effectiveness of treatment.

Despite the fidelity of our ABM framework, there remain several limitations which warrants further investigation. These include the static nature of lattice, timing and number of interactions between agents that may vary with location, type of activities, age, individuals' characteristics, and population demographics.[26] The inclusion of these factors will make the model more realistic, towards addressing the evolution of contact networks, but also more complex and may therefore demand more advanced computational techniques to maintain its efficiency and adaptability for implementation.

Acknowledgments

The authors would like to acknowledge the support of Mitacs and NSERC - Canada for this collaborative work. A.L.E. would like to thank the Agent-Based Modelling Laboratory at York University (Canada) for providing visiting fellowships which made this collaboration possible. A.S.M. would like to acknowledge the support of the Brazilian agency CNPq through the National Institute of Science and Technology of Complex Systems (Grant # 305738/2010-0). A.S.M would also like to acknowledge the support from the Brazilian agency CAPES.

References

1. C. D. Deangelis and A. Flanigin. Tuberculosis - a global problem requiring a global solution. *JAMA*, 293:2793–2794, 2005.
2. World Health Organization. Tuberculosis fact sheet nr. 104, November 2010.
3. S. M. Blower *et al.* The intrinsic transmission dynamics of tuberculosis epidemics. *Nature Medicine*, 8(1):815–821, 1995.
4. P. M. Small S. M.Blower and P. Hopewell. Control strategies for tuberculosis epidemics: new models for old problems. *Science*, 273:497–500, 1996.
5. Aquino L de Espíndola, Chris T. Bauch, Brenno Caetano Troca Cabella, and Alexandre Souto Martinez. An agent-based computational model of the spread of tuberculosis. *J. Stat. Mech.*, P05003, 2011.

6. Aquino L de Espíndola, Daniel Girardi, Thadeu Josino P. Penna, Chris T. Bauch, Alexandre Souto Martinez, and Brenno Caetano Troca Cabella. Exploration of the parameter space in an agent-based model of tuberculosis spread: Emergence of drug resistance in developing vs developed countries. *Int. J. Mod. Phys. C*, 23(6), 2012.
7. Luiz C. Mostaco-Guidolin *et al.* A software development framework for agent-based infectious disease modelling. *In: Biomedical Engineering, Trends, Researches and Technologies, In-Tech*, pages 641–664, 2011.
8. M. C. Gonzàlez, C. A. Hidalgo, and A. L. Barabási. Understanding individual human mobility patterns. *Nature*, 453(7196):779–82, Jun 2008.
9. N. M. Ferguson *et al.* Strategies for containing an emerging influenza pandemic in southeast asia. *Nature*, (437):209–214, 2005.
10. (U.S.) Committee on the Role of information Technology in Responding to Terrorism. Information technology for counterterrorism: Immediate actions and future possibilities. *National Research Council*, 2003.
11. D. S. Burke *et al.* Individual-based computational modeling of smallpox epidemic control strategies. *Academic Emergency Medicine*, 13:1142–1149, 2006.
12. Y. Meng, R. Davies, K. Hardy, and P. Hawkey. An application of agent-based simulation to the management of hospital-acquired infection. *Journal of Simulation*, 4(1):60–67, 2010.
13. N. Halder, J. K. Kelso, and G.J. Milne. Analysis of the effectiveness of interventions used during the 2009 a/h1n1 influenza pandemic. *BMC public health*, 10:168, 2010.
14. R. Paranjape and A. Sadanand. *Multi-Agent Systems for Healthcare Simulation and Modeling : Applications for System Improvement.* Hershey, New York: Medical Information Science Reference, IGI Global, 2010.
15. C. Dye and M. A. Espinal. Will tuberculosis become resistant to all antibiotics? *Proc. Biol. Sci.*, 268(1462):45–52, 2001.
16. C. Ozcaglar, A. Shabbeer, S. Vandenberg, B. Yener, and K. P. Bennett. Epidemiological models of mycobacterium tuberculosis complex infections. *Math Biosci.*, 236(2):77–96, April 2012.
17. S. M. Moghadas and M. E. Alexander. Exogenous reinfection and resurgence of tuberculosis: A theoretical framework. *Journal of Biological Systems*, 12(2):231–247, 2004.
18. J. A. Caminero *et al.* Exogenous reinfection with tuberculosis on an european island with moderate incidence of disease. *Am. J. Respir. Crit. Care Med.*, 163(17):720, 2001.
19. A. van Rie *et al.* Exogenous reinfection as a cause of recurrent tuberculosis after curative treatment. *N. Engl. J. Med.*, 341:1174–1179, 1999.
20. I. Sutherland, E. SvandovaLa, and S.E. Radhakrishna. The development of clinical tuberculosis following reinfection with tubercle bacilli. *Tubercle*, 62:255–268, 1982.
21. E. Vynnycky and P. E. M. Fine. The natural history of tuberculosis: The implications of age-dependent risks of disease and the role of reinfection. *Epidemiol. Infec.*, 119:183–201, 1997.

22. E. Vynnycky and P. E. M. Fine. The long-term dynamics of tuberculosis and other diseases with long serial intervals: Implications of and for changing reproduction numbers. *Epidemiol. Infec.*, 121:309–324, 1998.
23. C. Ford *et al.* Mycobacterium tuberculosis mutation rate estimates from different lineages predict substantial differences in the emergence of drug-resistant tuberculosis. *Nat. Genet.*, 45(7):784–90, 2013.
24. A. Pablo-Mendez *et al.* Global surveillance for antituberculosis-drug resistance, 1994-1997. *N. Engl. J. Med.*, 338(Chapter 33):1641–1649, 1998.
25. J. Alavez-Ramirez et al. Within-host population dynamics of antibiotic-resistant *M. Tuberculosis. Mathematical Medicine and Biology*, 24:35–56, 2006.
26. E. Pienaar *et al.* A Model of Tuberculosis Transmission and Intervention Strategies in an Urban Residential Area. *Comput Biol Chem.*, 34(2): 86–96, 2010.

SOME EXTENSIONS OF THE CLASSICAL EPIDEMIC MODELS*

FRED BRAUER
Department of Mathematics
University of British Columbia
Vancouver, BC V6T 1Z2, Canada
E-mail: brauer@math.ubc.ca

We consider two different extensions of the classical SIR epidemic model. The first (work joint with Z. Shuai and P. van den Driessche) is a model with two modes of transmission, direct and indirect, as in cholera. The second (joint work with S. Moghadas and Y. Xiao) is a model that includes treatment and the development of drug resistance. In each case we try to obtain a final size relation and to draw some epidemiological conclusions.

1. The Classical Kermack-McKendrick Epidemic Model

One of the early triumphs of mathematical epidemiology was the formulation of a simple model by Kermack and McKendrick in 1927 [29] whose predictions are very similar to the behavior, observed in countless epidemics, of diseases that invade a population suddenly, grow in intensity, and then disappear leaving part of the population untouched. The Kermack-McKendrick model is a compartmental model based on relatively simple assumptions on the rates of flow between different classes of members of the population. The SARS epidemic of 2002-3 revived interest in epidemic models, which had been neglected since the time of Kermack and McKendrick in favor of models for endemic diseases. More recently, the threat of spread of avian flu raised in 2005 and the H1N1 influenza A pandemic of 2009 have provided a continuing source of important modeling questions about epidemics. The first section of this paper is a summary of results for the Kermack-McKendrick epidemic model covered in more detail in [9]. The second and third sections describe extensions to two different situations.

*This work is supported by the Natural Sciences and Engineering Research Council and M-prime

1.1. *The simple Kermack-McKendrick epidemic model*

In the simplest form of the Kermack-McKendrick model, the population being studied is divided into three classes labeled S, I, and R. Individuals are either susceptible (S), that is, not (yet) infected at time t, or infectious (I), that is, infected and able to spread the disease by contact with susceptibles, or removed (R), that is, infected and then removed from the possibility of being infected again or of spreading infection. Removal is carried out either through isolation from the rest of the population, or through immunization against infection, or through recovery from the disease with full immunity against reinfection, or through death caused by the disease. We will use the terminology SIR to describe a disease which confers immunity against re-infection, to indicate that the passage of individuals is from the susceptible class S to the infective class I to the removed class R. Epidemics are usually diseases of this type, and in this paper we will consider only models in which infectives recover with full immunity against reinfection.

In the simple Kermack-McKendrick epidemic model we assume that an average individual makes a contacts sufficient to transmit infection in unit time and that there are no disease deaths, so that the total population size is a constant N. We assume also that on average the fraction of infected individuals who remain infective for at least a time τ is $e^{-\gamma\tau}$, so that the mean infective period is $1/\gamma$ and the rate of recovery at time t is $\gamma I(t)$. These asumptions lead to the simple Kermack-McKendrick model

$$\begin{aligned} S' &= -\frac{a}{N}SI \\ I' &= \frac{a}{N}SI - \gamma I, \end{aligned} \tag{1}$$

together with initial conditions

$$S(0) = S_0, \quad I(0) = I_0, \quad S_0 + I_0 = N.$$

The number of recovered members R in the population is governed by the equation $R' = \alpha I$, and is determined once S and I are known.

1.2. *The age of infection epidemic model*

The epidemic model described by Kermack and McKendrick in 1927 was much more general than (1). It included a dependence of infectivity on the time since becoming infected (age of infection). This allows inclusion of more complicated compartmental structure [19], [26], [27]. We let $\varphi(t)$ be the total infectivity at time t, defined as the sum of products of the number of infected members with each infection age and the mean infectivity for that infection age. We let $P(\tau)$ be the number of individuals who are still infected at infection age τ, and we let $\pi(\tau)$ with $0 \leq \pi(\tau) \leq 1$ be the mean infectivity at infection age τ. Then

$$A(\tau) = \pi(\tau)P(\tau),$$

is the mean infectivity of members of the population with infection age τ. We continue to assume that an average individual makes a contacts sufficient to transmit infection in unit time and that there are no disease deaths, so that the total population size is a constant N.

The age of infection epidemic model is

$$\begin{aligned} S' &= -a\frac{S}{N}\varphi \\ \varphi(t) &= \varphi_0(t) + \int_0^t a\frac{S(t-\tau)}{N}\varphi(t-\tau)A(\tau)d\tau \\ &= \varphi_0(t) + \int_0^t [-S'(t-\tau)]A(\tau)d\tau. \end{aligned} \tag{2}$$

For the model (2), it is known [7], [32] that

- $I(t) \to 0$ and $S(t) \to S_\infty > 0$ as $t \to \infty$.
- The basic reproduction number, defined as the number of secondary infections caused by introducing a single infective individual into an entirely susceptible population, is

$$\mathcal{R}_0 = a\int_0^\infty A(\tau)d\tau.$$

- There is a *final size relation*

$$\log\frac{S_0}{S_\infty} = \mathcal{R}_0\left[1 - \frac{S_\infty}{N}\right]. \tag{3}$$

In the special case (1) of (2)

$$\mathcal{R}_0 = \frac{a}{\gamma}.$$

The final size relation gives a relation between the basic reproduction number and the size of the epidemic. The final size of the epidemic, the number of members of the population who are infected over the course of the epidemic, is $N - S_\infty$. This is often described in terms of the *attack rate* $(1 - S_\infty/N)$. [Technically, the attack rate should be called an attack ratio, since it is dimensionless and is not a rate]. For the model (1), if

$S_0 \approx N$, it is easy to see that if $\mathcal{R}_0 > 1$, the number of infectives increases initially and we have an epidemic, while if $\mathcal{R}_0 < 1$, the number of infectives decreases from the start, and there is no epidemic. The same is true for the age of infection model (2) but the verification is more complicated. The basic reproduction number is a *threshold* quantity, distinguishing between an epidemic and a disease outbreak that does not spread. For models with complicated compartmental structure, there are methods to aid in the computation of the basic reproductive number [11], [45].

The final size relation assumes that all individuals who are infected at time $t = 0$ have infection age zero at time $t = 0$. If there are initial infectives with infection age greater than zero, the final size relation has the form

$$\log \frac{S_0}{S_\infty} = \mathcal{R}_0 \left(1 - \frac{S_\infty}{N}\right) - \Gamma$$

with

$$0 \le \Gamma \le \int_0^\infty (N - S_0)A(t)dt.$$

It is not difficult to prove [8] that there is a unique solution S_∞ of the final size relation (3), and that this solution satisfies the bound

$$S_\infty < \frac{S_0}{\mathcal{R}_0}.$$

Differentiation of (3) with repect to $\mathcal{R}_0$ gives

$$\left(\frac{\mathcal{R}_0}{N} - \frac{1}{S_\infty}\right) \frac{dS_\infty}{d\mathcal{R}_0} = 1 - \frac{S_\infty}{N} \ge 0.$$

Since $S_\infty < S_0/\mathcal{R}_0 \le N/\mathcal{R}_0$, it follows that

$$\frac{dS_\infty}{d\mathcal{R}_0} < 0.$$

Thus increasing the basic reproduction number decreases S_∞, which is equivalent to increasing the size of the epidemic.

1.3. *Models with disease deaths*

If there are disease deaths, so that the total population size decreases during the epidemic, it is necessary to make some changes in the model (1). The assumption in the model (1) of a constant rate of contacts per infective, called *standard incidence* is quite unrealistic, except possibly in the early stages of an epidemic in a population of moderate size. It is more realistic to assume a contact rate that depends on the total population size. If there are no disease deaths, so that the total population size remains constant, such a distinction is unnecessary. If there are disease deaths, so that the total population size decreases during the epidemic, we generalize the model (1) by assuming that an average member of the population makes $a(N)$ contacts in unit time. It is reasonable to assume $a'(N) \geq 0$. Also, we define

$$\beta(N) = \frac{a(N)}{N}.$$

and assume $\beta'(N) \leq 0$ to express the idea of saturation in the number of contacts.

If the total population size is not constant because of disease deaths, since the contact rate depends on total population size we must include an equation for total population size in the model. This forces us to distinguish between members of the population who die of the disease and members of the population who recover with immunity against reinfection. We assume that a fraction f of the members leaving the infective class at time t recover and the remaining fraction $(1-f)$ die of disease. For the simple Kermack-McKendrick model (1) we use S, I, and N as variables, with $N = S+I+R$. We now obtain a three-dimensional model

$$\begin{aligned} S' &= -a(N)\frac{S}{N}I \\ I' &= a(N)\frac{S}{N}I - \gamma I \\ N' &= -(1-f)\gamma I. \end{aligned} \tag{4}$$

Since N is now a decreasing function, we define $N(0) = N_0 = S_0 + I_0$. Then $\mathcal{R}_0 = a(N_0)/\gamma$ [17]. In deriving a final size relation, we obtain

$$\log\frac{S_0}{S_\infty} = \int_0^\infty \frac{a(N(t)}{N(t)}I(t)dt.$$

Since

$$\frac{a(N(t))}{N(t)} \geq \frac{a(N_0)}{N_0},$$

the final size relation becomes an inequality

$$\log\frac{S_0}{S_\infty} \geq \mathcal{R}_0\left[1 - \frac{S_\infty}{N}\right]$$

A similar extension of the age of infection model (2) may be made.

In the remainder of this paper it is assumed that there are no disease deaths, so that the final size relation is an equality and may be used to determine the extent of an epidemic. If there are disease deaths, it is necessary to integrate the model equations numerically to determine the extent, but if the disease death rate is small it is possible to show that the final size relation is an approximate equality.

In [9], we described some extensions of the Kermack-McKendrick model to situations with hetreogeneous mixing. In this paper, we will study two other kinds of extensions, namely to models for diseases in which there are two separate modes of transmission, and to models for diseases in which there is a possibility that treatment, such as with antiviral drugs, may lead to the development of drug resistance.

Some additional references for epidemic models and reproduction numbers are [3, 4, 10, 13, 15, 16, 24, 25, 23, 28, 33, 38, 39, 42].

2. Epidemic Models with Direct and Indirect Transmission

Some diseases may be spread in more than one manner. For example, cholera may be spread by person to person transmission but may also be transmitted indirectly through a pathogen released by infectives through a medium such as contaminated water [40, 41]. The work described in this section is joint with Z. Shuai and P. van den Driessche, and is contained in [12]. Other sources of information about cholera include [1, 5, 14, 22, 30, 40].

There is a theory of epidemic models for such diseases parallel to the theory of the preceding section. We begin with a simple model analogous to the simple SIR model (1).

2.1. *A simple indirect transmission epidemic model*

Consider an epidemic model with direct (person to person) and indirect (through a medium such as contaminated water) transmission. To a simple SIR model we add a pathogen B shed by infectives. We assume that the infectivity of the pathogen is proportional to its concentration, suggesting mass action transmission. The resulting model is

$$\begin{aligned} S' &= -a\frac{S}{N}I - \beta SB \\ I' &= a\frac{S}{N}I + \beta SB - \gamma I \\ R' &= \gamma I \\ B' &= rI - \delta B, \end{aligned} \tag{5}$$

with initial conditions

$$S(0) = S_0, \quad I(0) = I_0, \quad B(0) = B_0,$$

in a population of constant total size $N = S_0 + I_0$, with $R(0) = 0$. In general, $N = S + I + R$. In this model r represents the rate at which an infectious individual sheds pathogen and δ represents the rate at which the pathogen loses infectivity.

From the sum of the equations for S and I in (5) we see that

$$(S + I)' = -\gamma I.$$

This shows that $(S + I)$ decreases to a limit, and since $(S + I)$ is a smooth function it is possible to show that its derivative approaches zero, from which we may deduce that

$$I_\infty = \lim_{t \to \infty} I(t) = 0.$$

Also, integration of this equation gives

$$\gamma \int_0^\infty I(t)dt = N - \lim_{t \to \infty} S(t) = N - S_\infty. \tag{6}$$

This implies $\int_0^\infty I(t)dt < \infty$.

The next generation matrix approach of [43], viewing I, B as disease compartments, gives the basic reproduction number

$$\mathcal{R}_0 = \frac{a}{\gamma} + \frac{r\beta N}{\gamma\delta}.$$

In this expression, the first term represents secondary infections caused directly by a single infective introduced into a wholly susceptible population, infecting a susceptibles in unit time for a time period $1/\gamma$. The second term represents secondary infections caused indirectly through the pathogen since a single infective sheds a quantity r of pathogen in unit time for a time period $1/\gamma$ and this pathogen infects βN susceptibles in unit time for a time period $1/\delta$.

Integration of the equation for S in (5) gives

$$\log \frac{S_0}{S_\infty} = \frac{a}{N} \int_0^\infty I(t)dt + \beta \int_0^\infty B(t)dt. \tag{7}$$

Integration of the linear equation for B in (5) gives

$$B(t) = B_0 e^{-\delta t} + r \int_0^t e^{-\delta(t-s)} I(s)ds. \tag{8}$$

Next, we must show that

$$\lim_{t\to\infty}\int_0^t e^{-\delta(t-s)}I(s)ds = \lim_{t\to\infty}\frac{\int_0^t e^{\delta s}I(s)ds}{e^{\delta t}} = 0.$$

If the integral in the numerator of this expression is bounded, this is clear, and if the integral is unbounded, l'Hôpital's rule shows that the limit is $\lim_{t\to\infty} I(t)/\delta = 0$. Then (8) implies that

$$B_\infty = \lim_{t\to\infty} B(t) = 0.$$

But integration of (8) and interchange of the order of integration gives

$$\int_0^\infty B(t)dt = \frac{B_0 + r\int_0^\infty I(t)dt}{\delta}. \tag{9}$$

This implies $\int_0^\infty B(t)dt < \infty$.

Substitution of (9) into (7) gives

$$\log\frac{S_0}{S_\infty} = \left[\frac{a}{N} + \beta\frac{r}{\delta}\right]\int_0^\infty I(t)dt + \beta\frac{B_0}{\delta},$$

and now substitution of (6) gives the final size relation

$$\begin{aligned}\log\frac{S_0}{S_\infty} &= \left(\frac{a}{N} + \frac{\beta N r}{\gamma\delta}\right)\left[1 - \frac{S_\infty}{N}\right] + \beta\frac{B_0}{\delta}\\ &= \mathcal{R}_0\left[1 - \frac{S_\infty}{N}\right] + \beta\frac{B_0}{\delta}.\end{aligned} \tag{10}$$

This implies $S_\infty > 0$. The result here is a special case of the equation on the last line on p. 123 of [40].

2.2. *An age of infection model*

In order to cover such generalizations of the model (5) as multiple infective stages and arbitrary distributions of stay in a stage, we give an age of infection model

$$\begin{aligned} S'(t) &= -S(t)\left[\frac{a}{N}\varphi(t) + \beta B(t)\right]\\ \varphi(t) &= \varphi_0(t) + \int_0^t [-S'(t-\tau)]P(\tau)d\tau \\ B(t) &= B_0(t) + \int_0^t r\varphi(t-\tau)Q(\tau)d\tau. \end{aligned} \tag{11}$$

In this model, $\varphi(t)$ represents the total infectivity of individuals with age of infection t, $\varphi_0(t)$ represents the total infectivity at time t of individuals who were already infected at time $t = 0$, $B_0(t)$ represents the pathogen concentration at time t remaining from pathogen concentration that was already present at time $t = 0$, $P(\tau)$ represents the mean infectivity of individuals at age of infection τ, normally the product of the fraction of infectives still infective at age of infection τ and the relative infectivity at that infection age, and $Q(\tau)$ represents the fraction of pathogen remaining τ time units after having been shed by an infective. The function Q is monotone non-increasing with $Q(0) = 1$, $\int_0^\infty Q(\tau)d\tau < \infty$. Since infectivity of an individual may depend on the age of infection of the individual, the function P is not necessarily non-increasing, but we assume $\int_0^\infty P(\tau)d\tau < \infty$.

For the model (11), the basic reproduction number is

$$\mathcal{R}_0 = a\int_0^\infty P(\tau)d\tau + r\beta N\int_0^\infty P(\tau)d\tau\int_0^\infty Q(\tau)d\tau.$$

In this expression, the first term represents new infection transmitted directly by a single infectious individual inserted into a totally susceptible population, while the second term represents secondary infections caused by this individual indirectly through shedding of pathogen.

Integration of the equation for S in (11) gives

$$\log\frac{S_0}{S_\infty} = \frac{a}{N}\int_0^\infty \varphi(\tau)d\tau + \beta\int_0^\infty B(\tau)d\tau. \tag{12}$$

Routine calculations involving interchange of the order of integration give

$$\begin{aligned}\int_0^\infty \varphi(\tau)d\tau &= \int_0^\infty \varphi_0(\tau)d\tau + [S_0 - S_\infty]\int_0^\infty P(\tau)d\tau\\ \int_0^\infty B(\tau)d\tau &= \int_0^\infty B_0(\tau)d\tau + r\int_0^\infty \varphi(\tau)d\tau\int_0^\infty Q(\tau)d\tau\\ &= \int_0^\infty B_0(\tau)d\tau + r\int_0^\infty Q(\tau)d\tau\int_0^\infty \varphi_0(\tau)d\tau\\ &\quad + r[S_0 - S_\infty]\int_0^\infty P(\tau)d\tau\int_0^\infty Q(\tau)d\tau.\end{aligned}$$

Then substitution into (12) gives

$$\begin{aligned}\log\frac{S_0}{S_\infty} &= \mathcal{R}_0\left[\frac{S_0 - S_\infty}{N}\right] + \frac{a}{N}\int_0^\infty \varphi_0(t)dt\\ &\quad + r\beta\int_0^\infty Q(t)dt\int_0^\infty \varphi_0(t)dt + \beta\int_0^\infty B_0(t)dt.\end{aligned} \tag{13}$$

If all infections at time zero have infection age zero, then

$$\varphi_0(t) = [N - S_0]P(t), \quad \int_0^\infty \varphi_0(t)dt = [N - S_0]\int_0^\infty P(t)dt,$$

and if the entire pathogen concentration at time zero has infection age zero, then

$$B_0(t) = B_0 Q(t), \quad \int_0^\infty B_0(t)dt = B_0 \int_0^\infty Q(t)dt$$

with some constant B_0. In this case, the final size relation (13) takes the form

$$\log \frac{S_0}{S_\infty} = \mathcal{R}_0 \left[1 - \frac{S_\infty}{N}\right] + \beta B_0 \int_0^\infty Q(t)dt. \tag{14}$$

The final size relation has a term arising from an initial pathogen concentration that tends to decrease S_∞.

In general, because Q is monotone non-increasing,

$$\int_0^\infty B_0(t)dt \le B_0 \int_0^\infty Q(t)dt.$$

If P is monotone non-increasing,

$$\int_0^\infty \varphi_0(t)dt \le [N - S_0]\int_0^\infty P(t)dt.$$

If P is not monotone, which may occur, for example, if there is an exposed stage followed by an infectious stage with higher infectivity, this is not necessarily true. However, if there are no infectives initially, so that the epidemic is started by the pathogen, then $\varphi_0(t) = 0$ and $S_0 = N$. Then (13) remains valid without the need to assume that P is monotone.

The results in this section have been established only for a constant rate of pathogen shedding. If the rate of pathogen depends on the age of infection, the equation for B in the model (11) should be replaced by an equation

$$B(t) = B_0(t) + \int_0^t r(t-\tau)\varphi(t-\tau)Q(\tau)d\tau.$$

It is not possible to treat the corresponding model as an age of infection model, but we can view it as a staged progression model.

2.3. *A staged progression epidemic model*

The age of infection epidemic model (11) is very general, including models with multiple infective stages and treatment. In addition, it allows arbitrary distributions of stay in compartmnets. The drawback of the age of infection model is that it may be difficult to calculate the function $P(\tau)$. Some aids for this computation can be found in [11], [32], [45]. The staged progression epidemic model is a fairly general special case of the age of infection model, allowing multiple stages but allowing direct calculation of the function $A(\tau)$. Staged progression models for diseases with direct transmission only have been analyzed in [11], [26], [27].

We consider an epidemic with progression from S through k infected stages $I_1, I_2, \cdots, I_k$, as analyzed in [11], but with the addition of a pathogen. We assume that in stage j the relative infectivity is ε_j, the distribution of stay in the stage is given by P_j with $P_j(0) = 1, \int_0^\infty P_j(t)dt < \infty$, and P_j monotone non-increasing, so that the infectivity of an individual in stage j is $A_j(\tau) = \varepsilon_j P_j(\tau)$. There are no disease deaths and the total population size N is constant. We assume initial conditions

$$S(0) = 0, \quad I_1(0) = I_0, \quad I_2(0) = I_3(0) = \cdots = I_k(0) = 0, \quad R(0) = 0.$$

The total infectivity is given by

$$\varphi(t) = \sum_{j=1}^{k} \varepsilon_j I_j(t).$$

We let $B_j(t)$ be the quantity of pathogen shed by infectives in the stage I_j and let Q_j denote the distribution of stay of pathogen shed by infectives in this stage, with $Q_j(0) = 1, \int_0^\infty Q_j(t)dt < \infty$, and Q_j monotone non-increasing. We let r_j be the shedding rate in this stage. We also define the total quantity of pathogen,

$$B(t) = \sum_{j=1}^{k} B_j(t).$$

Then

$$B_j(t) = B_j^0(t) + \int_0^t r_j I_j(t-\tau) Q_j(\tau) d\tau. \tag{15}$$

A single infective introduced into a wholly susceptible population while in infection stage j causes a secondary infections in unit time directly for a period of $\int_0^\infty P_j(t)dt$. In addition, this individual sheds a quantity r_j of pathogen in unit time for a time period $\int_0^\infty P(t)dt$, and this pathogen causes βN infections in unit time for a time period $\int_0^\infty Q(t)dt$. This shows that the basic reproduction number is

$$\mathcal{R}_0 = a\sum_{j=1}^{k}\varepsilon_j\int_0^\infty P_j(t)dt + \beta N\sum_{j=1}^{k} r_j\int_0^\infty P_j(t)dt\int_0^\infty Q_j(t)dt.$$

We assume that all initial infectives are in the first stage with infection age zero at $t = 0$. Then the equation for I_1 in the model is

$$I_1(t) = I_1^0(t) + \int_0^t [-S'(t-\tau)]P_1(\tau)d\tau,$$

with $I_1^0(t) = I_0P_1(t)$. Then

$$\begin{aligned}\int_0^\infty I_1(t)dt &= I_0\int_0^\infty P_1(t)dt + \int_0^\infty\int_0^t [-S'(s-\tau)]P_1(\tau)d\tau dt\\ &= I_0\int_0^\infty P_1(t)dt + \int_0^\infty\left[\int_\tau^\infty [-S'(t-\tau)]dt\right]P_1(\tau)d\tau\\ &= I_0\int_0^\infty P_1(t)dt + [S_0 - S_\infty]\int_0^\infty P_1(t)dt = [N - S_\infty]\int_0^\infty P_1(t)dt.\end{aligned}$$

The same calculation made in [11] now gives

$$\int_0^\infty I_j(t)dt = [N - S_\infty]\int_0^\infty P_j(t)dt, \quad j = 1, 2, \cdots, k. \tag{16}$$

Integration of (15) using (16) gives

$$\begin{aligned}\int_0^\infty B_j(\tau)d\tau &= \int_0^\infty B_j^0(\tau)d\tau + r_j\int_0^\infty I_j(\tau)d\tau\int_0^\infty Q_j(\tau)d\tau\\ &= \int_0^\infty B_j^0(\tau)d\tau + r_j[N - S_\infty]\int_0^\infty P_j(\tau)d\tau\int_0^\infty Q_j(\tau)d\tau.\end{aligned}$$

Just as for the model (11) of the previous subsection,

$$S'(t) = -S(t)[\frac{a}{N}S(t)\varphi(t) + \beta B(t)] = -S(t)\left[\frac{a}{N}\sum_{j=1}^{k}\varepsilon_j I_j(t) - \beta\sum_{j=1}^{k}B_j(t)\right],$$

and integration gives

$$\log\frac{S_0}{S_\infty} = \frac{a}{N}\sum_{j=1}^{k}\varepsilon_j\int_0^\infty I_j(t)dt + \beta\sum_{j=1}^{k}\int_0^\infty B_j(t)]dt. \tag{17}$$

For simplicity, we assume that all individuals infected at time zero have infection age zero for $t = 0$, and also that there is a new quantity of pathogen

B_0 introduced at time zero, so that $B_0(t) = B_0 Q(t)$. Then, with the aid of (17) and (16) the relation (17) reduces to the final size relation

$$\begin{aligned}
\log \frac{S_0}{S_\infty} &= \frac{a}{N} \sum_{j=1}^{k} \varepsilon_j \int_0^\infty P_j(t)[N - S_\infty] \\
&\quad + \beta \sum_{j=1}^{k} r_j \int_0^\infty P_j(t)dt \int_0^\infty Q_j(t)dt[N - S_\infty] \\
&\quad + \beta B_0 \int_0^\infty Q(t)dt \qquad (18) \\
&= \mathcal{R}_0 \left[1 - \frac{S_\infty}{N}\right] + \beta B_0 \int_0^\infty Q(t)dt.
\end{aligned}$$

This is also equivalent to an expression in [40].

3. An Epidemic Model with Drug Resistance

Treatment of a virus with antiviral drugs raises a possibility of development of a drug-resistant strain of the virus. There is some experimental evidence that there are situations in which the treatment may cause development of more drug-resistant cases than it cures, so that treatment may become counter-productive. We formulate a model for an epidemic in which there is a drug-sensitive strain but treatment may cause some treated individuals to become drug-resistant. We divide a homogeneously mixing population into members who are susceptible (S), infected with the drug-sensitive strain (I_S), infected with the drug-sensitive strain under treatment (I_T), infected with the drug-resistant strain (I_R), and recovered individuals (R). Since treatment is ineffective against the drug-resistant infection, we do not distinguish between treated and untreated individuals in I_R. The work described in this section is joint with S. Moghadas and Y. Xiao, and is contained in [44]. References to the question of development of drug resistance include [2], [6], [21], [31], [35], [36], [37].

3.1. *A basic treatment model*

Many infectious pathogens can evolve and generate drug-resistant strains, generally less fit than the sensitive strains of the infectious pathogen. In the absence of treatment, resistant strains may be competitively disadvantaged compared to the sensitive strains, and may go extinct but treatment prevents the growth and spread of sensitive strains, and therefore induces a selective pressure that favors the resistant strain to replicate and restore its fitness to a level suitable for successful transmission. We assume that:

- the total population size is a constant N and there is a mean rate a of contacts sufficient to transmit infection.
- the population is divided into susceptibles (S), members with sensitive infections but not treated (I_S), members with sensitive infections being treated (I_T), members with resistant infections (I_R), and recovered members (R).
- the infectivity reduction factors are δ_T in I_T and δ_R in I_R.We assume $\delta_R < 1$.
- the recovery rates are $\gamma_S, \gamma_T, \gamma_R$ in I_S, I_T, I_R respectively.
- resistance develops as a result of treatment, and can then be transmitted.
- treatment is effective only against sensitive infections.
- the rate of developing resistance increases with time and the probability of developing resistance at time τ following the initiation of treatment is $(1 - e^{-\kappa\tau})$.

The resulting model is

$$\begin{aligned} S' &= -\frac{a}{N} S(I_S + \delta_T I_T + \delta_R I_R) \\ I_S' &= \frac{a}{N} S(I_S + \delta_T I_T) - (\gamma_S + \eta) I_S \\ I_T' &= \eta I_s - (\gamma_T + \kappa) I_T \\ I_R' &= \delta_R \frac{a}{N} S I_R + \kappa I_T - \gamma_R I_R, \end{aligned} \tag{19}$$

with initial conditions $S(0) = S_0$, $I_S(0) = I_0$, $I_T(0) = 0, I_R(0) \geq 0 = 0$, with $S_0 + I_0 + I_R(0) = N$.

We assume $I_T(0) = 0$, that is, that time is measured from the start of treatment. Since resistance generally emerges with compromised transmission fitness [18], we assume a lower transmissibility of the resistant infection compared to that of the sensitive infection without treatment ($\delta_R < 1$). We assume that treatment reduces the infectiousness, and therefore transmissibility, of the drug-sensitive infection ($\delta_T < 1$) [20]. Treatment may also shorten the infectious period ($\gamma_S \leq \gamma_T$) [36].

3.2. *Analysis of the model*

Using the next generation matrix method, as in [43], we calculate

$$\mathcal{R}_0 = \max[\mathcal{R}_S, \mathcal{R}_R],$$

where

$$\mathcal{R}_S = \frac{a}{\gamma_S + \eta} + \frac{\eta}{\eta + \gamma_S} \cdot \frac{\delta_T a}{\kappa + \gamma_T}, \quad \mathcal{R}_R = \frac{\delta_R a}{\gamma_R}.$$

Here, $\mathcal{R}_S$ is the number of secondary sensitive infections caused by a single sensitive infection in a wholly susceptible population, and $\mathcal{R}_R$ is the number of resistant infections caused by a single resistant infective in a wholly susceptible population. This calculation does not cover the

$$\frac{\eta}{\eta + \gamma_S} \cdot \frac{\kappa}{\kappa + \gamma_T} \cdot \frac{a\delta_R}{\gamma_R}$$

resistant infections that develop from treated sensitive infections when a sensitive infection is introduced into a wholly susceptible population. It is straightforward to show that there is an epidemic if and only if $\mathcal{R}_0 > 1$.

In the calculation of the final size relations for the model (19), we will use the notation $\hat{f}$ to denote $\int_0^\infty f(t)dt$ for an arbitrary non-negative integrable function. The sum of the equations of the model (19) is

$$(S + I_S + I_T + I_R)' = -(\gamma_S I_S + \gamma_T I_T + \gamma_R I_R) < 0.$$

Hence, $(S + I_S + I_T + I_R)$ is a decreasing function bounded below by zero, and therefore approaches a limit as $t \to \infty$. It is easy to verify that the system (19) is well-posed [10], and therefore S approaches a non-negative limit S_∞ as $t \to \infty$. Integration of this equation gives

$$N - S_\infty = \gamma_S \hat{I}_S + \gamma_T \hat{I}_T + \gamma_R \hat{I}_R. \tag{20}$$

Integration of the equation for I_T in (19) from 0 to ∞ gives, since by assumption $I_T(0) = 0$,

$$\eta \hat{I}_S = (\kappa + \gamma_T)\hat{I}_T, \tag{21}$$

and substituting this into (20) to eliminate $\hat{I}_T$, we obtain

$$N - S_\infty = \frac{\gamma_S\gamma_T + \gamma_S\kappa + \eta\gamma_T}{\kappa + \gamma_T}\hat{I}_S + \gamma_R \hat{I}_R. \tag{22}$$

Integration of the equation for S in (19) gives

$$\log \frac{S_0}{S_\infty} = \frac{a}{N}\left[\hat{I}_S + \delta_T \hat{I}_T + \delta_R \hat{I}_R\right],$$

and using (21) we reduce this to

$$\log \frac{S_0}{S_\infty} = \frac{a}{N}\left[\frac{\kappa + \gamma_T + \delta_T\eta}{\kappa + \gamma_T}\hat{I}_S + \delta_R \hat{I}_R\right]. \tag{23}$$

We view (22) and (23) as a pair of linear algebraic equations for I_S, I_R,

$$\frac{\gamma_S\gamma_T + \kappa\gamma_S + \eta\gamma_T}{\kappa + \gamma_T}\hat{I}_S + \gamma_R\hat{I}_R = N - S_\infty \tag{24}$$
$$\frac{a}{N}\frac{\kappa + \gamma_T\delta_T\eta}{\kappa + \gamma_T}\hat{I}_S + \frac{a}{N}\delta_R\hat{I}_R = \log\frac{S_0}{S_\infty}.$$

The solution of (24) is

$$\beta\Delta\hat{I}_S = \frac{a}{N}\delta_R(N - S_\infty) - \gamma_R\log\frac{S_0}{S_\infty}, \tag{25}$$
$$\beta\Delta\hat{I}_R = \left(\gamma_s + \gamma_T\frac{\eta}{\kappa + \gamma}\right)\log\frac{S_0}{S_\infty} - \frac{a}{N}\left(1 + \frac{\delta_T\eta}{\kappa + \gamma_T}\right)(N - S_\infty),$$

where

$$\Delta = \delta_R\left(\gamma_S + \frac{\gamma_T\eta}{\kappa + \gamma_T}\right) - \gamma_R\left(1 + \frac{\delta_T\eta}{\kappa + \gamma_T}\right).$$

It is easy to see that $\Delta(\eta) < 0$ is equivalent to

$$\frac{\delta_R}{\gamma_R} < \frac{\kappa + \gamma_S + \eta\delta_T}{\kappa\gamma_S + \gamma_S\gamma_T + \eta\gamma_T},$$

and if we define

$$\mathcal{R}^*(\eta) = a\frac{\kappa + \gamma_S + \eta\delta_T}{\kappa\gamma_S + \gamma_S\gamma_T + \eta\gamma_T},$$

this is equivalent to $\mathcal{R}_R < \mathcal{R}^*(\eta)$. The quantity $\mathcal{R}^*(\eta)$ has an epidemiological interpretation. If we write

$$p\mathcal{R}^*(\eta) = q,$$

with

$$p = \frac{\kappa\gamma_S + \gamma_S\gamma_T + \eta\gamma_T}{(\eta + \gamma_S)(\kappa + \gamma_T)}, \quad q = a\frac{\kappa + \gamma_S + \eta\delta_T}{\eta + \gamma_S)(\kappa + \gamma_T)},$$

then p is the probability that an individual imfected with the sensitive strain will not develop resistance, and q is the total number of new cases generated before the initial sensitive infection develops resistance. Thus $\mathcal{R}^*(\eta)$ is a treatment mediated reproduction number.

Returning now to (25), if $\Delta < 0$, that is, if $\mathcal{R}_R < \mathcal{R}^*(\eta)$, since the right side of each of the equations in (25) must be negative, we see that

$$\mathcal{R}_R\left[1 - \frac{S_\infty}{N}\right] < \log\frac{S_0}{S_\infty} < \mathcal{R}^*(\eta)\left[1 - \frac{S_\infty}{N}\right].$$

If $\Delta > 0$, then both inequalities are reversed.

Under the reasonable assumptions $\delta_T \gamma_S \leq \gamma_T$, $\delta_T < 1$ it is easy to verify that $\mathcal{R}^*(\eta)$ is a decreasing function of η. Thus

$$\mathcal{R}^*(0) = a\frac{\kappa + \gamma_S}{\gamma_S(\kappa + \gamma_T)} \geq lim_{\eta \to \infty} |\mathcal{R}^*(\eta) = \mathcal{R}^*(\infty) = a\frac{\delta_T}{\gamma_T}.$$

We distinguish three cases, corresponding to the location of $\mathcal{R}_R$ with respect to $\mathcal{R}^*(0)$ and $\mathcal{R}^*(\infty)$.

(i) $\mathcal{R}_R > \mathcal{R}^*(0)$, if

$$\delta_R \gamma_S(\kappa + \gamma_T) > \gamma_R(\kappa + \gamma_S).$$

In this case $\mathcal{R}_R > \mathcal{R}^*(\eta)$ for $0 \leq \eta < \infty$.

$$\mathcal{R}^*(\eta)\left[1 - \frac{S_\infty}{N}\right] < \log\frac{S_0}{S_\infty} < \mathcal{R}_R\left[1 - \frac{S_\infty}{N}\right].$$

(ii) $\mathcal{R}_R < \mathcal{R}^*(\infty)$, if

$$\delta_R \gamma_T < \gamma_R \delta_T.$$

In this case $\mathcal{R}_R < \mathcal{R}^*(\eta)$ for $0 \leq \eta < \infty$.

$$\mathcal{R}_R\left[1 - \frac{S_\infty}{N}\right] < \log\frac{S_0}{S_\infty} < \mathcal{R}^*(\eta)\left[1 - \frac{S_\infty}{N}\right].$$

(iii) $\mathcal{R}^*(0) > \mathcal{R}_R > \mathcal{R}^*(\infty)$, if

$$\delta_R \gamma_S(\kappa + \gamma_T) < \gamma_R(\kappa + \gamma_S), \quad \delta_R \gamma_T > \delta_T \gamma_R.$$

In this case $\mathcal{R}^*(\eta) > \mathcal{R}_R$ for small η and $\mathcal{R}^*(\eta) < \mathcal{R}_R$ for large η. Specifically, there is a value η_c such that $\mathcal{R}^*(\eta_c) = \mathcal{R}_R$, with

$$\eta_c = \frac{\gamma_R(\kappa + \gamma_S) - \delta_R \gamma_S(\kappa + \gamma_T)}{\gamma_T \delta_R - \delta_T \gamma_R}.$$

$$\mathcal{R}_R\left[1 - \frac{S_\infty}{N}\right] < \log\frac{S_0}{S_\infty} < \mathcal{R}^*(\eta)\left[1 - \frac{S_\infty}{N}\right] \quad (0 \leq \eta < \eta_c)$$

$$\mathcal{R}^*(\eta)\left[1 - \frac{S_\infty}{N}\right] < \log\frac{S_0}{S_\infty} < \mathcal{R}_R\left[1 - \frac{S_\infty}{N}\right] \quad (\eta_c < \eta < \infty).$$

In (22) and (23) we have two expressions for $\log(S_0/S_\infty)$ and $N - S_\infty$. In the analysis of the Kermack-McKendrick model (1) the corresponding expressions contained only one term of the form $\hat{I}$, and it was possible to eliminate that term from the two equations. Here, we have two terms $\hat{I}_S$

and $\hat{I}_R$, and elimination is not possible without some additional information or assumption.

In order to establish an equality relating the treatment rate and the epidemic final size, we will need to make additional assumptions. We have written $N - S_\infty$, the total number of members who are infected over the course of the epidemic, as the sum of two terms, one of which, $\gamma \hat{I}_R$ is the total number of drug-resistant infections. The other term,

$$\frac{\kappa\gamma_S + \gamma_S\gamma_T + \eta\gamma_T}{\kappa + \gamma_T}\hat{I}_S$$

is the number of cases of drug-sensitive infections that do not develop resistance. We consider the treatment rate η to be a parameter that can be controlled, with all the other parameters of the model fixed. Then the ratio of these two numbers is a function of η, and we define

$$\lambda(\eta) = \frac{\gamma(\kappa + \gamma_T)\hat{I}_R}{(\kappa\gamma_S + \gamma_S\gamma_T + \eta\gamma_T))\hat{I}_S}.$$

Simulations indicate that the effect of increasing the treatment rate η is to increase the number of resistant infections and to decrease the number of sensitive infections. If there is no transition from sensitive to resistant infections ($\kappa = 0$), the number of sensitive infections is a decreasing function of the treatment rate, while the number of resistant infections is independent of the treatment rate (except for a decrease because of the decrease in the number of sensitive infections). The effect of adding development of resistance at a given treatment rate is to decrease the number of sensitive infections because some sensitive infection would develop resistance and to increase the number of resistant infections. This makes it plausible that the ratio $\lambda(\eta)$ is an increasing function of the treatment rate, and therefore we make the assumption

$$\lambda'(\eta) \geq 0.$$

We write $\hat{I}_R$ as a function of $\lambda(\eta)$ and $\hat{I}_S$ and substitute into the expression for $N - S_\infty$, and then substitute into the expression for $\log(S_0/S_\infty)$. This leads to a pseudo-final size relation

$$\log\frac{S_0}{S_\infty(\eta)} = E(\eta)\Big(1 - \frac{S_\infty(\eta)}{N}\Big),$$

where

$$E(\eta) = \frac{\mathcal{R}^*(\eta) + \mathcal{R}_R\lambda(\eta)}{1 + \lambda(\eta)}.$$

For a final size equation of this form, determining S_∞ as a function of η, we have seen in Section 1.2 that the derivatives of $E(\eta)$ and S_∞ with respect to η have opposite sign. This implies that if

$$E'(\eta) = \frac{[1+\lambda(\eta)](R^*(\eta))' + \lambda'(\eta)[\mathcal{R}_R - R^*(\eta)]}{(1+\lambda(\eta))^2} > 0, \tag{26}$$

the effect of increasing the treatment rate η is to decrease S_∞, that is, to make the epidemic more severe. We see from (26), since $(R^*(\eta))' < 0$, $\lambda'(\eta) > 0$, that $E'(\eta) < 0$ if $\mathcal{R}_R < R^*(\eta)]$. However, it is possible to have $E'(\eta) > 0$ if $\mathcal{R}_R > R^*(\eta)]$, and this happens, at least for some values of η, in the cases (ii) $[\delta_R\gamma_T < \gamma_R\delta_T]$ and (iii) $[\delta_R\gamma_S(\kappa+\gamma_T) < \gamma_R(\kappa+\gamma_S)$, $\delta_R\gamma_T > \gamma_R\delta_T]$.

We have simulated the model (19) using parameter values for influenza infection that have been used previously. Simulations indicate that in both cases (ii) and (iii) epidemic final size may decrease to a minimum and then increases as the treatment rate is increased, so that there is a unique optimum treatment rate. In case (iii), where we have shown that there is a critical treatment rate η_c for which $\mathcal{R}^*(\eta_c) = \mathcal{R}_R$, we see from (26) that $E'(\eta_c) < 0$, and thus that the optimal treatment rate is greater than η_c.

The figure below shows the optimal treatment rates at which the final size is a minimum for two different levels of resistance transmission. We used parameter values

$$a = 0.45, \quad N = 10,000, \quad \gamma_S = \gamma_R = 1/4\ day^{-1}, \quad \gamma_T = 1/3\ day^{-1},$$
$$\kappa = 0.0002\ day^{-1}, \quad \delta_T = 0.4.$$

Initial values are

$$S_0 = 10^4 - 1, \quad I_S(0) = 1, \quad I_T(0) = I_R(0) = 0.$$

For a relatively low transmissibility of resistance $\delta_R = 0.7$, for which $\eta_c = 0.03$, the optimal rate is $\eta_0 = 0.238$ (red curves). As the transmissibility of resistance increases, the optimal treatment rate decreases. This is shown in the black curves for $\delta_R = 0.9$, which is in case (ii), with an optimal treatment rate $\eta = 0.139$. The solid curves are total infections; the dotted curves are the total number of infections without resistance, and the dashed curves are the total number of resistant infections.

4. Conclusions

The age of infection epidemic model is a very general form. We have considered extensions in two different directions. For diseases with direct and

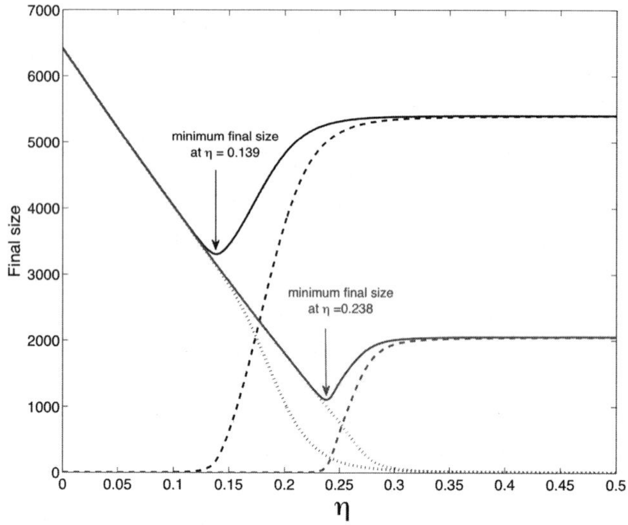

indirect modes of transmission it is possible to build a parallel development provided the rate of shedding pathogen does not depend on the time since infection. To allow a pathogen shedding rate that depends on the age of infection it is necessary to use a slightly less general epidemic model, namely a stage structured model.

For diseases in which there is a risk of development of drug resistance, experimental evidence indicates a possibility that increasing the treatment rate may be counter-productive. We have formulated a simple model that can exhibit such behavior, but analysis of the model depends on an assumption about the ratio of drug-resistant cases to drug-sensitive cases of disease. Although simulations indicate the validity of this assumption, we have not been able to establish its validity analytically. This remains an open problem, as does the question of describing an age of infection model for a disease with two strains.

A disease for which development of drug resistance is known to be possible is tuberculosis. Tuberculosis is a disease that acts on a very long time scale, and an epidemic model would be inappropriate. A model for tuberculosis must include demographic effects.

References

1. Alexanderian, A., M.K. Gobbert, K.R. Fister, H. Gaff, S. Lenhart, and E. Schaefer (2011) *An age-structured model for the spread of epidemic cholera: Analysis and simulation*, Nonlinear Anal. Real World Appl. **12**: 3483–3498.

2. Andersson, D., Levin, B. (1999) *The biological cost of antibiotic resistance*, Current Opinion in Microbiology **2**(5): 489–493.
3. Anderson, R.M. & May, R.M. (1979) *Population biology of infectious diseases I*, Nature **280**: 361–367.
4. Anderson, R.M. & R.M. May (1991) Infectious Diseases of Humans. Oxford University Press.
5. Andrews, J.R. & and S. Basu (2011), *Transmission dynamics and control of cholera in Haiti: an epidemic model*, Lancet **377**: 1248–1255.
6. Björkman, J., I. Nagaev, O.G. Berg, D. Hughes & D.I Andersson (2000) *Effects of environment on compensatory mutations to ameliorate costs of antibiotic resistance* Science **287**: 1479–1482.
7. Brauer, F. (2005) *The Kermack-McKendrick epidemic model revisited*, Math. Biosci. **198**: 119–131.
8. Brauer, F. (2008) *Age-of-infection and the final size relation*, Math. Biosc.& Eng. **5**: 681–690.
9. Brauer, F. (2013) *Compartmental age of infection epidemic models*, in BIOMAT 2012, R.P. Mondaini (ed), World Scientific Co., to appear.
10. Brauer, F. & C. Castillo-Chavez (2012), Mathematical Models in Population Biology and Epidemiology, 2nd Ed., Springer, New York.
11. Brauer, F., C. Castillo-Chavez & Z. Feng (2010) *Discrete epidemic models*, Math. Biosc. & Eng. **7**: 1–15.
12. Brauer, F., Z. Shuai, & P. van den Driessche (2013) *Dynamics of an age-of-infection cholera model*, Math. Biosc. & Eng. **10**, to appear.
13. Brauer, F., P. van den Driessche & J. Wu (eds.) (2008) *Mathematical Epidemiology*, Lecture Notes in Mathematics, Mathematical Biosciences subseries 1945, Springer.
14. C.T. Codeço,(2001) *Endemic and epidemic dynamics of cholera: the role of the aquatic reservoir*, BMC Infectious Diseases **1**: 1.
15. Diekmann, O., J.A.P. Heesterbeek & T. Britton (2012) *Mathematical Tools for Understanding Infectious Disease Dynamics* Princeton University Press, Princeton
16. Diekmann, O., J.A.P. Heesterbeek, and J.A.J. Metz (1990) *On the definition and the computation of the basic reproductive ratio mathcalR_0 in models for infectious diseases in heterogeneous populations*, J. Math. Biol., **28**: 365–382.
17. Dietz, K. (1982) *Overall patterns in the transmission cycle of infectious disease agents*, In: R.M. Anderson, R.M. May (eds), Population Biology of Infectious Diseases, Life Sciences Research Report **25**, Springer-Verlag, Berlin-Heidelberg-New York: 87–102.
18. Domingo, E. & J. Holland (1997) *RNA virus mutations and fitness for survival*, Ann. Rev. Microbiology **51**: 151–178.
19. Gumel, A., S. Ruan, T. Day, J. Watmough, P. van den Driessche, F. Brauer, D. Gabrielson, C. Bowman, M.E. Alexander, S. Ardal, J. Wu, & B.M. Sahai (2004) *Modeling strategies for controlling SARS outbreaks based on Toronto, Hong Kong, Singapore and Beijing experience*, Proc. Roy. Soc. London, **271**: 2223–2232.

20. Halloran, M., F. Hayden, Y. Yang, I. Longini & A. Monto (2006) *Antiviral effects on influenza viral transmission and pathogenicity: Observations from household-based trials*, Am. J. Epidemiology **165**: 212–221.
21. Hansen, E. & T. Day (2011) *Optimal antiviral treatment strategies and the effects of resistance*, Proc. Roy. Soc. London B **278**: 1082–1089.
22. Hartley, D.M., J.G. Morris Jr. & D.L. Smith (2006) *Hyperinfectivity: a critical element in the ability of V. cholerae to cause epidemics?* PLOS Med. **3**:63–69.
23. Heffernan, J.M., R.J. Smith? & L.M. Wahl (2005) *Perspectives on the basic reproductive ratio*, J. Roy. Soc. Interface,**2**: 281–293.
24. Heesterbeek, J.A.P. (1992) R_0, *Thesis*, CWI, Amsterdam.
25. Hethcote, H.W. (2000) *The mathematics of infectious diseases*, SIAM Review **42**: 599–653.
26. Hyman, J.M. & J. Li, *Infection-age structured epidemic models with behavior change or treatment*, J. Biol. Dyn., **1**: 109–131.
27. Hyman, J.M., J. Li & E. A. Stanley (1999) *The differential infectivity and staged progression models for the transmission of HIV*, Math. Biosci., **155**: 77–109.
28. Keeling, M. & P. Rohani (2008) Modelling Infectious Diseases in Humans and Animals, Princeton University Press, Princeton, N.J.
29. Kermack, W.O. & A.G. McKendrick (1927) *Contributions to the mathematical theory of epidemics*, Proc. Royal Soc. London, **115**: 700–721.
30. King, A.A. E.L. Ionides, M. Pascual, & M.J. Bouma (2008) *Inapparent infectious and cholera dynamics*, Nature **454**: 877–890.
31. Lipsitch, M., T. Cohen, M. Murray & B. Levin (2007) *Antiviral resistance and the control of pandemic influenza* PLoS Medicine **4**: 111–121.
32. Ma, J. & D.J.D. Earn (2006) *Generality of the final size formula for an epidemic of a newly invading infectious disease*, Bull. Math. Biol. **68**: 679–702.
33. May, R.M. & R.M. Anderson, R.M. (1979) *Population biology of infectious diseases II*, Nature **280**: 455–461.
34. Moghadas, S. (2011) *Emergence of resistance in influenza with compensatory mutations*, Mathematical Population Studies **18**: 106–121.
35. Moghadas, S. (2008) *Management of drug resistance in the population: influenza as a case study*, Proc. Roy. Soc. London B **275**: 1163–1169.
36. Moghadas, S., C. Bowman, G. Röst & J. Wu (2008) *Population-wide emergence of antiviral resistance during pandemic influenza*, PLoS One **3**(3), e1839.
37. Moghadas, S., C. Bowman, G. Röst, D. Fisman & J. Wu (2009) *Post-exposure prophylaxis during pandemic outbreaks*, BMC Medicine **7**(73), 1–10.
38. Murray, J.D. (2002) Mathematical Biology, Vol. I, Springer-Verlag, Berlin-Heidelberg-New York.
39. Murray, J.D. (2002) Mathematical Biology, Vol. II, Springer-Verlag, Berlin-Heidelberg-New York.

40. Shuai, Z. & P. van den Driessche (2011) *Global dynamics of cholera models with differential infectivity*, Math. Biosc. **234**: 118–126.
41. Tien, J.H. & D.J.D. Earn (2010) *Multiple transmission pathways and disease dynamics in a waterborne pathogen model*, Bull. Math. Biol. **72**: 1506–1533.
42. Thieme, H.R. (2003) Mathematics in Population Biology, Princeton University Press, Princeton, N.J.
43. van den Driessche, P. & J. Watmough (2002) *Reproduction numbers and sub-threshold endemic equilibria for compartmental models of disease transmission*, Math. Biosc. **180**: 29–48.
44. Xiao, Y., F, Brauer and S. Moghadas, it Can treatment increase the epidemic size?, to appear.
45. Yang, C.K. & F. Brauer (2008) *Calculation of $\mathcal{R}_0$ for age-of-infection models*, Math. Biosc. & Eng. **5**: 585–599.

INDEX